KU-777-846

Springer Series in Optical Sciences   Volume 26

Edited by Arthur L. Schawlow

ANDERSONIAN LIBRARY
★
WITHDRAWN
FROM
LIBRARY
STOCK
★
UNIVERSITY OF STRATHCLYDE

Springer Series in Optical Sciences

Editorial Board: J. M. Enoch   D. L. MacAdam   A. L. Schawlow   T. Tamir

bcn .015840

Lasers and Applications

Proceedings of the Sergio Porto Memorial Symposium
Rio de Janeiro, Brasil, June 29 – July 3, 1980

Editors:
W.O.N. Guimaraes, C.-T. Lin, and A. Mooradian

With 200 Figures

ANDERSONIAN LIBRARY
★
WITHDRAWN
FROM
LIBRARY
STOCK
★
UNIVERSITY OF STRATHCLYDE

Springer-Verlag Berlin Heidelberg New York 1981

Professor WLADIMIR O.N. GUIMARAES
Instituto de Fisica, Unicamp
13.100 Campinas Sp, Brasil

Professor CHHUI-TSU LIN
Instituto de Quimica, Unicamp
13.100 Campinas Sp, Brasil

Dr. ARAM MOORADIAN
Massachusetts Institute of Technology,
Lincoln Laboratory
Lexington, MA 02173, USA

Editorial Board

JAY M. ENOCH, Ph. D.
School of Optometry
University of California
Berkeley, CA 94720, USA

DAVID L. MACADAM, Ph. D.
68 Hammond Street,
Rochester, NY 14615, USA

ARTHUR L. SCHAWLOW, Ph. D.
Department of Physics, Stanford University
Stanford, CA 94305, USA

THEODOR TAMIR, Ph. D.
981 East Lawn Drive, Teaneck,
NJ 07666, USA

D
621·366
LAS

ISBN 3-540-10647-2 Springer-Verlag Berlin Heidelberg New York
ISBN 0-387-10647-2 Springer-Verlag New York Heidelberg Berlin

This work is subject to copyright. All rights are reserved, whether the whole or part of the material is concerned, specifically those of translation, reprinting, reuse of illustrations, broadcasting, reproduction by photocopying machine or similar means, and storage in data banks. Under § 54 of the German Copyright Law, where copies are made for other than private use, a fee is payable to "Verwertungsgesellschaft Wort", Munich.

© by Springer-Verlag Berlin Heidelberg 1981
Printed in Germany

The use of registered names, trademarks, etc. in this publication does not imply, even in the absence of a specific statement, that such names are exempt from the relevant protective laws and regulations and therefore free for general use.

Offset printing: Beltz Offsetdruck, Hemsbach/Bergstr. Bookbinding: J. Schäffer oHG, Grünstadt.
2153/3130-543210

Preface

The International Conference on Lasers and Applications was held in Rio de Janeiro, Brazil from 29 June to 3 July 1980. This conference was held to commemorate the memory of Professor Sergio Porto who died suddenly about one year earlier while attending a laser conference in the Soviet Union. The subject matter covered the active areas of laser devices, photochemistry, nonlinear optics, high-resolution spectroscopy, photokinetics, photobiology, photomedicine, optical communication, optical bistability, and Raman spectroscopy.

The conference was attended by over 150 people including scientists from Japan, France, England, West Germany, Norway, Italy, Brazil, Chile, Argentina, India, Canada, and the United States. A memorial session attended by members of the Porto family and ranking Brazilian government dignitaries preceded the start of the conference.

The location of the conference in Rio de Janeiro, Brazil, was chosen because it was in the homeland of Sergio Porto and provided an opportunity for his friends, colleagues, and countrymen to pay homage to him. The setting on Copacabana Beach afforded access to the lovely beaches, restaurants, and nightlife of one of the most beautiful and exciting cities of the world. There were tours of the city together with a banquet that featured a performance by one of the best Samba Schools in Rio.

Financial support from many sponsors in Brazil and the United States is gratefully acknowledged in making this working conference a fitting tribute to the memory of Professor S.P.S. Porto.

January, 1981

W.O.N. Guimaraes
C.T. Lin
A. Mooradian

Contents

Part VI. *Picosecond Bistability*

Part I

Raman Spectroscopy

Surface Brillouin Scattering

R. Loudon

Physics Department, Essex University, Colchester CO4 3SQ, UK

1. Introduction

During the past 20 years, since the invention of the laser, light scattering has become an increasingly powerful means of measuring the spectra of systems in thermal equilibrium. Sergio Porto played a leading role in the development of techniques to take maximum advantage of the properties of laser light in its application to light scattering, and he and his collaborators made many of the first observations of the spectra of excitations which thereby became accessible to this kind of measurement. The power spectra of the thermal fluctuations of almost all dynamic variables in solids and liquids can be studied by light scattering over part of the range of frequency and wavevector.

Most light scattering work of the past two decades has been concerned with excitations in bulk materials but there has been a parallel interest in the spectra of surface excitations. For example, in the case of light scattering by thermally-excited surface waves on a liquid, where experimental observations date back to 1913 [1], the first measurements of the ripplon frequency spectra were made in 1967 and 1968 [2,3], followed by more detailed investigations [4]. The measurements are made on flat surfaces; we consider only this case.

It is possible to distinguish two effects that arise in the presence of a surface. Firstly, the fluctuations associated with the bulk modes of excitation are modified close to the surface because of the boundary conditions imposed on the dynamic variable concerned. Secondly, there are usually new modes of excitation whose fluctuations have significant amplitude only at or very close to the surface. For a liquid, the longitudinal acoustic waves are bulk modes while the ripplons are surface modes.

2. Surface Fluctuations

The total surface fluctuation in a dynamical variable is calculated, as in the bulk case [5], from the energy of a static displacement of the variable concerned. Fig. 1 shows the co-ordinates and geometry to be used throughout, with the z=0 plane as the surface of the undisturbed medium, and the excitation wavevector taken as $(Q, 0, q)$. Consider a static ripple on a liquid, with vertical displacement

$$u^z(0) = u_o \cos Q x \qquad\qquad u_o Q \ll 1. \qquad\qquad (1)$$

To lowest order in u_o, the increase in area of a large section of the surface of area A is

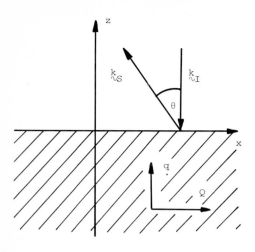

<u>Fig.1.</u> Surface light-scattering geometry

$$\Delta A = \frac{1}{4} u_0^2 \ Q^2 \ A \ . \tag{2}$$

The energy of the ripple is given by a sum of the gravitational and surface-tension contributions,

$$\Delta E = \frac{1}{4} u_0^2 \ A(\rho g + \alpha Q^2) \ , \tag{3}$$

where α is the surface tension and ρ is the density. The classical thermal excitation factor $\exp(-\Delta E/k_B T)$ then leads to a mean-square displacement

$$\langle u^z(0)^2 \rangle_Q = \frac{1}{2} \ \langle u_0^2 \rangle = k_B T/A(\rho g + \alpha Q^2) \ . \tag{4}$$

This is the integrated surface displacement spectrum for ripples with a given direction of surface wavevector Q.

The total mean-square surface displacement obtained by summation over all wavevectors is

$$\langle u^z(0)^2 \rangle = (A/4\pi^2) \int_0^{Q_m} \langle u^z(0)^2 \rangle_Q \ 2\pi \ Q \ d \ Q$$
$$\approx (k_B T/4\pi \ \alpha) \log(\alpha \ Q_m^2/g\rho) \ . \tag{5}$$

If the maximum wavevector is taken to be that for which the viscous damping rate equals the ripple frequency, then numerical values for mercury at room temperature give

$$Q_m \approx 3 \times 10^9 \ m^{-1} \qquad \langle u^z(0)^2 \rangle^{\frac{1}{2}} \approx 1.4 \times 10^{-10} \ m \ , \tag{6}$$

a root-mean-square displacement of about 3 Bohr radii. The wavevectors Q accessible to light-scattering spectroscopy lie typically in the range

$$10^5 \ m^{-1} < Q < 10^7 \ m^{-1} \ , \tag{7}$$

where the surface-tension energy greatly exceeds the gravitational, and (4) can be written

$$\langle u^z(0)^2 \rangle_Q = k_B T/A \; \alpha \; Q^2 \; . \tag{8}$$

The results for other kinds of surface fluctuation are obtained in a similar manner. Thus for surface ripples on an isotropic solid, where the restoring force results from the elastic stiffness expressed in terms of the Lamé parameters λ and μ, the mean-square surface displacement is

$$\langle u^z(0)^2 \rangle_Q = k_B T(\lambda + 2\mu)/2A\mu(\lambda + \mu)Q \; . \tag{9}$$

The root-mean-square displacement obtained by summation over all wavevectors gives

$$\langle u^z(0)^2 \rangle^{\frac{1}{2}} \approx 2.3 \times 10^{-11} \; m \tag{10}$$

for aluminium, equal to about one half the Bohr radius. Note that for a liquid where the shear stiffness μ is zero, the fluctuations become unrestrained unless surface tension is included.

Electromagnetic fluctuations can be treated similarly. The perpendicular component of the electric field at the surface of a dielectric with static relative permittivity κ_0 has a mean-square fluctuation

$$\langle E^z(0)^2 \rangle_Q = k_B T \; Q/\varepsilon_0 \; A \; \kappa_0(1 + \kappa_0) \; . \tag{11}$$

The fluctuations may be associated with surface plasmons or polaritons. Finally, for a ferromagnet with applied field H_0 and spontaneous magnetization M_0 parallel to the y axis, the magnetization perpendicular to the surface has a mean-square fluctuation

$$\langle M^z(0)^2 \rangle_Q = k_B T \; \gamma \; M_0/\mu_0 \; A \; D^{\frac{1}{2}}(\gamma \; H_0 + D \; Q^2)^{\frac{1}{2}} \; , \tag{12}$$

where γ is the gyromagnetic ratio and D is the exchange stiffness.

The varying dependence of the strength of the fluctuations on Q has important consequences for light-scattering experiments. For ripples on liquids, where the surface tension energy proportional to Q^2 is confined to the surface itself, the same factor Q^2 appears in the denominator of (8). For ripples on solids the elastic energy in a layer of given z-co-ordinate is also proportional to Q^2. However, the distortion now penetrates some distance into the bulk material with a spatial dependence that includes a term of the form

$$u^z(z) = u_0 \; exp(Qz) \; cos \; Qx \; . \tag{13}$$

The penetration depth is thus proportional to $1/Q$, and the volume-integrated elastic energy is proportional to Q, giving the factor Q in the denominator of (9). Similar qualitative remarks apply to the other examples. We note that the spatial dependence (13) occurs in the static limit for any excitation whose total wavevector is proportional to a positive power of the frequency.

The common thermal factor $k_B T$ in the fluctuations is a consequence of the classical statistics, and the results are incorrect when the main contributions occur at angular frequencies comparable to or larger than $k_B T/\hbar$. The main contributions to the electric-field fluctuations associated with surface polaritons usually occur at angular frequencies of order 10^{13} Hz, when (11) is invalid. Such frequencies correspond to Raman scattering rather than

Brillouin scattering and we consider this case no further. For comparison, with a surface wavevector Q of order 10^7 m^{-1}, the main acoustic-wave contributions to the surface ripple spectra occur at angular frequencies of order 10^{10} Hz for both solids and liquids, and the ferromagnetic spin wave frequencies are also in this region. The liquid ripplon frequencies are of order 10^8 Hz. The classical statistics are valid in these cases.

3. Kinematics and Techniques

For the scattering of light by an excitation of surface wavevector Q and frequency ω, the incident and scattered optical wavevectors and frequencies are related by

$$k_I^X = k_S^X \pm Q \tag{14}$$

$$\omega_I = \omega_S \pm \omega, \tag{15}$$

where the upper (lower) signs refer to the Stokes (anti-Stokes) components of the scattered light.

Suppose for example that the incident light beam is normal to the sample surface and that scattered light is collected at angle θ to the normal, as shown in Fig. 1. Then (14) and (15) give

$$\omega/\omega_I = \pm 1 - (cQ/\omega_I) \csc \theta. \tag{16}$$

Fig. 2 shows the scans across the ωQ plane for various scattering angles. The excitation wavevectors that contribute to the anti-Stokes spectrum are slightly larger than those that contribute to the Stokes spectrum for a given θ, but the differences are negligible in a typical Brillouin experi-

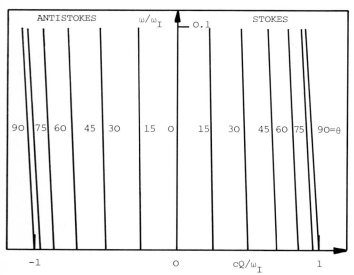

Fig.2. Experimental scans for various scattering angles

ment where ω/ω_T is of order 10^{-4} or less. The most important feature is the change in ^{1}sign of the wavevector of the excitation with which the light interacts between the two sides of the spectrum. Thus although the wavevector transfer Q is the same for the entire spectrum, the Stokes side corresponds to creation of a quantum of excitation of wavevector Q and the anti-Stokes side to destruction of a quantum with wavevector -Q. This can lead to assymetry in the measured spectra around $\omega=0$ in systems that are not invariant under time reversal, for example ordered magnetic materials or media with an externally generated flow of acoustic waves.

Two main experimental techniques are used to measure surface Brillouin spectra. For ripplons, with frequency

$$\omega_R = (\alpha Q^3/\rho)^{\frac{1}{2}} \tag{17}$$

and a scattered intensity obtained from (8) proportional to $1/Q^2$ there are advantages in achieving small Q, of order 10^5 m^{-1} in practice, by scattering close to the direction of specular reflection of the incident light. Then ω_R is typically of order 10^5 Hz and the scattered light is resolved by optical heterodyne spectroscopy. For the acoustic wave and magnetic spectra, where the dependences of intensity on Q given by (9) and (12) are less rapid and the frequencies are generally higher, the spectra are best measured by Fabry-Perot interferometry. Brillouin scattering measurements in this regime have been dominated by the work of John Sandercock using progressively refined multipass interferometers.

Theoretical analysis of the measurements requires the calculation of fluctuation spectra, and not merely their integrated values discussed in §2. The required spectra are readily obtained by linear response theory [6] in terms of the susceptibility of the excitation variable to a suitable applied force of frequency ω. The imaginary part of the susceptibility (or linear response function or Green function) determines the frequency spectrum of the excitation via the fluctuation-dissipation theorem.

4. Liquid Fluctuation Spectra

The spectrum of the mean-square surface displacements on a viscous liquid calculated by linear response theory [7] is

$$\left\langle u^z(0)^2 \right\rangle_{Q,\omega} = \frac{2k_B T}{\pi A \omega} \text{ Im } \frac{i\, q_L(q_T^2 + Q^2)^2}{(q_T^2 - Q^2)^2 \rho \omega^2 + q_L Q^2\{4\rho\, \omega^2\, q_T + i\alpha(q_T^2 + Q^2)^2\}} \tag{18}$$

where

$$q_L = \{(\omega/V_L)^2 - Q^2\}^{\frac{1}{2}} \tag{19}$$

$$q_T = \{(i\rho\omega/\eta) - Q^2\}^{\frac{1}{2}} \tag{20}$$

are the components of the longitudinal and transverse wavevectors perpendicular to the surface, V_L is the longitudinal acoustic velocity, and η is the shear viscosity. Fig. 3 shows the calculated spectrum for mercury for $Q = 10^7$ m^{-1}. With increasing frequency, the first spectral contribution comes from the surface modes, centred on the frequency (17) and with a modest width $4\eta Q^2/\rho$. The second much broader contribution, commencing at a

8

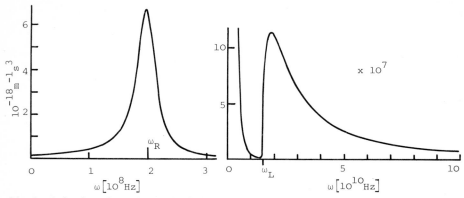

Fig.3. Calculated surface-ripple spectrum of mercury

frequency

$$\omega_L = V_L \, Q \, , \tag{21}$$

comes from the bulk modes, the longitudinal acoustic waves. The bulk modes give a broad continuous distribution because for a given Q in (19) the wave-vector component $q_\perp$ perpendicular to the surface can take all values from 0 to ∞, producing a range of frequencies ω. The intrinsic damping caused by the viscosity has a negligible broadening effect on the continuum.

The ripplon part of the spectrum measured experimentally [8,9] is much broader than expected from the ordinary value of the viscosity, but the measured acoustic wave part [10] is in close agreement with the calculated spectrum. More recent work has studied spectra of fluids covered by monolayers [11] and there remain various problems in the spectra of more complex fluid surfaces.

5. Solid Fluctuation Spectra

The spectrum of the mean-square surface displacements on an isotropic solid has a calculated form almost identical to (18). However, the presence of shear restoring forces in the solid causes changes in the interpretation of some of the symbols. There are now propagating transverse acoustic waves with velocity V_T given by

$$\rho V_T^2 = \mu \, , \tag{22}$$

and their wavevector z component is

$$q_T = \{(\omega/V_T)^2 - Q^2\}^{\frac{1}{2}} \, , \tag{23}$$

replacing (20). The longitudinal velocity is expressed in terms of the Lame parameters by

$$\rho V_L^2 = \lambda + 2\mu \, . \tag{24}$$

The surface displacements are restored by the elastic stiffness forces, much stronger than the surface tension in a liquid, and α can be set equal to zero in (18).

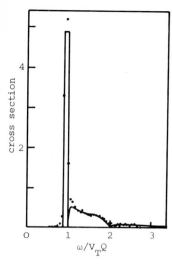

Fig.4. Calculated spectrum and measured points for surface ripples on aluminium (from [15])

Figure 4 shows the measured [12] and calculated [13-16] surface fluctuation spectra of polycrystalline aluminium. Because there is now only one kind of force in the system, all of the spectrum occurs in the same frequency region. The surface mode is the Rayleigh acoustic wave; the theory does not include the anharmonic forces that generate its damping, and the rectangle in Fig. 4 shows the integrated area of this contribution. The continua to higher frequencies come from the bulk transverse and longitudinal acoustic waves.

In the spectra described above, where the media are metals, the coupling of incident and scattered light occurs via the surface distortion. The cross section is derived by a generalization of ordinary reflectivity theory, where the flat surface is perturbed by a travelling ripple [15,17,18]. In less opaque materials, where the incident light penetrates further, the ripple mechanism is augmented by the elasto-optic mechanism responsible for the normal bulk scattering, where the coupling of incident and scattered light occurs by way of the acoustic modulation of the relative permittivity. The elasto-optic mechanism contributes for excitations where the influence of the surface extends some distance into the bulk material. The surface Brillouin spectra of several solids show the effects of simultaneous scattering by the two mechanisms [19-22].

Other kinds of surface excitation produce no mechanical distortion of the surface and their light-scattering spectra result solely from the elasto-optic mechanism. For a solid film mounted on a solid substrate, there are various modes that have an oscillatory spatial dependence in the film but decay exponentially with distance into the substrate. With the co-ordinates of Fig. 1, there are Sezawa waves polarized in the zx plane that produce a surface rippling and have been observed experimentally [23,24], and there are Love waves polarized in the y direction that scatter only by the elasto-optic mechanism [25] and have not so far been observed. This collection of waves originally discovered in seismic studies is completed by the Stoneley waves that propagate along the interface between two media and scatter by both mechanisms [26].

Magnetic excitations do not of course produce any significant surface ripple, and their light scattering occurs entirely by the magneto-optical modulation of the relative permittivity. The cross section is thus determined not merely by the fluctuations in magnetization at the z=0 surface but also by their spatial dependence. The bulk and surface mode fluctuations have different spatial dependences and their relative contributions to the scattering are sensitive to the optical penetration depth. Surface effects should occur for all kinds of ordered magnetic material but most work to date is concerned with ferromagnets.

The surface mode frequency of a ferromagnet in the magnetostatic limit is

$$\omega_M = \gamma(H_0 + \frac{1}{2} M_0) \ , \tag{25}$$

and the associated fluctuations have a mean-square magnetization component [27]

$$\left\langle M^z(z)^2 \right\rangle_{Q,\omega} = (k_B T \ \gamma \ M_0 \ Q/\mu_0 \ A \ \omega_M) \ \exp \ (2Qz) \ \delta \ (\omega \pm \omega_M) \ . \tag{26}$$

The most intriguing aspect of the magnetic surface spectra is the lack of symmetry between the Stokes and anti-Stokes sides. Thus in (26), the positive sign is to be taken in the delta function when the wavevector transfer of magnitude Q is directed parallel to the x-axis, but the negative sign is to be taken when the wavevector transfer is directed antiparallel to the x axis. With a given experimental geometry the surface mode therefore contributes a peak either to the Stokes or to the anti-Stokes side of the spectrum, but not to both sides.

Figure 5 shows the calculated spectrum [28,29] for Fe when the wavevector transfer is positive. Note that the bulk modes contribute more conventionally with symmetrically placed continua. The calculated magnetic Brillouin spectra give good interpretations of the measurements [30,31]. There is potential for further studies of bulk and surface magnetic modes in ordered samples by Brillouin spectroscopy, particularly to determine the role of surface perturbations [32].

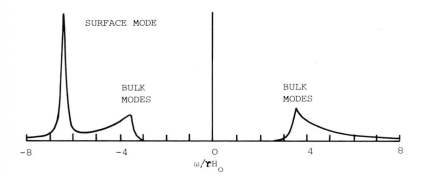

Fig.5. Calculated surface magnetic spectrum of iron (from [22])

6. Conclusions

Most of the work reviewed above is very recent and the field of surface Brillouin spectroscopy promises several years of fruitful development. The experiments carried out to date have used samples with the most perfect available surfaces in order to obtain spectra interpretable in terms of known properties of the materials. There have not so far been many surprises and the theory has largely involved techniques that were already available [33]. Brillouin scattering can now be used to study less perfect surfaces, for example surfaces with some controlled degree of roughness or surfaces wholly or partially covered by layers of a different substance. There are possibilities for investigating the effects of structural phase transitions on surface spectra [34]. Finally, although the light-scattering wavevectors Q are relatively small, it is possible to study some aspects of the microscopic surface structure, for example in the effects of monolayers and in the surface pinning of magnetic excitations.

References

1. L.I. Mandelshtam: Ann. Physik 41, 609 (1913)

2. R.H. Katyl, K.U. Ingard: Phys. Rev. Lett. 19, 64 (1967)

3. R.H. Katyl, K.U. Ingard: Phys. Rev. Lett. 20, 248 (1968)

4. See for example M.A. Bouchiat, J. Meunier: J. Phys. (Paris) suppl. 33, C1, 141 (1972)

5. L.D. Landau, E.M. Lifshitz: Statistical Physics (Pergamon Press, Oxford 1969)

6. See for example D. Forster: Hydrodynamic Fluctuations, Broken Symmetry and Correlation Functions (Benjamin, New York 1975)

7. R. Loudon: Proc. Roy. Soc., in press

8. M. Bird, G. Hills: In Lasers in Chemistry ed. by M.A. West (Elsevier, Amsterdam 1977) p.18

9. M. Bird, G. Hills: In Physicochemical Hydrodynamics ed. by D.B. Spalding (Advance Publications, London 1977) p. 609

10. J.G. Dil, E.M. Brody: Phys. Rev. B14, 5218 (1976)

11. D. Byrne, J.C. Earnshaw: J. Phys. D12, 1145 (1979)

12. J.R. Sandercock: Solid St. Commun. 28, 547 (1978)

13. R. Loudon: Phys. Rev. Lett. 40, 581 (1978)

14. N.L. Rowell, G.I. Stegeman: Phys. Rev. B18, 2598 (1978)

15. R. Loudon, J.R. Sandercock: J. Phys. C13, 2609 (1980)

16. V.R. Velasco, F. Garcia-Moliner: Solid St. Commun. 33, 1 (1980)

17. A.M. Marvin, F. Toigo, V. Celli: Phys. Rev. B11, 2777 (1975)

18. G.S. Agarwal: Phys. Rev. B15, 2371 (1977)

19. K.R. Subbaswamy, A.A. Maradudin: Phys. Rev. B18, 4131 (1978)

20. A.M. Marvin, V. Bortolani, F. Nizzoli: J. Phys. C13, 299 (1980)

21. A.M. Marvin, V. Bortolani, F. Nizzoli, G. Santoro: J. Phys. C13, 1607 (1980)

22. D.L. Mills, K.R. Subbaswamy: Progress in Optics, in press

23. N.L. Rowell, G.I. Stegeman: Phys. Rev. Lett. 41, 970 (1978)

24. V. Bortolani, F. Nizzoli, G. Santoro, A.M. Marvin, J.R. Sandercock: Phys. Rev. Lett. 43, 224 (1979)

25. E.L. Albuquerque, R. Loudon, D.R. Tilley: J. Phys. C13, 1775 (1980)

26. E.L. Albuquerque: J. Phys. C13, 2623 (1980)

27. M.G. Cottam: J. Phys. C12, 1709 (1979)

28. R.E. Camley, D.L. Mills: Phys. Rev. B18, 4821 (1978)

29. R.E. Camley, D.L. Mills: Solid St. Commun. 28, 321 (1978)

30. P. Grunberg, F. Metawe: Phys. Rev. Lett. 39, 1561 (1977)

31. J.R. Sandercock, W. Wettling: IEEE Trans. Magn. 14, 442 (1978)

32. R.C. Moul, M.G. Cottam: J. Phys. C12, 5191 (1979)

33. F. Garcia-Moliner: Annls. Phys. 2, 179 (1977)

34. R.T. Harley, P.A. Fleury: J. Phys. C12, L863 (1979)

Momentum Transfer in Surface Brillouin Scattering

A.F. Khater

Instituto de Física, Universidade Federal Fluminense
Niteroi - RJ - CEP 24000 - Brasil

1. Introduction

The scattering of light in surfaces is of recent interest. The experimental
techniques depend on a coherent and powerful radiation source such as a laser.
Sergio Porto played a leading role in the development and use of laser sources
in light scattering measurements, and this work is dedicated to his memory.
The surface Brillouin spectra give information on the bulk excitations in the
neighbourhood of the material boundary, and on surface excitations proper
[1,2]. Light scattering in surfaces does not obey the wavevector selection
rule familiar for scattering in transparent materials. Rather, the incident
radiation couples to all the modes, which results in opacity broadening.
The theory of Brillouin scattering from the acoustic excitations in the
surfaces of opaque materials involves the question of the momentum transferred
from the radiation fields to the material when incident light (ω_i, k_i) is
scattered into a measurable spectra (ω_s, k_s). The spectral peak is given in
theory at a pseudo-momentum transfer of $2\bar{n}k_i$ for example in [1]. In another
instance this is given by $2|n|k_i$ in [3], where $n = \eta + i\kappa$ is the complex
refractive index which characterises the opaque material. These results are
obtained from a rule of addition of wavevectors, sometimes complex [3].
The question of what the momentum transfer is when light scatters in surfaces
of opaque materials needs to be answered, and takes us into a longstanding
debate as to the momentum of light in a refracting medium [4-10].

2. Momentum

The momentum of light in a material has been measured by boundary experiments,
mechanically [5,6]. A convenient starting point for theoretical study is then
the form of a mechanical force which acts on the material when the electromagnetic radiation propagates in it. The only such force which is consistent
with the requirements of special relativity [7] is the volume force

$$F_\alpha = \frac{\partial T_{\alpha\beta}}{\partial x_\beta} - \frac{\eta^2}{c^2} \frac{\partial}{\partial t} (\underline{E} \times \underline{H})_\alpha \ . \tag{1}$$

$T_{\alpha\beta}$ is the Maxwell stress tensor. $\underline{E}$ and $\underline{H}$ are the electric and magnetic field
vectors, respectively. α and β run over cartesian coordinates. It is conven-
tional to put $\eta = 1$ for free space.

Consider the material and free space to occupy the regions $z > 0$ and $z < 0$,
respectively. The light is incident in free space perpendicular to the material surface at $z = 0$. For steady state conditions integrate (1) to give

$$\overline{p}_\alpha = \overline{p}_\alpha(0^+) + \int F_\alpha \; \Pi_\beta \; dx_\beta \; dt \; . \tag{2}$$

The bars denote time averages. $\overline{p}_\alpha$ is the mechanical momentum which is experienced by the material due to the radiation. $\overline{p}_\alpha(0^+)$ in this representation is the mechanical momentum experienced by the material in the surface boundary when the light just enters, at $t = 0^+$, $z = 0^+$. The expression (2) is convenient for it separates any boundary effects from those due to the volume force F_α inside the material.

In a transparent material for which $n \equiv \eta$, the time average of the integral in (2) can be evaluated using (1), and is zero. $\overline{p}_\alpha(0^+)$ is not zero, however. The theoretical debate gives different results for the momentum of light in a transparent material [4]. The experiments on the other hand consistently give the same result [5,6,8]. Significantly these experiments are boundary experiments, a fact which has been commented on [4]. It seems reasonable therefore to identify $\overline{p}_\alpha(0^+)$ of (2) with the result of experiments

$$\overline{p}_\alpha(0^+) = \eta k_i \; . \tag{3}$$

Any energy-momentum exchange between the radiation fields and the material must satisfy the requirements of the conservation of total energy and total momentum, in all inertial frames. This leads in transparent materials to Abraham's form $(E \times H)/4\pi c$ for the momentum density, rather than Minkowski's $(D \times B)/4\pi c$ [9]. In opaque materials: (i) E and H, rather than D and B are the field variables taken to interact with the dynamic variables of the material, and further (ii) any energy-momentum exchange between the fields and the material must still satisfy the above requirements [9], which again leads to a symmetric energy-momentum tensor. (i) and (ii) imply that Abraham's remains the appropriate form for momentum density in opaque materials, as in metals [10]. The only way this result can be consistent with the representation of mechanical momentum in (2) is to suppose that (2) applies formally for transparent and opaque materials alike, but that only the variation of $(E \times H)/4\pi c$ needs be considered to evaluate energy-momentum exchange between the fields and the material.

The time average of the integral in (2) is now calculated for an opaque material in which the incident monochromatic light is completely absorbed. The result can be written using (3) as

$$\overline{p} = (\; \eta - \frac{2\kappa^2}{(1 + \eta)^2} \;) \; k_i \; . \tag{4}$$

The subscript α is dropped for perpendicular incidence.

3. Surface Brillouin Scattering

Consider that $\overline{p}$ characterises an excited state $|\overline{p},0,0>$ for the material, and that in Brillouin scattering this state decays into another $|0,q,\eta k_s>$. The quantities q and ηk_s denote in the material the momentum of the acoustic excitation and of the scattered light, respectively. The scattering Hamiltonian at a depth z in the material can be written as

$$H(z) = E_s \; (\; \sum_q P_q \;) \; E_i \; \exp(-\kappa k_i z) \; . \tag{5}$$

E_i and E_s are the amplitudes of incident and scattered electric fields.

E_s is homogeneous for $z > 0$. P_q is the $\underline{q}$th component of the material polarisation.

The total scattering cross-section is

$$\sigma = \int_0^\infty dz \; (A \; \text{Im} \; \sum_{\mp \underline{k}_s, \overline{p}, \underline{q}} \; |< \eta \underline{k}_s, \underline{q}, 0 | H | \overline{p}, 0, 0 >_{Q^z_{\pm}, \omega} |^2) . \tag{6}$$

A is an appropriate constant. The principle of conservation of momentum gives

$$Q^z_{\pm} = \overline{p} + \eta k_s^{\;z} \pm q^z \tag{7}$$

along the z axis, where $\pm$ determine the two possible ways of phonon creation (annihilation) in Stokes (Anti-Stokes) processes.

For opaque materials for which the elasto-optic scattering mechanism is dominant, the differential scattering cross-section is obtained from (6) and appropriate H [3] in the form

$$d^2\sigma/d\Omega \; d\omega = F \quad \kappa(s^2 + 4\kappa^2 + x^2)/[(s^2 - 4\kappa^2 - x^2) + 16s^2\kappa^2] . \tag{8}$$

where $s = \overline{p} + \eta$, and $x = q^z/k_i$. F is a constant of the order of magnitude of f_s in [3]. The peak x_m occurs in (8) when

$$x_m^2 = - (s^2 + 4\kappa^2) \pm 2s \; (s^2 + 4\kappa^2)^{1/2} . \tag{9}$$

For opaque materials where $\kappa \lesssim \eta$ and $s \approx 2\eta$, (9) yields a peak position $x_m \approx 2\eta(1 - \lambda)$ where $\lambda = (\kappa^2/2\eta^2)^2$. The discrepancy between this x_m and 2η in [1] when measuring the speed of sound in opaque materials is of the order of a few percent, and thus can serve as a test for the theory.

References

1. J.R. Sandercock: Solid St. Commun. 28, 547 (1978)

2. R. Loudon: "Surface Brillouin Scattering", in this Volume

3. A. Dervisch and R. Loudon: J. Phys. C9, L669 (1976)

4. Sir R. Peierls: Proc. Roy. Soc. A347, 475 (1976)

5. R.V. Jones and J.C.S. Richards: Proc. Roy. Soc. A221, 480 (1954)

6. A. Ashkin and M. Dziedzic: Phys. Rev. Lett. 30, 139 (1973)

7. W.K.H. Panofsky and M. Phillips: *Classical Electricity and Magnetism* (Addison-Wesley, London 1955) p. 191

8. R.V. Jones and B. Leslie: Proc. Roy. Soc. A360, 347 (1978)

9. M.G. Burt and Sir R. Peierls: Proc. Roy. Soc. A333, 149 (1973)

10. L.D. Landau and E.M. Lifshitz: *Electrodynamics of Continuous Media* (Pergamon, Oxford 1975) p. 242

High-Resolution Studies of Phase Transitions in Solids

P.A. Fleury and K.B. Lyons

Bell Laboratories, Murray Hill, NJ 07974, USA

Abstract

Progress in the study of solid-state phase-transition dynamics using high-resolution light scattering is reviewed. Techniques to enhance contrast include multipass interferometers and resonant reabsorption of stray light. Critical dynamic central peaks arising from several different mechanisms have been observed in the spectra of several solids. Specific examples discussed include $Pb_5Ge_3O_4$, $TbVO_4$, $BaMnF_4$, and $TaSe_2$.

1. Introduction

The invention of the laser brought with it entirely new types of spectroscopy as well as substantial improvement in both the sensitivity and resolving power achievable with conventional optical spectroscopies. This conference is concerned with both aspects of laser related improvements in the ability to probe matter and its excitations. This paper is devoted to those improvements in spontaneous light scattering spectroscopy made possible by the laser and the computer which have opened to study physical phenomena in regimes previously inaccessible to solid state spectroscopy. These include, for example, very low frequency excitations related to instabilities which signal the onset of continuous phase transitions and dynamic critical phenomena. The original applications of laser scattering spectroscopy to solid state phase transitions [1] focussed upon the temperature evolution of the so-called "soft-mode" frequency. The need to follow the critical slowing down ever closer to zero frequency was frustrated because the finite contrast of conventional spectroscopic instruments and strong parasitic or stray scattered light often prevented observation of very low frequency (but dynamic) spectral features. This difficulty distorted the nearly ideal match which scattering spectroscopy provides to the fundamental dynamic quantity of interest for phase transitions, namely the autocorrelation function of the order parameter fluctuations.

Conventional Raman and Brillouin spectroscopy covers the $1-10^4$ cm^{-1} spectral range, but it is the frequency region below 1 cm^{-1} which represents the final stages of dynamic evolution and which is often of most interest. Two methods have been developed which significantly increase the combined contrast and resolving power of spectrometers: 1) the multipass interferometer and 2) the tandem interferometer used in combination with computer-assisted resonant reabsorption techniques. Although we will briefly discuss both in the next section we shall focus dominatly in the examples to be covered in Sect. 3 upon the second method. These examples include dynamics of commensurate phase transitions in lead germanate and terbium vanadate, and incommensurate transitions in barium manganese fluoride and in members of the transition metal dichalcogenide family. The latter embraces the subject of reflection scattering from opaque materials — in which the problems of parasitic or stray light can be particularly severe.

2. Experimental Techniques and Apparatus

The generic spectroscopic problem to be addressed here is the combined effect of intense elastically scattered light and the finite contrast of the ordinary spectrometer. This impairs the experimenter's ability to distinguish, for example, between two very different dynamic behaviors near the critical temperature of a solid-state phase transition.

One experimental strategy seeks to narrow the inaccessible region by greatly increased contrast and improved finesse over that available in a conventional Fabry-Perot interferometer. By arranging to pass the scattered light through the instrument several times, one may achieve contrasts of nearly 10^{10}, approximately seven orders of magnitude greater than is available in a single-pass instrument of the same finesse [2]. To a rough approximation the instrumental response of an N pass interferometer is an N-fold self-convolution of the single-pass response. The advantages of such an instrument are obvious in permitting observation of weak, low-frequency components in the presence of a very much stronger elastic component. Since the instrument is scanned by the relative motion of two interferometer plates in N pass as well as in single-pass operation, there is no problem with spectral registration of the various passes or "stages". However, the instrument does not reject the elastic or stray light but merely compresses it into a narrower instrumental region around zero frequency, so that quasi-elastic spectral features remain difficult to separate from the elastic response. The multipass interferometer does not discriminate against order overlap contributions to the spectrum, and does not permit the free spectral range and the instrumental resolution to be chosen independently. These restrictions may be substantially circumvented by operating in tandem two multipass interferometers of different free spectral ranges. SANDERCOCK [3] has recently overcome the considerable difficulties associated with synchronously scanning the plates of the two instruments. Although some of the above-mentioned disadvantages remain with the tandem multipass interferometer, it has certainly opened a number of previously impossible situations to experimental study. In particular thermally excited surface acoustic and magnetic waves on metals and opaque semiconductors have been successfully investigated [4].

An alternative scheme to increase the effective contrast and resolving power is based on resonant reabsorption of the elastically scattered light using a molecular iodine filter placed between the sample and the interferometer [5]. The elastically scattered light as well as the quasi-elastic light within ±300 MHz of the exciting laser frequency may be attenuated by as much as a factor of 10^7 in such a cell. Most of the quasi-elastic and inelastic spectrum is passed to the detector (although somewhat distorted by subsidiary absorptions in the molecular iodine). The distortions can be quantitatively removed by computer-assisted normalization procedures, provided that the laser frequency, the iodine cell temperature, and the interferometer are sufficiently stable. To achieve high stability, to minimize order overlap, and to obtain high finesse and synchronous scanning we employ a pressure-scanned tandem incommensurate Fabry-Perot interferometer system [5]. Effective finesse approaching 1,000 and contrast greater than 10^9 have been achieved with this instrument when combined with the iodine cell.

Figure 1 shows a schematic diagram of this apparatus which permits not only computer assisted normalization, but computer controlled operation of both the tandem Fabry-Perot and the Spex double monochromator. The ability to observe both Raman and Brillouin spectra simultaneously is particularly useful in the study of solid-state critical dynamics. Figure 2 shows a high-resolution spectrum of $TbVO_4$ taken with the iodine-tandem interferometer combination before and after computerized normalization. Clearly the quantitative recovery of rather complex line shapes for frequency shifts greater than 0.01 cm^{-1} is achievable — even with virtually complete suppression of stray light. This system has permitted the observation of several new light scattering phenom-

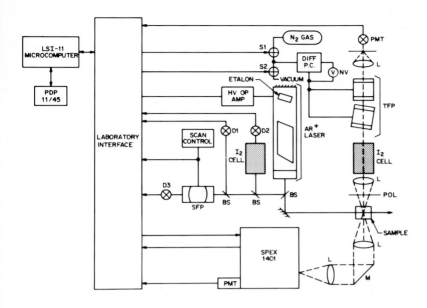

Fig. 1. Apparatus schematic for computer-assisted high-resolution light scattering experiments. The single-mode argon laser frequency is stabilized to the I_2 absorption and monitored using a spherical Fabry-Perot (SFP). For Brillouin studies the tandem Fabry-Perot (TFP) is pressure scanned under computer control. Raman spectra may be simultaneously gathered using the Spex double monochromator

ena, including all of the phase-transition examples discussed in the remainder of this paper.

3. Central-Peak and Soft-Mode Interactions in Structural Phase Transitions

The simple quasi-harmonic soft-mode description of critical slowing down in structural phase transitions fails to account for observed light scattering spectra in several respects [6]. First, since light scattering is a constant momentum transfer ($\hbar q$) process and for right angle scattering the excitations probed have $q \sim 10^5$ cm^{-1}, virtually every spectrum will contain Brillouin scattering from the acoustic phonons whose frequencies lie in the 0.1 to 2 cm^{-1} range. Coupling between these acoustic phonons and any soft optic mode or other critical dynamic excitation will complicate the spectral response and must be understood if one is to extract properly the parameters of the phase-transition dynamics from the spectral line shape. The detailed manifestations of such mode-coupling effects depend upon the symmetry of the system and the

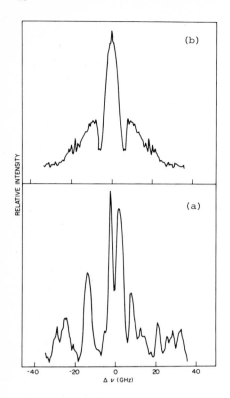

Fig. 2 a,b. Spectrum of TbVO$_4$ at 34.5 K with q at 15° to X axis taken with apparatus of Fig. 1. (a) raw data as received through the I$_2$ cell, TFP combination; (b) computer-normalized data which corrects for distortions of I$_2$ subsidiary absorptions

precise experimental probe used. Considerable care must be exercised in comparing the results, for example, of light scattering, neutron scattering, spin resonance, and acoustic experiments, since each of these experimental techniques couples differently to the order parameter and to the acoustic phonon system. As derived more fully elsewhere [7], the quantitative description of the light-scattering spectrum S (ω) for a coupled-mode system may be expressed as shown in (1),

$$S(\omega) = IM \sum_{i,j} F_i F_j x_{ij} (\omega) \quad . \tag{1}$$

Here the F_i represent the scattering strengths of the uncoupled modes and the x_{ij} are expressed in terms of x_i, the uncoupled dynamic susceptibilities, in the following way:

$$X_{ii} = \frac{x_i}{1-A^2 x_i x_j} \quad ; \quad X_{ij} = \frac{A x_i x_j}{1-A^2 x_i x_j} \tag{2}$$

where A^2 measures the coupling strength between modes i and j.

In (2) the dynamic susceptibility represented by x_i can be expressed generally to account for all of the other interactions and self-energy effects associated with a given dynamic variable. The particular form of the individual x_i will depend of course upon the dynamical behavior of the uncoupled degree of freedom which it describes. For example, a simple phonon will be described by the quasi-harmonic Lorentzian form

$$x_s = [(\omega^2 - \omega_s^2) + i2\Gamma_s\omega]^{-1} \quad . \tag{3}$$

The simple soft-mode theory predicts that $\omega_s^2 = a(T-T_c)$. The second inadequacy of the simple soft-mode theory is its failure to account for the appearance of the often observed "central peak" close to T_c. In many cases this can be done by adding [7] to Γ_s a term $^c\Sigma(\omega)$ of the form given in (4). This describes the coupling of the phonon mode to a relaxation process characterized by a relaxation time, t, and a coupling strength, δ.

$$\Sigma(\omega) = \frac{\delta^2 t}{1 - i\omega t} \quad . \tag{4}$$

The self-energy term, Σ, introduces a second characteristic time (besides ω_s^{-1}) into the problem which may complicate the critical dynamics. Until recently there was insufficient experimental data on the values of t, δ and their dependences upon wave vector, temperature, and symmetry to ascribe these terms unambiguously to any particular physical process. Table 1 lists several theoretical mechanisms for the so-called central-peak complication. These include entropy fluctuations, two phonon difference processes, impurity- or defect-related phenomena, phasons in incommensurate systems, and electronic transitions. Each of these mechanisms implies distinctive dependence of the central peak upon wave vector, polarization selection rules, temperature, electric field, and impurity content. In the remainder of this section we will illustrate some of these mechanisms by considering specific phase transitions.

Table 1. Central peak mechanisms

Mechanism	Linewidth	Material
Entropy Fluctuations	$\Gamma = D_{TH}q^2$	KDP
Phasons	$\Gamma = (D_{||}\cos^2\phi + D_{\perp}\sin^2\phi)q^2$	$BaMnF_4$
Phonon Density Fluctuations	$\Gamma = \left(\vec{v}_g\right)\cdot\vec{q}$	$Pb_5Ge_3O_{11}$
Static Defects	$\Gamma = 0$	$Pb_5Ge_3O_{11}$
Mobile Defects	$\Gamma = D_m q^2$; etc.	?
Static Domain Walls	0	?
Solitons	$\vec{v}_s\cdot\vec{q}$	?
Defect Local Modes	?	?
Electronic Degeneracy	t_e^{-1}	$TbVO_4$

3.1 Ferroelectric Transition: Lead Germanate

Consider first the displacive ferroelectric transition in lead germanate at 451 K. Order parameter measurements [8] have shown the phase transition to be continuous and nearly mean-field-like. Theoretical considerations place this transition in the uniaxial dipolar class expected to exhibit at most, logarithmic deviations from mean field behavior. Light scattering studies [7] have shown that the dynamics evolve in three regimes. For $(T-T_c) > 40$ K the soft-mode frequency collapses in a mean-field fashion, becoming over damped at $(T-T_c) = 40$ K. The overdamped soft-mode response then continues to narrow in accordance with simple mean-field behavior. This persists down to $[(T-T_c)/T_c]$ ~ 0.01 whereupon a change is observed. The overdamped soft-mode wing stabilizes at a frequency width of about 2.5 cm^{-1} and a central peak appears near zero frequency. As shown in Figure 3, this central peak grows up dramatically between the two longitudinal acoustic Brillouin components at ±20 GHz.

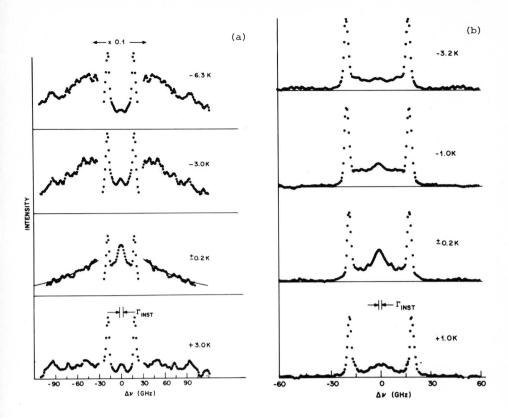

Fig. 3 a,b. Light scattering spectra of $Pb_5GE_3O_{11}$ near the ferroelectric $T_C =$ 451 K. Temperature expressed as $[T_C-T]$. (a) intermediate resolution, showing soft-mode wing for $|\Delta\nu| > 30$ GHz; with $10\times$ gain change for $|\Delta\nu| < 30$ GHz. (b) high-resolution scans emphasize central peak - LA phonon interactions

Under higher resolution (Fig. 3b) the central-peak line shape is shown to be not only broader than the instrumental resolution but also of unusual shape, indicative of considerable interaction with the acoustic phonons. Thus the dynamic response of lead germanate contains a singular dynamic central peak which interacts strongly with both a soft optic phonon and the acoustic phonons of the proper symmetry. A coupled mode analysis based upon (1-4) above produced the line-shape fits shown in Figure 4. The nonsingular value for t^{-1} required to fit all of the spectra is 29 GHz. The persistence of the dynamic central peak slightly above T_C evident in Figure 4, is in apparent violation of mean-field theory predictions but consistent with the predictions of renormalization group theory. Similarly the integrated intensity of the

24

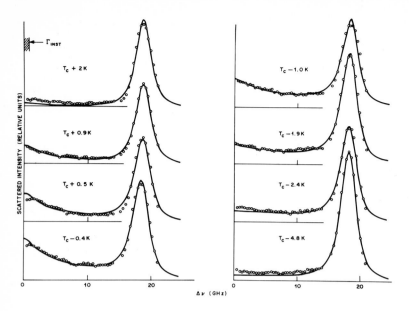

Fig. 4. Solid curves represent least-squares fits to observed spectra (ooo) in $Pb_5Ge_3O_{11}$ using theory described in the text

inelastic central peak exhibjts a weak but significant maximum at T_c consistent with expected deviations from mean-field behavior[c] [7].

As to the central-peak's microscopic origin, experiments [7] varying externally imposed electric field, scattering wave vector, and polarization symmetry have eliminated several of the mechanisms shown in Table 1. The overall uncoupled line width of the central peak, and its lack of dependence upon q or electric field are consistent with the phonon density fluctuation (or two-phonon difference) mechanism. This process is perhaps most simply viewed as the simultaneous creation of a phonon at wave vector k on a given branch while a phonon at wave vector (k + q) on the same branch is destroyed. Anharmonic lattice dynamical interactions in systems such as lead germanate are sufficiently complex that a first-principles calculation of the two-phonon difference spectrum has not yet proved feasible. However, a recent high-resolution study [9] of the simple fcc solid rare gas (Xe) has also revealed a dynamic central peak attributable to two-phonon difference processes. This feature has been reproduced in molecular dynamics calculations [10] of the two-phonon response in such a simple crystal. These observations suggest that two-phonon difference processes are a relatively common source of dynamic central peaks in solids.

In lead germanate excellent spectra may be obtained even without the use of the iodine absorption cell. Although not of sufficient contrast to permit quantitative measurements on the dynamic central peak such spectra reveal an additional *static* singular central peak whose intensity exhibits a power law divergence near T_c and whose origin is probably static symmetry breaking defects. A combination of spectroscopic techniques [7,11] has set an upper limit of 10 Hz for the frequency of this 'static' peak. Data are still insufficient to identify either the number or the nature of the symmetry breaking defects responsible for this static singular feature.

To date there have been at least half a dozen other systems in which defect scattering has been identified as the cause of a singular central peak. As discussed more thoroughly elsewhere [12] it seems likely that the use of high-resolution spectroscopy for the study of interactions between critical fluctuations and defects in model systems will experience significant increase in the future.

3.2 Cooperative Jahn-Teller Transition: TbVO$_4$

The cooperative Jahn-Teller transition represents the best understood class of structural transitions from a microscopic point of view. Not only is the mechanism for the transition itself quantitatively attributable to a specific electron-phonon interaction, this same interaction provides a microscopic description of the soft-mode and central-peak dynamics. The low-lying electronic levels of the Tb^{3+} ion within the crystal field exhibit a degenerate doublet interposed between two closely spaced singlets [13]. As the temperature is lowered the thermal population of this quartet of levels is redistributed until it becomes energetically favorable for the crystal to distort. This structural distortion lifts the degeneracy of the doublet so that the energy cost of the structural distortion is repaid by a lowering of the overall electronic energy.

The quartet of levels are more strongly split as the temperature is lowered into the ordered phase. Indeed the singlet-singlet energy difference can be directly observed using electronic Raman scattering to provide a direct measurement of the order parameter. The accompanying structural distortion is isomorphic to a transverse acoustic mode polarized in the basal plane, which is the appropriate soft mode for this transition. The associated elastic anomaly has been thoroughly studied by ultrasonic [14] and small angle Brillouin scattering [14] experiments.

A complete dynamic theory of the cooperative Jahn-Teller transition [15] had predicted in addition to a soft mode, the existence of a singular dynamic central peak arising from transitions within the degenerate electronic doublet.

The symmetries of the soft acoustic phonon and the electronic excitations are such that these two degrees af freedom are essentially uncoupled when the acoustic wave propagates in the (110) direction, whereas coupling is maximal for propagation in the (100) direction. Furthermore, the coupling parameter is predicted theoretically [16] to vary in a prescribed way with wave vector direction, thereby providing an experimental means of turning on smoothly the electron phonon coupling. In Figure 5a is shown the uncoupled spectrum [17] measuring in detail the dynamic central peak associated with the un-

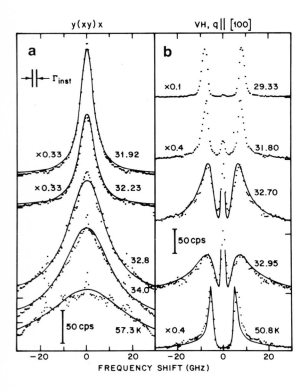

Fig. 5 a,b. Observed spectra (...) near $T_c = 32.6$ K in TbVO$_4$ in (a) uncoupled and (b) fully coupled geometries. Solid curves are theoretical line shapes for $T > T_c$

coupled electronic degrees of freedom. The strong temperature dependence of this line width is evident. By contrast Figure 5b shows a temperature sequence of spectra in the fully coupled geometry [17]. In both cases the solid dots represent the normalized experimental data while the solid lines represent theoretical fits. The highly temperature dependent line shape complexities observed are indicative of strong interactions and significant interference effects between the various terms contributing to the sum in (1). In $TbVO_4$ the individual terms contributing to the sum in (1) are nearly two orders of magnitude larger than the sum, reflecting nearly perfect cancellation of individual contributions in various regions of the spectrum. This accounts for the extreme sensitivity of the spectral line shape to temperature and q direction. A more striking indication of the strong interference effects is provided in Figure 6 where spectra taken at three temperatures in the vicinity of T_c are displayed for different orientations of the q vector. All of these consequences are quantitatively accounted for by the theory. The partic-

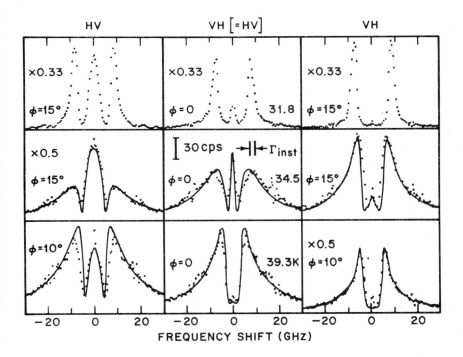

Fig. 6. Line shape dependence of $TbVO_4$ spectra upon temperature and geometry. ϕ denotes the angle between $\overline{q}$ and the X axis

ular dynamic response functions for the electronic and phonon degrees of freedom have been derived above T_c by HUTCHINGS et al. [15] and have the following
forms:

$$X_e = X_e^0[1-JX_e^0]^{-1}$$

$$X_e^0 = \{\sinh(\beta\epsilon) \cdot 4\epsilon[4\epsilon^2 - \omega^2 - i\omega\Gamma_1]^{-1} + \beta\Gamma_2[\Gamma_2 - i\omega]^{-1}\}[1+\cosh(\beta\epsilon)]^{-1}$$

$$X_a = 2\omega_a[\omega_a^2 - \omega^2 - i\omega\nu_a]^{-1}$$

where $\beta^{-1} = kT$; $2\epsilon(= 18$ cm^{-1}) and $\Gamma_1 = 10$ cm^{-1} are the frequency and width of
the singlet-singlet transition-measurable in the Raman spectrum. The electronic
central-peak width Γ_2 observed in the uncoupled geometry together with T_c determine the maximal coupling, A. The temperature-dependent value of Γ_2 found
from study of the uncoupled geometry is used in constructing the fits to the
data in the coupled geometry (see, in particular, Figure 6, solid lines).
Both the temperature dependence and the angular dependence of the ratio Fa/F$_e$
are in agreement with theoretical expectations.

These results represent the first time that any dynamic central peak has
been quantitatively described on the basis of a microscopic theory for a
structural phase transition. The agreement between theory and experiment where
a theory exists $(T > T_c)$ is essentially perfect. Equally good data have been
obtained below the transition as well and should provide considerable stimulus
for the development of a dynamical theory for fluctuations in the ordered
state $(T < T_c)$ of the Jahn-Teller system.

3.3 Incommensurate Transition: BaMnF$_4$

In an incommensurate phase transition the spatial periodicity of the resulting order is not a simple multiple of the parent unit cell dimension. In this
case the order parameter may be written in the plane wave approximation in
terms of a phase and an amplitude. The spectrum of the order parameter fluctuations will have two branches: one representing fluctuations in the amplitude of the order parameter and the other representing phase fluctuations.
The amplitude mode is expected to behave as an ordinary soft optic mode. But
because a *uniform* change of phase for the order parameter (at constant amplitude) can be made without cost in energy, the phase mode (or phason) is expected to have zero frequency at zero wave vector. Further the phason in the in-

commensurate phase is not expected to exhibit singular behavior near T_i, the normal-to-incommensurate transition. It is, however, expected to change character at T_c, the lock-in transition, which signals the onset of a new commensurately ordered phase. Although several incommensurate transitions have been studied direct observation of the phason in the incommensurate phase has been reported only in BaMnF$_4$ [18]. These experiments have shown a temperature-dependent critical central peak whose line width varies strikingly with the scattering wave vector as indicated in Figure 7, but *not with temperature*. The behavior of the observed central peak is consistent with the interpretation that it results from an overdamped phason mode. The line width can be fit to the expression

$$\Gamma_p = q^2(D_{||} \cos^2\phi + D_\perp \sin^2\phi) \tag{6}$$

where $D_{||}$ and $D_\perp$ are constants and ϕ is the angle between q and the a axis (direction of incommensuration). Equation (6) is based on theoretical expectations [19] although a priori values for $D_{||}$ and $D_\perp$ are not available.

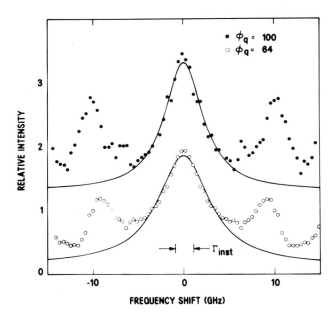

Fig. 7. $90°$ scattering in incommensurate BaMnF$_4$ at 247 K, showing q-dependent central peak and TA Brillouin doublet. Solid curves represent deconvolved Lorentzian widths of 4.0 ($\bullet\bullet\bullet$) and 6.1 GHz (ooo) respectively. The angle ϕ_q is that formed by the X axis and $\vec{q}$, with the latter in the XZ plane

Figure 7 shows the central peak (plus a Brillouin doublet) at $T = 247$ K for two different values of ϕ. Similar data over a range of ϕ are described by (6) using $D_{||} = 0.98$ cm^2/s and $D_\perp = 0.145$ cm^2/s. By contrast the measured thermal diffusivity [20] is $D_{TH} = 0.005$ cm^2/s, so that any central peak due to entropy fluctuations would be much narrower than is observed here.

While these points argue strongly that the observed BaMnF$_4$ central peak is due to phasons, several aspects of this transition remain unexplained. First, the observed temperature dependence of the central peak intensity — reaching a maximum apparently ~7K below T_i — is not understood. Second, a singular *static* central peak has also been observed whose intensity reaches a maximum ~15K below $T_i = 254$ K. This is the same temperature at which the pyroelectric coefficient in BaMnF$_4$ reaches a maximum. Finally the acoustic anomalies near T_i are observed to be different for TA and LA phonons [18,21]. While this in itself is not surprising, the detailed connections between these anomalies and the phason behavior remains to be understood. For example, the observed dispersion in the TA phonon velocity has been interpreted [21] as due to coupling with the phason, but the assumed linearity of this coupling has been called into question [18].

3.4 Charge Density Wave Transitions

Much of the current interest in incommensurate transitions was spawned by the discovery of charge density wave transitions in the layered transition metal dichalcogenides [22]. In these metallic or semimetallic compounds the structural instability is thought to arise from phonon coupling to conduction electrons with the Fermi momentum, k_F. As with BaMnF$_4$, the incommensurate CDW structures are expected to sustain phason excitations. Depending upon the phason lifetime, one expects either a central peak or an additional acoustic-like doublet in the dynamic scattering response. However, the opacity of these metallic materials has until recently precluded the kinds of high-resolution reflection spectroscopy required to examine the critical dynamics with light scattering. Recently, both the multipass and the resonant reabsorption techniques have been brought to bear on these materials. Both thermally excited surface acoustic waves [23] (in seven members of the MX$_2$ family) and a dynamic central peak [24] (in TaSe$_2$) have been observed. We shall conclude this paper with a brief discussion of these preliminary results.

Shown in Figure 8 are room-temperature surface spectra from both the Rayleigh waves and near surface bulk transverse and longitudinal phonons in some transition metal dichalcogenide layered materials taken with a triple pass interferometer. By varying the scattering geometry it is possible to probe selectively thermal surface waves of varying wave vector and thereby to map out over at least a limited range the surface wave dispersion relation. All of the features shown in Figure 8 have acoustic phononlike dispersion relations. These spectra [26] illustrate the utility which light scattering might have as a probe of thermally excited surface waves, in materials where the conventional acoustic techniques prove difficult or impractical. They further reveal that in semi-opaque materials of moderate-to-high reflectivity, both of the proposed mechanisms for the coupling between photons and surface excitations may be of comparable strength. The surface ripple mechanism is primarily responsible for the peaks labeled R in Figure 8 and the photoelastic mechanism is primarily responsible for the peaks labeled A and B. Experiments are presently underway to assess the possible influence of bulk phason modes upon the surface spectra in the incommensurate phase of these materials.

While no definitive observations have yet been made on surface phasons, an anomalous temperature-dependent central peak has been reported [27] in $2H\text{-}TaSe_2$. Complete studies on the linewidth as well as the wave vector, temperature, and symmetry dependence of this feature remain to be done, however. Nevertheless, it appears encouraging that light scattering might provide dynamic information of similar utility and detail for low frequency excitations and phase transitions in metals and other opaque materials that it has already provided in transparent materials.

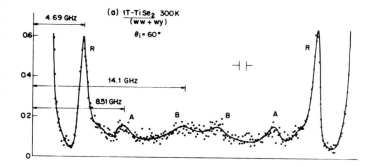

Fig. 8. Reflection spectrum at 300 K of $1T\text{-}TiSe_2$, obtained with a triple pass FP and no I_2 cell. R labels the surface Rayleigh wave. A and B correspond approximately to the fast shear and the longitudinal bulk phonons, and are not observed in similar materials with higher opacity

References

1. See, for example, G.B. Wright (ed.): *Light Scattering Spectra of Solids*, Proc. Int. Conf., New York Univ., 1968 (Springer, Berlin, Heidelberg, New York 1969)
2. J.R. Sandercock: In *Light Scattering in Solids*, ed. by M. Balkanski (Flammarron Sciences, Paris 1972) p. 9
3. J.R. Sandercock: Solid State Commun. *26*, 547 (1978)
4. W. Senn et al.: Physics of Semiconductors, Inst. Phys. Conf. Series *43*, 709 (1978)
5. K.B. Lyons, P.A. Fleury: J. Appl. Phys. *47*, 4898 (1976)
6. See, for example, T. Riste˙ (ed.): *Anharmonic Lattices, Structural Transitions and Melting* (Noordhoff, Leiden 1974)
7. K.B. Lyons, P.A. Fleury: Phys. Rev. *B17*, 2403 (1978)
8. H. Iwasaki et al.: J. Appl. Phys. *43*, 4970 (1972)
9. K.B. Lyons, P.A. Fleury, H.L. Carter: Phys. Rev. *B21*, 1653 (1980)
10. B.J. Alder et al.: Physica *83B*, 249 (1976)
11. D.J. Lockwood et al.: Solid State Commun. *20*, 703 (1976)
12. P.A. Fleury: Third Europhysics Conf. on Defects in Ionic Crystals. J. de Physique (1980)
13. R.J. Elliott et al.: Proc. R. Soc. London *A328*, 217 (1972)
14. J.R. Sandercock et al.: J. Phys. *C5*, 3126 (1972)
15. M.T. Hutchings, R. Scherm, S.R.P. Smith: A. I. P. Conf. Proc. 29, Magnetism and Magnetic Materials, 372 (1975)
16. M.C. Marques: J. Phys. *C13* (to be published, 1980)
17. R.T. Harley, K.B. Lyons, P.A. Fleury: J. Phys. *C13* (to be published, 1980)
18. K.B. Lyons, T.J. Negran, H.J. Guggenheim: J. Phys. *C13* (to be published, 1980)
19. R.N. Bhatt, W.L. McMillan: Phys. Rev. *B12*, 2042 (1975)
20. T.J. Negran: Ferroelectrics (to be published, 1980)
21. D.W. Bechtle, J.E. Scott, D.J. Lockwood: Phys. Rev. *B18*, 6213 (1978)
22. See F.J. DiSalvo: In *Electron-Phonon Interactions and Phase Transitions*, ed. by T. Riste (Plenum, New York 1977) p. 107
23. R.T. Harley, P.A. Fleury: J. Phys. *C12*, L863 (1979)
24. R. Sooryakumar, D.G. Burns, M.V. Klein: In *Light Scattering in Solids*, ed. by J.L. Birman, H.Z. Cummins, K.K. Rebane (Plenum, New York 1979) p. 347

A Statistical Analysis of Trends in Research on Laser Raman Spectroscopy[*]

R.S. Krishnan and R.K. Shankar

Indian Institute of Science, Bangalore 560 012, India

1. Introduction

In 1921, Professor C.V. Raman started a series of experimental studies on
the scattering of light in the Indian Association for the Cultivation of
Science at Calcutta. Working over a period of seven years Raman and his
students established the various laws of scattering of light. Continuing
his researches in the same field he also discovered a new scattering phenomenon of a fundamental character now known as "Raman Effect", in the early
part of 1928. In this experiment it was found that when a beam of monochromatic light was incident on a medium, whether it be a gas, liquid or
solid, part of the light was scattered in all directions with a change of
frequency. Quantum mechanically the phenomenon is visualized as inelastic
scattering of light in which an incident photon is absorbed and a scattered
photon of higher or lower frequency compared to that of the absorbed one is
emitted. The energies of the incident and scattered photons differed by an
amount corresponding to the energy difference between the quantum-mechanical
states or levels of the scattering medium. The fundamental and most important
aspect of Raman scattering is that it provided an easily accessible tool for
the spectroscopic investigation of energy levels of systems not usually accessible by the usual absorption and emission techniques. Depending on the
range of energy levels and the agencies responsible for this mechanism, the
Raman scattering process could be classified under different connotations.
This aspect became very significant and real after the use of lasers for
excitation of Raman spectra, i.e., after 1963.

2. Prelaser Era

For the first thirty-five years since the discovery in 1928, i.e., till 1963,
studies on Raman effect were mainly confined to vibrational, rotational, and
anisotropic (Rayleigh wing) spectra of molecules in the gaseous and liquid
state and to the internal and lattice spectra of mostly clear and transparent
crystals. These studies were more concerned with the establishment of quantum-
mechanical principles and supplying information about the structure of molecules and crystals with practical applications in analytical chemistry and

---

[*] The late Professor Sergio Porto was one of the two pioneers who initiated
research on laser Raman spectroscopy and had contributed substantially to
its progress. It was therefore considered appropriate to present the results
of this survey at the International Conference on Lasers and Applications
arranged to commemorate Professor Sergio Porto. One of us (R.S.K) had the
privilege of working with him for a short period.

for understanding the nature of many chemical problems such as association, dissociation, polymerisation, molecular interaction, nature of the chemical bond, hydrogen bonding, isomerism, molecular rotations, solvent effect, spectra-structure correlation, etc. The highlight of this era lay in the fact that there was very little innovation in the technique of study. It was simple but exotic and laborious involving the use of mercury radiations for excitation, ordinary spectrographs and prolonged photographic exposures. In spite of the limitations of the technique of study investigations of a fundamental character were carried out on second-order Raman spectra of simple crystals like diamond, alkali halides, etc., which led to a series of theoretical investigations on the lattice dynamics of crystals. Even some studies on electronic Raman spectra, resonance (Raman effect), and Raman spectra of phase transitions in ferroelectric and nonferroelectric crystals were carried out with some success.

3. Laser Era

The most important and decisive discovery in physics which brought in a revolution in the studies on Raman spectroscopy was the laser by Maiman in 1960. He was the first to observe stimulated emission in a ruby crystal at 694.3 nm. Schawlow gave a lecture on this topic at the National Research Council in Ottawa on this subject in January 1961. During the discussion that followed Herzberg and Stoicheff suggested the use of ruby laser as a light source for Raman spectroscopy. At the Columbus Meeting on Molecular Spectroscopy in August 1961 Porto and Wood, working in the Bell Telephone Laboratories, reported the recording of the Raman spectra of carbon-tetrachloride and benzene excited by the radiations from a ruby laser. Almost at the same time, Stoicheff was also successful in recording the Raman spectrum using a laser. He presented his results at the Xth Colloquium Spectroscopicum Internationale, Washington, the proceedings of which were published by E.R. Lippincott and M. Margoshes in 1963. Although the first few laser Raman spectrograms taken in those days were very inferior to those obtained with excitation using conventional mercury radiations and recorded photoelectrically, the pioneering experiments of Porto and Stoicheff laid the foundation for the first renaissance in Raman spectroscopy which followed immediately thereafter. Since 1963 with the rapid development of gas lasers (helium-neon, argon ion and helium-cadmium) and photomultiplier scanning monochromators excitation of Raman spectra by mercury radiations was slowly but steadily replaced by laser excitation. The change-over was complete by 1968. The development of more sensitive photomultiplier detecting and electronic recording devices and computer interfaces for integrating Raman data from weakly scattering substances revolutionised the practice of Raman spectroscopy. With the progressive innovations in the techniques of excitation and detection using lasers of low and high power, accurate data on frequency shifts and degree of polarisation of the electronic, vibrational, rotational, lattice and internal Raman spectra of many compounds and crystals including those of coloured and opaque substances like metals have been collected. The volume of sample required for study by Raman spectroscopy shrank from cc's to milli or even micro cc's and it became easy to record the spectra of solutions, single crystals, powders, films, fibres, etc. maintained at any desired temperature and pressure. The introduction of spinning or flowing sample arrangements extended the utility of Raman spectroscopy still further by minimizing sample heating and consequent decomposition. This quantum jump in the versatility of the technique constituted the first laser revolution; thereby, studies on Raman effect increased by leaps and bounds with regard to the range of coverage for tackling both fundamental and applied problems. This was the first renaissance in the history of Raman spectroscopy after a lapse of nearly forty years after its discovery.

The most startling developments during the laser era were the discoveries of new types of Raman excitations evisaged in the quantum theory, but not realised in the prelaser era. They are scattering by polaritons, magnons, plasmons, plasmaritons, Landau levels, excitons, soft modes connected with phase transitions, spin flip transitions, etc. and nonlinear scattering processes such as stimulated, hyper, inverse, coherent anti-Stokes and coherent Stokes Raman scattering and other higher order Raman spectral excitations. Continued technological innovation and the development of more powerful lasers, tunable dye lasers, time-resolved and space-resolved Raman spectrometers, multichannel analysers and Raman microprobes have led to a second renaissance in Raman effect studies in the seventies.

During the first thirty-five years from 1928 to 1962 nearly 6,000 papers were published on Raman effect, while after the induction of laser excitation in 1963, more than 19,000 papers have appeared up to date. These include original contributions, reviews and reports on all aspects of Raman scattering. An exhaustive year-wise bibliography of the 25,000 and odd papers published on Raman effect since the discovery up to 1979 has been prepared. It is hoped to publish the same in the near future with comprehensive author and subject indices. The collection of data for 1979 is not yet complete and hence the present analysis is restricted to the period 1962-1978.

4. Statistical Analysis

Table 1 gives an year-wise distribution of the number of papers published during each year arising from research on Raman effect. In the case of scientific journals, the official year of the respective volume is taken as the year of publication although in the case of some journals the publication has been delayed. In the case of papers presented at conferences and symposia, the year of publication of the proceedings is taken as the year for statistical analysis and for preparing year-wise bibliographies.

Table 1. List of papers published on Raman effect

Year	No.	Year	No.
1962	275	1971	1249
1963	298	1972	1354
1964	362	1973	1362
1965	403	1974	1516
1966	454	1975	1616
1967	528	1976	2184
1968	736	1977	1722
1969	756	1978	2136
1970	911	1979	~1700

In the first five years beginning from 1963, studies using laser excitation accounted for only 10, 40, 60, 100 and 155 papers out of the total 298, 362, 403, 454 and 528, respectively. The exceptional intensity, the high mono-chromaticity and the perfectly polarised nature of the laser radiations en-abled one to get very accurate data concerning the parameters of Raman lines for many of the substances already investigated in the prelaser era. Besides, during the first few years of the laser era a major part of the research on laser Raman spectroscopy was directed towards the study of stimulated Raman effect (SRS) which was first observed accidentally by Woodbury and Ng in 1962 while working with a Q-switched ruby laser. The correct interpretation of their observations was given a year later by Echard, et al. This type of excitation could be easily demonstrated with the aid of a powerful laser source such as a ruby laser or Nd-impregnated YAG or glass laser and hence this formed the subject of study for many in the first few years of the laser era. The number of yearly publications was of the order of 10, 30, 45, 65, 60 and 70 in the period from 1963 to 1968. Scientists working in USA and USSR laboratories had more or less a monopoly in this field. During this period this type of work was started in a small way in a few more countries.

The sudden spurt in the number of publications from 1968 onwards should be attributed to the availability of laser Raman monochromators with fast photon counting and electronic recording devices and with the added facility of computerisation of data. From 1968 many more organic and inorganic simple and complex compounds and a whole gamot of coloured and even opaque compounds were studied by laser Raman spectroscopy. International conferences on light scattering spectra of solids organised in 1968 and 1971 were also partly responsible for the increase in the number of publications in these years. The development of tunable dye lasers and the time-resolved and space-resolved Raman spectroscopy opened up a new vista of research on resonance Raman scat-tering in absorbing substances and biomolecules and on nonlinear phenomena such as CARS, HRE, IRE, etc. and the number of publications began to pile up very rapidly since 1971. This has led to second renaissance in the field of Raman spectroscopy during the period 1972-74. International Conferences on Raman spectroscopy held at Freiburg in 1976 and in Bangalore 1978 where about 300 papers and review articles were presented and the simultaneous publication of the Proceedings in the same years are responsible for the number of publi-cations exceeding 2,100 in these two years. In the year 1979 preliminary estimates indicate that the number of publications may be the same order as in 1977. This trend is expected to continue for some more years to come.

Another aspect of interest is the geographical distribution of research on Raman effect in the laser era. Starting with nine countries where work on Raman effect was started in 1928, the work was continued in 17 countries during the first year of the laser era (1963). It rose to 27 countries in 1968, to 39 in 1973 and over 60 in 1978. The same nine countries of the world still continue to lead in the programmes of research on laser Raman spectros-copy. The number of publications from the top 23 countries expressed as a percentage of the total for each year is listed in Table 2. As is to be ex-pected the United States of America is far ahead of others in the matter of research on Raman effect from 1965 onwards. This is because lasers and photo-electric recording Raman spectrometers were developed and manufactured in USA in a big way and were easily available to scientists working in this field in that country. France which occupied the second place in the pre-laser era was pushed to the third place during the laser era. France, Fed. Rep. of Germany and United Kingdom are nearly equal now in research output on Raman effect. India, where the discovery was made, was leading the coun-tries in the first few years after the discovery. It still continued to be one of the leading countries in the prelaser era and had many important achieve-ments to its credit. But with the sophistication in technique and the rising

Table 2. Country-wise distribution of papers in percentage

Year	1962	1963	1964	1965	1966	1967	1968	1969	1970	1971	1972	1973	1974	1975	1976	1977	1978
USA	15.3	22.1	28	33	29.9	32.9	35.4	32.5	36.4	39.4	33.8	32.3	29.4	27.8	29.1	28.1	27
USSR	32.8	34.9	34	22.3	22.2	21.2	17.2	16.1	11.6	10.9	11.2	14.9	19.3	17.4	13	15.5	13.3
France	5.8	8.1	7.5	8.0	7.5	10.4	7.4	7.5	11.1	11.6	10.2	8	8.1	10.7	10.6	8.4	11.4
Fed. Rep. of Germany	12.4	6.7	8.6	8	10.3	10	7.2	7.6	6.6	6	7.5	8.4	9.2	8.9	10.6	9.6	9.1
U.K.	7.6	8.4	4.7	8.3	7.8	9.6	15	14.6	10.4	10.9	13	9	8.4	9.9	10.1	9.3	8.9
Japan	7.27	1.68	4.15	3.75	4.83	3.25	3	2.63	3.08	2.8	3.77	4.7	5.73	5.22	5.75	7.44	6.48
Canada	6.55	3.0	3.04	1.75	1.98	3.25	4.63	5.13	8.35	6.89	5.1	5.07	5.46	4.03	4.32	3.93	4.33
India	7.27	7.38	3.97	7	5.5	1.89	3.27	3.55	2.62	3.67	3.55	3.61	2.67	2.11	1.79	1.91	3.41
Italy	1.09	1.34	1.1	0.5	1.76	1.89	1.63	1.45	1.21	1.04	1.55	1.54	1.2	1.72	1.97	1.67	1.75
Holland	0.73	-	0.55	0.5	0.22	0.19	0.4	1.18	1.21	1.96	1.11	1.47	1.32	1.32	1.19	1.56	1.41
Australia	0.36	1.34	0.83	0.5	0.22	0.75	0.95	1.45	0.77	0.8	0.22	0.95	0.53	0.53	0.41	0.23	0.62
Austria	-	-	-	0.75	-	-	-	0.39	0.55	0.16	0.29	0.37	0.20	0.46	0.32	0.23	0.73
Belgium	0.36	1.00	-	0.75	0.22	0.75	0.13	0.39	0.44	0.08	0.74	0.88	1.4	1.19	0.55	0.58	0.68
Brazil	-	-	-	-	-	-	0.27	0.13	0.11	0.08	0.22	0.59	0.92	0.79	1.39	1.09	1.36
Czechoslovakia	0.72	1.68	0.83	1.00	-	0.38	0.4	0.13	0.22	0.24	0.22	0.29	0.33	0.4	0.37	0.06	0.28
Denmark	-	-	-	0.25	-	1.9	0.4	0.13	0.33	0.4	0.59	0.51	0.66	0.86	0.69	0.62	1.17
G.D.R.	-	-	-	-	0.22	0.19	-	0.52	-	0.14	0.66	0.51	0.67	0.46	0.55	1.21	0.88
Finland	-	1.33	-	0.25	-	-	-	-	0.33	0.8	0.15	0.59	0.27	0.26	0.46	0.34	0.39
Israel	-	-	-	-	-	0.38	0.4	0.79	0.11	0.72	1.03	1.03	0.92	1.19	1.15	1.15	0.78
Norway	-	-	-	-	-	-	0.55	0.52	1.09	0.72	0.89	1.03	0.8	0.79	0.50	0.41	0.49
Poland	1.08	0.67	1.66	0.75	0.66	1.14	0.27	0.66	0.55	0.24	0.66	0.66	0.2	0.99	1.33	1.56	1.61
Switzerland	0.73	0.33	-	0.5	0.22	-	0.4	-	0.33	0.48	0.66	0.29	0.26	0.4	0.73	0.68	0.49
Yugoslavia	-	-	-	-	-	-	0.13	-	0.22	0.32	0.37	0.29	0.46	0.33	0.69	0.52	0.49

cost of laser Raman spectrometers India lagged behind to the 6th place in the sixties and 8th place in the seventies. This should be attributed to the nonavailability of sophisticated instruments to the Indian Raman spectroscopists.

From the statistical point of view the next aspect to be considered is the nature of the subjects covered by investigations using Raman spectroscopy. The subject-wise distribution of papers published during the period 1968 to 1978 is listed in Table 3. The abbreviations used in the table have the following connotations: *SRS* = Stimulated Raman Scattering; *RRS* = Resonance Raman Scattering; *SORS* = Second Order Raman Scattering; *CARS* = Coherent Anti-Stokes Raman Scattering; *ERS* = Electronic Raman Scattering; *HRE* = Hyper Raman Effect. Under the broad title "Organic" the following topics are included - aliphatic compounds, aromatic compounds, organic molecular crystals, organic hydrogen bonds, organic halides, organic boron, silicon, germanium, phosphorus, arsenic, sulphur and selenium compounds, organic metal halide complexes and other organometallic complexes. Under "Inorganic" the following categories of compounds are covered: inorganic general (gases, liquids and solids), inorganic crystals (pure, doped, with defects, and irradiated), inorganic halides, mixed halides, mixed and complex crystals, crystal melts (simple and complex), inorganic ions, alkali halides (pure, with defects, doped and irradiated), superconducting crystals, minerals, inorganic molecular crystals, elements, metals and hydrogen-bonded inorganic crystals. Under the head "Applications" are included gas diagnostics in flow and combustion chambers, atmospheric pollution, remote sensing, Raman lidar, chemical analysis and processes. Many papers have been enumerated under different key words or subjects.

Before the advent of lasers the subjects covered by studies using Raman effect could be broadly classified under twenty categories. In 1978 this number increased sixfold. Only twenty-seven topics which form the subjects of major activity on Raman effect are listed in the Table. The use of powerful laser beams for excitation paved the way for an unexpected expansion of the vista of research on Raman spectroscopy. Many new types of Raman excitations and nonlinear and higher order effects were discovered during the period 1964-1966 mainly from laboratories in the United States along with the development of lasers and detecting techniques. These include scattering by polaritons, magnons, plasmons, plasmaritons and spin flip and SRE, HRE, and CARS. In the case of resonance Raman spectroscopy leading theoretical and experimental contributions were from German and Russian laboratories. The extensive application of RRS to biomolecules was initiated in American laboratories.

In many subjects there is a progressive increase in the number of publications over the period analysed. This does not apply to subjects like SRS, ERS, HRE, RRS, polariton, magnon, spin flip and soft mode. Work in these fields started with the use of laser excitation and the number of publications began to increase up to the year 1970 and later on the activity remained nearly constant over the period from 1971 to 1978. Spectacular increases are noticed in the case of RRS, biomolecules, RRS of biomolecules, phase transition, ferroelectric crystals, applications and adsorption. Water and ice continue to be of special interest of study using Raman effect. At least a dozen papers appear every year. It is the nearest known universal solvent and it exhibits a high degree of intermolecular coupling due to hydrogen bonding and plays an important role in the biological systems.

The present exciting fields are surface scattering and scattering from adsorbed molecules. Work on these topics and also on resonance Raman spectra of biomolecules and polymers and on technological applications such as temperature and concentration measurements in combustion chambers, in jet flows and in polluted atmospheres, medical diagnostics, etc. are expected

Table 3. Subject-wise classification

Subject	1968	1969	1970	1971	1972	1973	1974	1975	1976	1977	1978
Inorganic	297	293	410	400	440	497	485	505	735	565	675
Organic	244	237	295	328	398	391	395	435	551	515	612
Semi Conductor	27	28	28	65	57	56	81	68	131	65	95
SRS	72	51	51	60	59	46	59	72	64	75	58
RRS	9	11	24	32	52	70	89	126	235	141	221
Biomolecules	2	3	20	19	36	75	52	84	129	93	145
Phase Trans.	9	12	15	27	19	24	27	45	110	58	73
Ferro-electric	13	13	20	33	28	32	35	39	65	28	45
Polymer	13	17	23	28	24	33	25	33	38	44	63
Temperature	19	22	22	12	23	38	47	53	67	55	54
Pressure	5	6	7	15	16	24	25	31	54	36	49
SORS	17	9	9	9	9	20	24	19	52	24	32
CARS	-	-	-	-	4	3	13	8	39	46	64
ERS	13	9	11	4	12	10	12	13	16	17	20
HRE	-	-	-	3	8	1	5	6	11	12	20
Polariton	5	11	12	14	28	19	30	33	46	21	21
Magnon	12	6	14	19	13	16	13	9	29	15	11
Adsorption	1	-	7	4	5	4	7	6	22	18	44
Surface	4	-	-	-	3	-	-	16	23	14	10
Spin Flip	2	4	5	16	20	10	16	13	17	6	14
Matrix Isomers	-	1	2	14	18	14	8	15	31	18	27
Soft Mode	4	2	5	-	-	-	17	20	30	13	22
Liquid Crystal	1	1	2	8	9	10	10	13	13	7	14
Review	23	31	33	62	70	65	43	61	91	102	120
Theory	52	46	56	83	74	79	100	89	152	90	111
Applications	5	6	18	20	37	36	70	62	75	71	93

to increase rapidly in the near future. The nonlinear coherent scattering phenomena such as *CARS* (Coherent Anti-Stokes Raman Spectroscopy), *CSRS* (Coherent Stokes Raman Spectroscopy), *RIKES* (Raman-Induced Kerr Effect) and *HORSES* (Higher Order Raman Excitation Studies) which involve basically four photon parametric mixing processes, are expected to get increasing attention. Of these, work on CARS and its application would open up a new vista of investigations. Time-resolved and space-resolved Raman spectroscopy with the pulsed sources of picosecond duration and the Raman microprobe developed by the French group are expected to play an important role in the study of fast chemical reactions, photochemical processes, phase transitions in solids, analysis of heterogeneous samples, biological processes, medical diagnostics, etc.

The number of journals in which papers on Raman effect appeared in 1928 was 15 covering mainly physics and chemical physics. It rose to 72 covering nearly the same subjects in 1962. In 1978 the papers on Raman effect appeared in over 140 journals covering all branches of science. More than 80 books have been published so far.

Although laser Raman spectroscopy has become an important and essential tool for solving problems in many branches of science and technology, one finds that studies using modern Raman spectroscopic techniques have become very costly over the years and are therefore beyond the means of most laboratories in the developing countries.

Acknowledgement

The authors are grateful to the Department of Science and Technology, Government of India for financial assistance.

Relaxation Mode in SrTiO$_3$:
A Mode to Test Melting Models?

G.A. Barbosa and J.I. Dos Santos[1]

Departamento de Física - ICEx, Universidade Federal de Minas Gerais
Belo Horizonte, MG - Brazil

Among the problems which Professor S.P.S. Porto had a special delight in dis-
cussing with his characteristically contagious enthusiasm, were those related to perovskites crystals, with their anomalous properties in the dielectric constant and Raman spectra.

These discussions cropped up quite frequently in his laboratories and
almost everyone of his students and co-workers had a bit of participation
in them. The difficulties inherent to those problems forced Porto to present
them in a quasi-cyclic fashion, as in Gabriel García Márquez's "Cien Años de
Soledad" — a mark of his perseverance.

And the wheel still revolves.

Abstract

In this work, Raman spectra are presented from room temperature up to 800°C.
Each spectrum is seen as a contribution of two processes — a "one-phonon"
disordered process and a relaxation mode. The relaxation time associated with
this mode shows a linear decrease with increasing temperature, going to zero
as the melting point is approached.

Raman Scattering is recognized as an effective probe to reveal disorder
manifestations in crystals as the wave vector $\vec{K}$ conservation rule is relaxed
producing deviations of the expected group theory predictions for the normal
modes.

In the perovskites, strong bands appear in the Raman spectra in all phases.
For BaTiO$_3$ and SrTiO$_3$ with a phase transition tetragonal-cubic, these bands
show no discontinuity in intensity [1] as the phase transition temperature
is crossed.

FONTANA and LAMBERT [2] called upon the disorder model proposed by COMES
et al. [3] as an explanation for another anomalous behavior of these bands
in BaTiO$_3$; they seem to present the scattering intensity decreasing with increasing temperature above 150°C. This behavior is unexpected for two-phonon
or one-phonon processes [4]. They took these bands as "first-order" phonons
produced by the disorder, these phonons being forbidden for an ordered crystal in the O$_h$ cubic phase. They also proposed that above 300°C, thermal agi-
tation destroys the disordered configuration forcing the bands, which are
now forbidden to disappear.

Other authors [5] called attention to the fact that in these crystals at
high temperatures care should be taken to compensate for light absorption

---

[1] Present address: Dept. de Física, Universidade Federal de Goiás, GO,
Brazil

processes inside the crystals, Indeed, they showed that for a particular BaTiO$_3$ sample up to 300°C and for another one of SrTiO$_3$ up to 580°C, after corrections for light absorption, no intensity decrease was manifested.

Further measurements [6] in BaTiO$_3$ supported the main results presented in [2].

In this work Raman spectra were taken in SrTiO$_3$ from room temperature up to 800°C. Several precautions were taken: 1) the light-absorption coefficients used were obtained for same sample [7] used in the Raman Scattering experiments; 2) a fixed geometry was used and a careful calculation compensates for the laser and scattered light absorptions; 3) a backward scattering geometry was chosen to achieve good scattered light levels even at high temperatures.

A Krypton laser was used as the exciting source ($\lambda = 6471$Å). A Spex double monochromator and photon counting techniques were used. The furnace had a temperature control within 3°C. At high temperatures the furnace light emission was determined by turning off the laser and running the spectrometer in the same wavelength range and these results subtracted form the obtained Raman spectra.

The obtained spectra are presented in Figs. 1a and b. Curve A in Fig. 2 shows the total integrated intensity as function of temperature showing a clear decrease above 600°C, thus supporting FONTANA and LAMBERT's idea [2] also for SrTiO$_3$.

Furthermore, the spectra shape at the higher temperatures strongly indicates the existence of a relaxation process as a background for the "one-phonon" bands (see dashed curves in Fig. 1). Curve B and C in Fig. 2 shows the obtained intensities below and above the dashed curves of Fig. 1. In this interpretation it is seen that the "one-phonon" process shows an almost monotonously decreasing intensity with increasing temperature. The relaxation

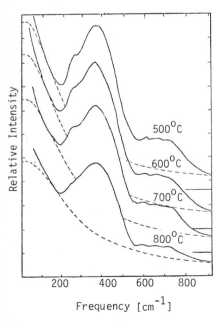

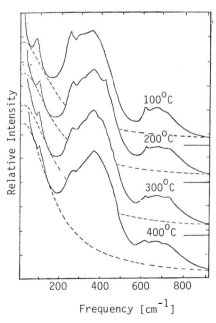

Fig. 1 (a). Raman Spectra up to 400°C

Fig. 1 (b). Raman Spectra up to 800°C

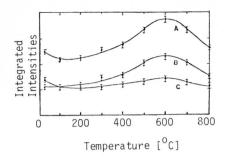

Fig. 2. Integrated intensities in function of temperature, compensated for
the light absorption

Temperature [°C]

intensity increases up to 600°C and then decreases. Fig. 3 shows the relaxation time as function of temperature.

It is then proposed that, in this crystal, the disorder manifests itself
with the forbidden "phonons" as well as in this relaxation process. This
Debye-like feature might be the evolution of the broad wing related by FIR-
STEIN, BARBOSA and PORTO on the 110°K phase transitions of $SrTiO_3$ [8] and
later by LYONS and FLEURY [9].

The linear variation of the relaxation time τ with temperature can be
fitted for the measured interval as $\tau \cong (4.7 - 2.4 \times 10^{-3}\ T) \times 10^{-14}$ s showing,
by extrapolation, $\tau \to 0$ as the crystal goes to the melting temperature.

As single electrons cannot be thermally excited at these temperatures to
produce excitations [10] with such short lives, the physical interpretation
for this mode should rely on the motion of heavy ions. A rattling motion in
a loose potential, with increasing amplitude as the temperature increases,
would produce a relaxation mode with an increasingly faster decay due to the
interactions with the surrounding ions. The indication $\tau \to 0$ as $T \to T_M$ very
closely associates this mode with a melting mechanism.

Acknowledgments. This research was supported by FINEP (Financiadora de Estudos e Projetos) and CNPq (Conselho Nacional de Desenvolvimento Científico e
Tecnológico).

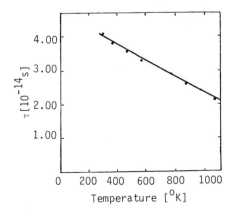

Fig. 3. Relaxation time in function of temperature

$\tau \simeq (4.7 - 2.4 \times 10^{-3}\ T) \times 10^{-14}$ s

References

1. W.G. Nilsen, J.G. Skinner: J. Chem. Phys. *48*, 2240 (1968)
2. M.P. Fontana, M. Lambert: Sol. State Commun. *10*, 1 (1972)
3. R. Comes, M. Lambert, A. Guinier: J. Phys. Soc. Jpn. *28*, 195 (1970)
4. H. Poulet, J.P. Mathieu: *Spectres de Vibration et Symétrie des Cristaux* (Gordon & Breach, London 1970);
   A.S. Barker, R. Loudon: Rev. Mod. Phys. *44*, 18 (1972)
5. G.A. Barbosa, A.S. Chaves, S.P.S. Porto: Sol. State Commun. *11*, 1053 (1972)
6. A.M. Quittet, M. Lambert: Sol. State Commun. *12*, 1053 (1953)
7. G.A. Barbosa, R.S. Katiyar, S.P.S. Porto: J. Opt. Soc. Am. *68*, 610 (1978)
8. L.A. Firstein, G.A. Barbosa, S.P.S. Porto: Proceedings of the Third International Conference on Light Scattering in Solids (1975), ed. by M. Balkansky, R.C.C. Leite, S.P.S. Porto (Flammarion, Paris 1976) p. 866
9. K.B. Lyons, P.A. Fleury: Phys. Rev. Lett. *37*, 161 (1976); J. Appl. Phys. *47*, 4898 (1976); Sol. State Commun. *23*, 477 (1977)
10. P.M. Platzman: Phys. Rev. *139*, A379 (1965);
    P. Mooradian: Phys. Rev. Lett., 1102 (1968)

Raman Scattering in Superconductors

M.V. Klein

Department of Physics and Materials Research Laboratory
University of Illinois at Urbana-Champaign
Urbana, IL 61801, USA

1. Introduction

Two types of Raman scattering are possible in a superconductor--excitation
of optical phonons and excitations of quasi-particles across the superconducting energy gap. Phonon Raman scattering, which can also be observed
in normal metals, is of interest because of the information it gives about
the electron-phonon interaction, including possible effects of that interaction to produce "anomalous dispersion" of optical phonons near zero wave
vector [1].

The energy gap Δ plays a key role in the modern theory of superconductivity [2]. For a superconductor at zero temperature an energy of at
least 2Δ is necessary to create an excitation, in this case a pair of quasi-
particles. The most common probe of these pair excitations is probably
far infrared absorption, which shows a continuous rise from zero at a
frequency of 2Δ. For almost 20 years there have been theories of electronic Raman scattering from such excitations. They assume the inequality
$2\Delta \ll qv_f$, where q is the wave vector transferred and v_f is the Fermi
velocity. At zero temperature the scattering intensity is predicted to
rise from zero at a frequency shift of 2Δ either discontinuously [3-6]
or continuously [7], depending upon the assumed details of the coupling
of the electrons to light. The intensity should then asymptotically
approach the value expected for free electrons, which is proportional to
the frequency shift ω.

There is a reported observation of such an effect by PORTO's group made
on Nb_3Sn under extremely difficult conditions of surface quality [8].
To my knowledge this result has not been reproduced in almost ten years,
and I think it fair to say that there has not yet been a clean observation
of the superconducting energy gap via direct coupling of light to the
electrons.

Our group at Illinois has made a clean observation of the gap by Raman
scattering in the layered compound $2H\text{-}NbSe_2$ [9]. The experimental facts
[10], including some new ones to be mentioned here, suggest strong coupling of the Raman gap excitations to a Raman-active optical phonon induced
by a charge-density-wave phase transition. The results can be understood
qualitatively in terms of a model in which the phonon has all the Raman
activity and "lends it" to the gap excitations via electron-phonon coupling [11].

Because optical phonons and their coupling to electrons are so important in understanding the superconducting gap excitations, I shall first discuss Raman scattering from phonons and why the predicted anomalous dispersion has not been observed.

2. Phonons

An interesting example of phonon Raman scattering in a complex metal, which happens to be a high temperature superconductor, is provided by the A15 compound V_3Si [12]. This cubic material owes its superconducting properties at least in part to the high density of states of d electrons from the transition metal atoms, which form nearly 1-dimensional chains along the three cube axes. Of the several $q \approx 0$ optical phonon modes, two are Raman active, but only the E_g mode should couple to the d electrons [13]. Figure 1 shows the Raman spectrum for this mode. Its asymmetric line shape

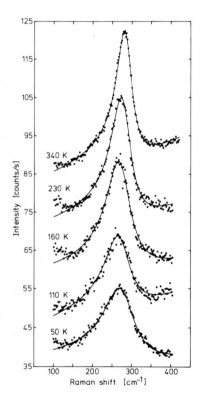

Fig.1 The E_g optical phonon in V_3Si [12]; curves are shifted vertically for clarity

is unusual, as is the increasing width with decreasing temperature and the slight softening following by hardening, which may be related to a martensitic transformation at 20K. The solid lines represent fits to a Lorentzian multiplied by an antiresonance factor of the form $(\omega^2 - \omega_a^2)^2$, where ω_a is the antiresonance frequency. We believe that the phonon is

broadened and renormalized by electron-phonon coupling and that the antiresonance is due to simultaneous scattering of light from the phonon and from
a continuum of electronic excitations. The electron-phonon coupling
constant obtained from these fits can quantitatively predict the superconducting transition temperature if this phonon is assumed to be a "typical"
phonon [14].

We do not see evidence for "anomalous dispersion" in these data or in
similar results in other metals. The signature for this effect is an
unusual line shape that changes with the penetration depth and hence with
the wavelength of the laser light used to excite the spectra. In a Raman
experiment in a metal, q_z, the component of the wave vector transferred
normal to the surface, is not well defined, and one must average what would
otherwise be the spectrum for fixed q_z over a distribution of q_z characterized by the inverse of the optical penetration depth δ [15]. This will
affect the measured spectrum if the phonon frequency ω_0 has dispersion
over the wave-vector range from zero to δ^{-1}. In metals with a simple Fermi
surface at small q, namely when $q < \omega_0/v_f$, electron-phonon vertex corrections
are necessary for the calculation of the renormalized phonon frequency,
resulting in strong dispersion [1].

The Fermi surface of V_3Si is not simple. The d-electron bands are quite
flat, and they are forced by symmetry to become degenerate at high-symmetry
points of the Brillouin zone, such as the center of the face or the zone
corner [16]. Whereas in a simple metal an optical phonon can give up
energy only by intraband "Landau damping", in V_3Si and other complex metals
damping can occur by interband electronic transitions. For them there is
no anomaly when the wave vector of the phonon goes to zero.

This reasoning also explains the discrepancy found by GRANT et al [17]
between the Raman linewidth observed in hexagonal close packed (hcp)
metals like Mg and Zn and estimates made using the anomalous dispersion
theory of IPATOVA et al [1]. Anomalous dispersion predicted a line width
considerably wider than the measured width. The hcp structure forces the
energy bands to be degenerate in pairs at the top and bottom of the zone
(if spin-orbit coupling is neglected), and the Fermi surface passes through
these faces [18]. The Raman-active optical phonon has E_{2g} symmetry, and
it can mix the nearly degenerate bands near the intersection of the Fermi
surface with the zone faces, thus providing interband decay channels for
the phonon.

The phonon Raman spectra of 2H-NbSe$_2$ are shown in Fig.2 [19]. This
layered compound has the same space group as that of the hcp lattice. The
Raman selection rules for light polarized in the basal place are (xx) for
A_{1g} and both (xx) and (xy) for E_{2g}. The A_{1g} spectrum is therefore obtained
by subtracting the (xy) spectrum from the (xx) spectrum. The Nb atoms
form an hcp lattice with each plane sandwiched between two planes of Se
atoms. The small E_{2g} peak at 28 cm^{-1} is the "interlayer mode" in which
alternating Se-Nb-Se sandwiches vibrate against each other in the basal plane.
The peaks near 40 cm^{-1} are associated with charge-density-waves and will be
discussed below. For the E_{2g} mode at 248 cm^{-1} the Nb atoms in alternating
layers vibrate in the basal plane against each other and also against the
Se atoms in the same layer. For the A_{1g} mode at 233 cm^{-1} the Nb atoms
remain at rest at the center of each sandwich, whereas the outer Se layers
of each sandwich vibrate against each other along the c axis. This latter
mode will not split any electronic degeneracies required by symmetry. The

E_{2g} modes will split the degeneracies at the top and bottom of the Brillouin zone and hence will have more interband decay channels. This may explain why the 248 cm^{-1} mode is broader than the 233 cm^{-1} mode (although in 2H-TaSe$_2$ the analogous modes have the same width [20]).

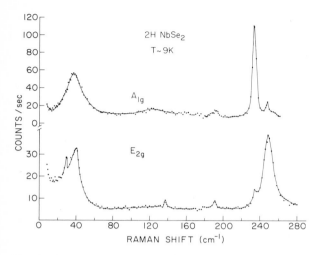

Fig.2 Raman spectrum of a high-quality sample of 2H-NbSe$_2$ at 9K

3. Charge Density Waves

Neutron diffraction studies by MONCTON et al [21] show that 2H-NbSe$_2$ undergoes a phase transition from a normal lattice to one with an incommensurate charge-density-wave (CDW) at the onset temperature T_d of 33K. Three wave vectors Q_j are simultaneously present in the CDW, lying 120° apart in the basal plane along the ΓM symmetry directions. Their length is about 0.98 of the value $[(2/3)(\Gamma M)]$ that would give a commensurate $3a_0 \times 3a_0$ superlattice. Elastic modulus measurements show that the incommensurate phase persists at least as low as 1.3K [22]. Above T_d longitudinal acoustic phonons of wave vector Q_j are observed to soften. Below T_d these phonons "condense" into a static distortion, about which new quasi-harmonic vibrations are possible. The peaks near 40 cm^{-1} in Fig.2 are "amplitude modes" of the CDW wherein the amplitude of the components of the distortion at each of the three Q_j varies either symmetrically in all of the Q_j (A_{1g}) or antisymmetrically in, say, Q_1 and Q_2 (E_{2g}). Modulations in phase ("phasons") are also possible. In the harmonic limit, they have zero frequency at zero wave vector for an incommensurate distortion. In higher order this is no longer true [23]. In the extreme anharmonic (soliton) limit there is both a zero frequency phason branch and one at finite frequency. We are apparently not observing any phasons in 2H-NbSe$_2$, and will ignore them in the following discussion.

4. Gap Excitations in Superconducting NbSe$_2$

Below 7.2K 2H-NbSe$_2$ becomes a highly anisotropic type II superconductor [24].
Raman spectra when the sample used for Fig.2 is immersed in superfluid He
are shown in Fig.3 [9,10]. Two new peaks are seen: one of A_{1g} symmetry at
19 cm^{-1} and an E_{2g} peak at 15.5 cm^{-1}. Their weighted average agrees with
the position of the peak in far infrared transmission observed by CLAYMAN
and FRINDT [25].

We have studied approximately 10 samples of 2H-NbSe$_2$. The one used
for Figs.2 and 3 gave the strongest and narrowest CDW Raman peaks. Other
samples gave broader CDW peaks, particularly in the E_{2g} spectrum, but the
peaks at the gap were only slightly affected [10]. Changes in the technique
of crystal growth can vary the presence of non-magnetic impurities,
which tend to inhibit the formation of CDW's [16], whereas they have a
small effect on superconductivity [27]. Raman measurements have been made
on several samples studied by HUNTLEY and FRINDT [16] which have no CDW's
but which still are superconductors below 7K [28]. The spectra below
60 cm^{-1} show no CDW phonon peaks and no superconducting gap peaks.
This last and very recent result had been predicted theoretically by
BALSEIRO and FALICOV [11] before the measurements were made.

Somewhat older results proving coupling between CDW phonons and gap
excitations are presented in Fig.4 [10], which shows that a magnetic field
completely suppresses the gap peak and transfers its strength to the CDW
phonon. The upper critical field H_{c2} for H perpendicular to the layers is
about 40kG at 2K [24]. It is seen that suppression is complete for
$H \approx H_{c2}/3$. This result holds for both Raman symmetries and for H parallel
and perpendicular to the layers.

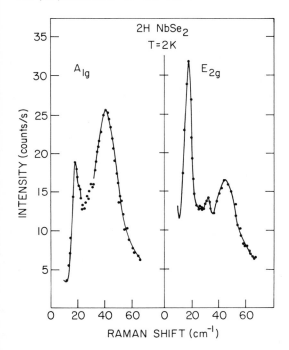

Fig.3 Raman spectrum of the
sample in Fig.2 when immersed
in superfluid He

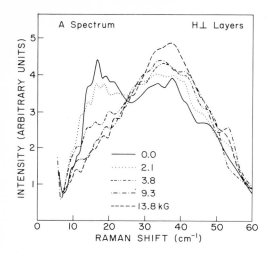

Fig.4 The A_{1g} Raman spectrum of a $NbSe_2$ sample (different from that in Figs.2, 3) at 2K as a function of an applied magnetic field perpendicular to the layers

5. Theory of the Raman-Active Gap Excitations

The shape of the observed gap peaks and their coupling to CDW phonons suggests that previous theories [3-7] of Raman scattering by gap excitations do not apply in 2H-$NbSe_2$. The direct scattering of light by the electrons is too weak to be observed in these experiments. Indirect coupling is assumed in the theory of BALSEIRO and FALICOV [11]. There are four implicit or explicit assumptions in their theory: Assumption 1: A Raman-active phonon lies close in energy to the superconducting gap 2Δ. Assumption 2: The phonon couples to the electrons via the usual type of electron-phonon coupling

$$H' = \sum_{k,\sigma} g_k (b_q + b^+_{-q}) c^+_{k+q,\sigma} c_{k,\sigma} \cdot \tag{1}$$

Here g_k is the electron-phonon coupling constant, q is the phonon wave vector, σ is the spin index, b_q and $c_{k,\sigma}$ are phonon and electron destruction operations, respectively. Assumption 3: The renormalized phonon spectrum is calculated using the electron "vacuum polarization" diagram for BCS electrons. This leads to a phonon self-energy for frequency ω at zero temperature proportional to

$$\Sigma_k \, g_k^2 \left(1 - \frac{\varepsilon\varepsilon' - \Delta^2}{EE'} \right) \left(\frac{1}{\omega^+ - E-E'} - \frac{1}{\omega^+ +E+E'} \right) \tag{2}$$

where $\omega^+ = \omega + i0^+$, $E = (\varepsilon^2 + \Delta^2)^{1/2}$, $E'= (\varepsilon'^2 + \Delta^2)^{1/2}$, $\varepsilon = \varepsilon_k$, and $\varepsilon' = \varepsilon_{k+q}$, the ε's being normal electron excitation energies relative to the Fermi surface. Assumption 4: The small q limit holds, i.e.,

$$\underline{q} \cdot \underline{v}_k \ll 2\Delta \tag{3}$$

where $v_k = \partial\varepsilon/\partial k$. Then $\varepsilon = \varepsilon'$ and $E = E'$, and the phonon self-energy becomes

$$\omega_o^2 <g^2\rho> \Delta^2 \int_{-\infty}^{\infty} \frac{d\varepsilon}{\sqrt{\Delta^2 + \varepsilon^2}} \quad \frac{1}{[(\omega^+/2)^2 - \Delta^2 - \varepsilon^2]} . \qquad (4)$$

Here $<g^2\rho>$ is the average of g_k^2 times the local density of states averaged over the Fermi surface. If g_k is assumed to be a constant g we find

$$<g^2\rho> = g^2\rho_o \qquad (5)$$

where ρ_o is the usual density of normal electrons per spin at the Fermi level. This is the form assumed by BALSEIRO and FALICOV.

The real part of (4) has an inverse square root singularity when ω approaches 2Δ from below. If the bare phonon is assumed to be undamped and have the frequency ω_o, this leads to a pole in the phonon Green's function at a frequency $\omega_p < 2\Delta$ that is the solution of [11]

$$\omega_p^2 = \omega_o^2 - \frac{16\,\omega_o\Delta^2<g^2\rho>}{\omega_p[4\Delta^2-\omega_p^2]^{\frac{1}{2}}} \tan^{-1} \frac{\omega_p}{[4\Delta^2-\omega_p^2]^{\frac{1}{2}}} \qquad (6)$$

A solution exists for all values of the coupling constant g. The self-energy (4) also leads to phonon damping when $\omega \sim \omega_o > 2\Delta$ and renormalization of the phonon peak near ω_o to a higher value.

An example of the resulting phonon spectral density is shown in Fig.5 [11] for a coupling constant $<g^2\rho> = 0.12 \; \hbar\omega_o$. From the residue of the pole (6) the strength of the "gap mode" below 2Δ can be determined. It is indicated in Fig.5 along with that of the renormalized phonon.

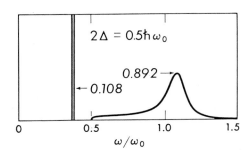

$2\Delta = 0.5\hbar\omega_0$

$0.892 \rightarrow$

$\leftarrow 0.108$

0

0.5

1.0

1.5

ω/ω_0

Fig.5 Phonon spectral density for the coupled modes [11]

The CDW phonon will have an intrinsic width in the absence of electron-phonon interactions due to anharmonicity. It may also be inhomogeneously broadened due to interactions of impurities with the CDW. BALSEIRO and FALICOV chose an inhomogeneous distribution of ω_0's to fit the A_{1g} spectrum of Fig.2. The fit is shown in Fig.6(a). They then calculated the resulting phonon spectral function in the superconducting state. The result for $<g^2\rho> = 0.08 \; \hbar\omega_0$ is shown in Fig.6(b). The experimental A_{1g} spectrum of Fig.3 is reproduced in Fig.6(c). Qualitatively there is good agreement.

A similar qualitative result is obtained if lifetime broadening is assumed for the bare phonon.

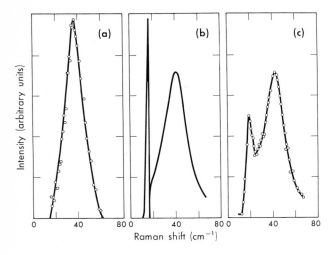

Fig.6 Phonon spectral density for the coupled modes when the phonon is inhomogeneously broadened to fit the data of Fig.3 [11]

The small q limit (3) warrants further discussion. In the pseudo-back-scattering geometry used for the experiments q will have two components: a fixed $q_\parallel \approx 1.2 \cdot 10^5$ cm^{-1} in the basal plane equal to the parallel component of the incident photon wave vector and a variable q_z, as discussed above, of order of the inverse skin depth $\delta^{-1} \approx 5.10^5$ cm^{-1} [29]. Calculations of the conduction band give it a width of about 1 eV [30,31], leading to an estimate of $v_\parallel$, the Fermi velocity in the basal plane of $1.6 \cdot 10^7$ cm/s and an estimate of $q_\parallel v_\parallel$ of about $2 \cdot 10^{12} \approx 10$ cm^{-1}. In the other direction one might estimate $q_z v_z \approx 60$ cm^{-1}. These represent overall averages. From (2) it can be seen that one needs to know the Fermi surface properties where the coupling constant g_k is large. For charge-density-wave phonons this will be near the intersections of the high-temperature Fermi surface with itself shifted by the wavevectors Q_j. These regions depend on the extent of "nesting" and are not really well known. For instance the band structure calculated in [30-31] is unable to account for the observed Landau quantum oscillations in 2H-NbSe$_2$ [32]. As with latter measurements, CDW effects are sensitive to subtle changes in the Fermi surface. We conclude that not enough is known to prove the inequality (3), but it is probably not grossly violated.

A calculation of the phonon spectral density has been made by SCHUSTER in the other limit, namely $qv_f \gg 2\Delta$ [33]. In this limit and with a simple Fermi surface the inverse square-root singularity of the real part of the phonon self-energy in (6) becomes merely a logarithmic singularity. Strong modifications of the spectrum are obtained only when 2Δ is close to ω_o; for $2\Delta \sim \omega_o/2$ they are essentially negligible (1% effects). It appears that the experiments require the small q limit if this model is to work.

6. Final Remarks and Conclusions

In many normal metals with complex band structures the electron-optical phonon interactions are in a sense simpler than predicted by quantum field theory; the complexity of anomalous dispersion and vertex corrections is prevented by the complexity of the Fermi surface. This remark may also apply to the renormalization of the CDW phonons in $NbSe_2$ by superconducting electrons.

The superconducting gap has probably not yet been observed directly in a Raman experiment, but it is clearly seen in $NbSe_2$ via coupling to CDW phonons. Other low-lying Raman-active phonons should show the same effect provided electron-phonon coupling is strong enough [11], and provided kinematical conditions give the small q limit. For the 28 cm^{-1} E_{2g} interlayer mode in Figs.2 and 3 we see no evidence of coupling to superconducting electrons--apparently at least one of these latter two conditions is not met.

Acknowledgments The Raman experiments on $NbSe_2$ were done in collaboration with R. Sooryakumar, who showed patience, energy, and skill in obtaining the data and in helping to explain it. We thank S. F. Meyer, F. J. DiSalvo, R. V. Coleman, and R. F. Frindt for providing crystals and J. P. Wolfe for the use of a cryostat for the magnetic field-dependent studies. Thanks go to W. L. McMillan and John Bardeen for theoretical discussions, and to L. M. Falicov for many conversations, for his patient explanation of the theory of Ref.11 and for permission to present it here in Figs.5 and 6. This work was supported by the National Science Foundation under MRL Grant DMR-77-23999.

References

1. I. P. Ipatova, A. V. Subashiev, A. A. Maradudin: In Light Scattering in Solids, ed. by M. Balkanski (Flammarian Sciences, Paris, 1971), p. 86; I. P. Ipatova, A. V. Subashiev: Zh. Eksp. Teor. Fiz. 66, 722 (1974) [Sov. Phys. JETP 39, 349 (1974)].
2. J. Bardeen, L. N. Cooper, J. R. Schrieffer: Phys. Rev. 108, 1175 (1975).
3. A. A. Abrikosov, L. A. Falkovskii: Zh. Eksp. Teor. Fiz. 40, 262 (1961) [Sov. Phys. JETP 13, 179 (1961)].
4. S. Y. Tong, A. A. Maradudin: Mat. Res. Bulletin 4, 563 (1969).
5. D. R. Tilley: Z. Phys. 254, 71 (1972); J. Phys. F 38, 417 (1973).
6. A. A. Abrikosov, V. M. Genkin: Zh. Eksp. Teor. Fiz. 65, 842 (1973) [Sov. Phys. JETP 38, 417 (1974)].
7. C. B. Cuden: Phys. Rev. B13, 1993 (1976); Phys. Rev. B18, 3156 (1978).
8. L. M. Fraas, P. F. Williams, S. P. S. Porto: Solid State Commun. 8, 2113 (1970).
9. R. Sooryakumar, D. G. Bruns, M. V. Klein: In Light Scattering in Solids, ed. by J. L. Birman, H. Z. Cummins, and K. K. Rebane (Plenum, N.Y., 1979), p. 347.
10. R. Sooryakumar, M. V. Klein: Phys. Rev. Letters, to be published.
11. C. A. Balseiro, L. M. Falicov: to be published.
12. H. Wipf, M. V. Klein, B. S. Chandrasekhar, T. H. Geballe, J. H. Wernick: Phys. Rev. Letters 41, 1752 (1978).
13. L. J. Sham: Phys. Rev. B6, 3584 (1972); J. Noolandi, L. J. Sham: Phys. Rev. B8, 2468 (1973).
14. R. Merlin, S. Dierker, M. V. Klein: unpublished work.

54

15. D. L. Mills, A. A. Maradudin, E. Burstein: Ann. Phys. (N.Y.) 56, 504 (1970).
16. B. M. Klein, L. L. Boyer, D. A. Papaconstantopoulos, L. F. Mattheis: Phys. Rev. B18, 6411 (1978).
17. W. B. Grant, H. Schulz, S. Hüfner, J. Pelzl: Phys. Stat. Sol (b) 60, 331 (1973).
18. J. B. Ketterson, R. W. Stark: Phys. Rev. 156, 751 (1967); W. Harrison: Phys. Rev. 118, 1190 (1960).
19. R. Sooryakumar, M. V. Klein: unpublished work.
20. John A. Holy: Ph.D. Thesis, University of Illinois at Urbana-Champaign (1977), unpublished.
21. D. E. Moncton, J. D. Axe, F. J. DiSalvo: Phys. Rev. B16, 801 (1977).
22. M. Barmatz, L. R. Testardi, F. J. DiSalvo: Phys. Rev. B12, 4367 (1975).
23. W. L. McMillan: Phys. Rev. B16, 4655 (1977); A. D. Bruce, R. A. Cowley: J. Phys. C11, 3609 (1978).
24. P. deTrey, Suso Gygax, J. P. Jan: J. Low Temp. Phys. 11, 421 (1973).
25. B. P. Clayman, R. F. Frindt: Solid State Commun. 9, 1881 (1971).
26. D. J. Huntley, R. F. Frindt: Can. J. Phys. 52, 861 (1974).
27. D. J. Huntley: Phys. Rev. Lett. 36, 490 (1976); J. R. Long, S. P. Bowen, N. E. Lewis: Solid State Commun. 22, 363 (1977).
28. R. Sooryakumar, M. V. Klein, R. F. Frindt: unpublished work.
29. R. T. Harley, P. A. Fleury: J. Phys. C12, L863 (1979).
30. L. F. Mattheis: Phys. Rev. B8, 3719 (1973).
31. G. Wexler, A. M. Wooley: J. Phys. C9, 1185 (1976); N. J. Doran, B. Ricco, D. J. Titterington, G. Wexler: J. Phys. C11, 685 (1978).
32. J. E. Graebner, M. Robbins: Phys. Rev. Lett. 36, 422 (1976).
33. H. G. Schuster: Solid State Commun. 13, 1559 (1973).

Enhanced Raman Scattering of Molecules Adsorbed on Ag, Cu and Au Surfaces

R.K. Chang, R.E. Benner,[1] R. Dornhaus,[2] and K.U. von Raben

Yale University, Department of Engineering and Applied Science
New Haven, CT 06520, USA, and

B.L. Laube

United Technologies Research Center, East Hartford, CT 06108, USA

Introduction

Raman spectroscopy has proved to be a valuable in situ technique for chemical speciation under conditions which are not amenable to the use of other diagnostic tools. S. P. S. PORTO recognized the potential of the method and completed pioneering work in this field by determining Raman cross sections and molecular symmetry for numerous molecules [1-4]. Until recently, however, spontaneous Raman spectroscopy was thought to lack sufficient sensitivity for chemical identification of monolayer adsorbates because of the smallness of the Raman cross sections under nonresonant conditions. Thus, the initial report of intense Raman scattering from pyridine adsorbed on Ag in an electrochemical cell [5] and the subsequent estimate [6,7] of the enhancement factor of $10^5 - 10^6$ compared to pyridine in the electrolyte stimulated great excitement. Recent investigations on surface enhanced Raman scattering (SERS) have been aimed at determining the physical mechanisms responsible for SERS, as well as at establishing the generality of the effect and the extent to which it is limited by adsorbates, substrates, and the surrounding environment. Since several reviews on SERS have been published [8-12], this paper summarizes, in the context of current theoretical models and experimental findings from other laboratories, our results on SERS from metal-cyanide complexes adsorbed on metal electrodes (Ag, Pt, Cu, and Au) and on Au colloids, as well as from isonicotinic acid adsorbed on continuous and discontinuous Ag films evaporated on a prism substrate.

Electrochemical Environment

The electrochemical configuration consisted of a mechanically polished (0.3 μm Linde B abrasive) polycrystalline metal (Ag, Pt, Cu, or Au) working electrode, a Pt wire loop counter electrode, and a saturated calomel reference electrode (SCE). A voltammogram recorded at 50 mV/sec for the Ag electrode immersed in an aqueous electrolyte (pH = 11) containing 0.1 M K_2SO_4 and 0.01 M KCN is shown in Fig.1. Figure 2 shows the spectral evolution of Raman scattering from a freshly polished Ag electrode placed in the electrolyte at −0.88 V as detected with an optical multichannel analyzer [13] during the first oxidation and reduction cycle (see corresponding voltammogram in Fig.1). During the first oxidation cycle (−0.88 → 0.55 V), no SERS was observed even though $Ag(CN)_{x+1}^{x-}$ complexes and AgCN were produced near the

---

[1] Present address: Sandia Laboratories, Livermore, CA 94550, USA.
[2] Present address: I. Physikalisches Institut der RWTH Aachen,
51 Aachen 1, West Germany.

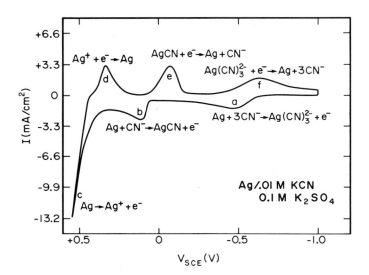

Fig.1 Voltammogram at 50 mV/sec with Ag working electrode in aqueous 0.1 M K_2SO_4 and 0.01 M KNC. Electrochemical reactions are indicated and labeled as peaks a through f

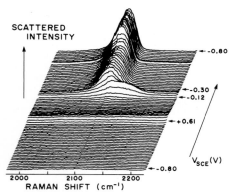

Fig.2 Development of SERS for cyanide complexes on Ag electrode with λ_i = 514.5 nm during the first oxidation-reduction cycle. The corresponding voltammogram is shown in Fig.1

Ag electrode (peaks a and b in Fig.1, respectively). Roughening of the electrode surface was begun by the oxidation of Ag (peaks a, b, and c in Fig.1). Only after the creation of freshly formed Ag sites on the electrode (e.g., Ag adatoms [14,15] or microscopic sized Ag sites) by the reduction of Ag^+ (after peak d in Fig.1) did a broad continuum of inelastic radiation and a discrete Raman peak from AgCN at 2165 cm^{-1} appear [16]. Upon further reduction, CN^- ions--either from solution or freed by the reduction of AgCN (peak e in Fig.1)--complexed with the remaining AgCN to form $Ag(CN)_2^-$. This gave rise to a Raman peak at 2140 cm^{-1}, accompanied by the disappearance of the AgCN Raman peak (see Fig.2). Additional CN^- ions became available at the electrode surface during the reduction of $Ag(CN)_2^-$ [indistinguishable from the reduction of $Ag(CN)_3^{2-}$ at peak f in Fig.1] and complexed with $Ag(CN)_2^-$ to form $Ag(CN)_3^{2-}$ and $Ag(CN)_4^{3-}$. These complexes gave rise to a strong Raman signal at 2110 cm^{-1} while the 2140 cm^{-1} peak of $Ag(CN)_2^-$ disappeared. Coulometry based on the geometric area of a polished electrode indicated the oxidation of the equivalent of 166 monolayers and the reduction of 69 monolayers of Ag during the complete voltage cycle (with 50 mV/sec ramp rate) as shown in Fig.1. During the second and subsequent oxidation cycles, the SERS signal

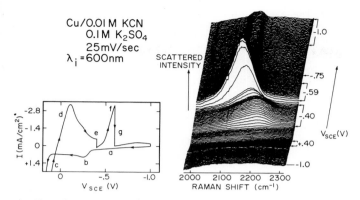

Fig.3 Development of SERS for cyanide complexes on Cu electrode
with λ_i = 600 nm during the first oxidation-reduction cycle, consisting of a ramp-hold activation procedure shown in the associated
voltammogram. The ramping rate is 25 mV/sec with voltage holds at
points c, e, g, and -1.0 V

progression reversed from $Ag(CN)_4^{3-}$ and $Ag(CN)_3^{2-}$ (2110 cm^{-1}) to $Ag(CN)_2^-$
(2140 cm^{-1}) to AgCN (2165 cm^{-1}), and to the disappearance of all SERS signals when the Ag sites were oxidized. Good correlation was found between
solution data and the Raman shifts of the species at the electrode surface.

Similar experiments were performed at a Pt working electrode but SERS
from Pt-CN$^-$ complexes could not be observed. The nobility of the Pt electrode may have precluded electrochemically induced surface roughening and
the formation and reduction of Pt-CN$^-$ complexes. Alternatively, the dielectric properties of roughened Pt, in contrast to those of Ag, may not have
enhanced the local electromagnetic (EM) fields in the blue-green region.
However, the evolution of the SERS peaks associated with $Ag(CN)_{x+1}^{x-}$ was
observed during cyclic voltammetry at a Pt electrode immersed in an aqueous
solution of 0.01 M $K_2Ag(CN)_3$ and 0.1 M K_2SO_4 [16]. The voltage cycling deposited several layers of Ag on the Pt surface in a roughened state. Once
the Ag sites on the Pt surface were established, the SERS from $Ag(CN)_2^-$ was
observed at 0.55 V and transformed to $Ag(CN)_3^{2-}$ and $Ag(CN)_4^{3-}$ upon further
reduction. During subsequent oxidation half cycles, the evolution of the
SERS peaks reversed. Therefore, the creation of Ag roughness by the electrochemical reduction of Ag$^+$ in solution was requisite to the observation of
SERS from $Ag(CN)_{x+1}^{x-}$ complexes. Furthermore, the effect did not require that
roughened Ag be deposited on a continuous Ag substrate.

The growth and decay of SERS from cyanide complexes adsorbed on a Cu electrode in an electrolyte of 0.1 M K_2SO_4 and 0.01 M KCN were also correlated
with the electrochemical reactions indicated by the cyclic voltammetry results shown in Fig.3 [17].

The peaks in the voltammogram represent:

peak b: $Cu + CN^- \rightarrow CuCN + e$

peak c: $Cu \rightarrow Cu^{2+} + 2e$

peak d: $Cu^{2+} + 2e \rightarrow Cu$

peak f: $CuCN + e \rightarrow Cu + CN^-$
$CuCN + xCN^- \rightarrow Cu(CN)_{x+1}^{x-}$.

58

Generation of a strong SERS signal from the Cu electrode required a voltage ramp and hold procedure rather than the uninterrupted voltage ramping used for the Ag electrode. Prerequisite to the first occurrence of SERS at 2170 cm^{-1} from CuCN on a freshly polished Cu electrode was oxidation of the Cu surface to Cu^{2+} ions and CuCN during the potential hold at 0.4 V (peak c in Fig.3) followed by the reduction of Cu^{2+} ions to form fresh Cu sites on the Cu surface (peak d in Fig.3). Upon further decrease in the voltage (peak f in Fig.3), CuCN was reduced, making CN^- ions available to form $Cu(CN)_{x+1}^{x-}$ at the fresh Cu sites. These complexes gave rise to a Raman peak at 2100 cm^{-1}, accompanied by the disappearance of the CuCN peak at 2170 cm^{-1} (see Fig.3). With better spectral resolution, the broad peak centered at 2100 cm^{-1} was resolved into at least two peaks consistent with the assignment to a mixture of $Cu(CN)_{x+1}^{x-}$ complexes.

With λ_i = 600 nm, SERS from Cu was found to be only slightly less intense than that from Ag after identical ramp-hold electrochemical processing. For λ_i below 575 nm, the discrete Raman peaks of the Cu-cyanide complexes sharply decreased in intensity relative to the inelastic continuum emission intensity. Figure 4 shows a portion of the inelastic spectra obtained with λ_i = 514.5 nm during the ramp-hold electrochemical procedure shown in Fig.3. The discrete peaks of the Cu-cyanide complexes which were pronounced at λ_i > 575 nm are marginally discernible. Without the addition of KCN to the electrolyte, the continuum was present but reduced in intensity. With λ_i = 457.9 nm, the continuum was present but no indication of discrete SERS peaks was detected. The analogous continuum emission from Ag electrodes has been found to exhibit no Raman gain in time resolved experiments [18]. Based on these observations, we ascribe this voltage dependent inelastic continuum emission to a surface roughness enhanced electron-hole pair radiative recombination [8].

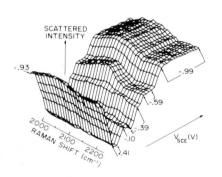

Fig.4 The voltage dependent continuum of inelastic radiation on Cu electrode with λ_i = 514.5 nm during the ramp-hold procedure shown in the voltammogram of Fig.3

A SERS spectrum with λ_i = 600 nm was obtained from a Au electrode in an aqueous electrolyte containing 0.1 M K_2SO_4, 0.01 M $KAu(CN)_2$, and 0.01 M KCN by a series of voltage step-holds at -0.9, -3.0, and -0.9 V (see Fig.5). The hold at -3.0 V produces a roughened surface by the reduction of $Au(CN)_2^-$ to Au accompanied by the H_2 evolution. A SERS peak at 2110 cm^{-1} which developed at -0.9 V was significantly weaker than the cyanide peaks from Ag and Cu electrodes. EM models for SERS could not predict this result since the dielectric properties of Au are intermediate between those of Ag and Cu. However, Au also tends to be chemically inert. The IR absorption of solid AuCN occurred at 2261 cm^{-1} [19] and the Raman sprectrum from a solution of $Au(CN)_2^-$, the only complex formed in the Au-cyanide system, has a peak at 2164 cm^{-1} [20], significantly displaced from the 2110 cm^{-1} peak observed in this work. The explanation for this difference is still under investigation. Analogous to the results with the Cu electrodes, no discrete SERS peak was detectable with incident photon energy sufficient to excite d-band electrons in Au, e.g., with λ_i = 514.5 nm. However, a voltage dependent continuum emission was easily observed for such photon energies.

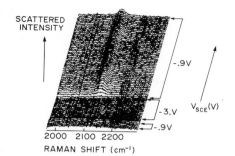

SCATTERED
INTENSITY

2000 2100 2200
RAMAN SHIFT (cm⁻¹)

$V_{SCE}(V)$

−.9V

−3.V

−.9V

Fig.5 Development of SERS for
Au(CN)$_2^-$ on Au electrode with
λ_i = 600 during the first series
of voltage step-holds at −0.9,
−3.0, and −0.9 V

The appearance of the discrete SERS peaks at Ag, Pt, Cu, and Au electrodes
in electrolytes containing cyanide ions has been correlated with the electrochemical reactions occurring during cyclic voltammetry. The electrochemically
induced surface roughness is important for both discrete Raman peaks and the
associated inelastic continuum. The wavelength dependence of the SERS emission requires further study to explain the physical mechanisms. Isolation of
the EM and chemical contributions to the overall SERS is of particular importance.

Metal Films

Recent explanations for SERS suggest that the random roughness of a planar
surface can be characterized as a statistical distribution of gratings having
different spacings, amplitudes, and orientations [8,9,21,22]. The gratings
couple volume EM waves to surface plasmon polaritons (SPP). Another way of
coupling volume EM waves to SPP, using smooth Ag film, is through internal
reflection in a KRETSCHMANN configuration [23]. We have investigated the
contribution of SPP to SERS of molecules adsorbed on Ag films evaporated on
a hemicylindrical prism. For λ_i = 514.5 nm, nearly optimal coupling is
achieved with a 57 nm thick Ag film, which results in a 75x enhancement of
the electric field intensity at the Ag-air interface [24,25]. The SERS
spectrum from isonicotinic acid adsorbed on the 57 nm Ag film is shown in
Fig.6 [26]. Coupling to SPP did increase the Raman intensity by approximately
two orders of magnitude when the Raman radiation was collected on the air
side of the Ag film. Based on our results with fluorescent dye molecules in
contact with the Ag film, another order of magnitude increase could be
obtained by collecting the Raman radiation at the prism side with the detection angle set at either of the two SPP angles for the scattered wavelength
λ_s [27]. At these two angles, the reradiation efficiency is significantly
increased because of the coupling of the near-field radiation of the fluorescent or Raman active molecules to the SPP which, in turn, are transformed to
volume EM waves emanating from the prism side of the Ag film.

For a 5 nm thick Ag film, no coupling of the volume EM waves to SPP on the
planar Ag surface is expected [25]. Thus, the electric field intensity at
the Ag-air interface should not be enhanced for the incident angle used
with the 57 nm thick Ag film. Contrary to initial expectations, the Raman
intensity from the isonicotinic acid adsorbed on the 5 nm thick Ag film
was an order of magnitude greater than that for the 57 nm'thick Ag film,
which was an optimal thickness for SPP coupling (see Fig.6; note difference in incident power for the two curves, 40 vs 100 mW). In contrast to
the 57 nm thick Ag film results, the Raman signal from the 5 nm thick

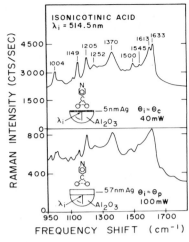

Fig.6 Comparison of SERS for isonicotinic acid chemisorbed on a
5 nm Ag island film with SPP excitation (top) and chemisorbed on a
57 nm Ag more homogeneous film
with SPP excitation. Note laser
power scale change, 40 vs 100 mW

Ag film was not sensitive to either the incident angle or the incident
polarization. SEM photographs of the 5 nm Ag film revealed a much more
pronounced surface structure (contiguous islands about 10nm in diameter)
compared to the more homogeneous 57 nm Ag film [26].

Several explanations for the much larger Raman intensity from the 5 nm Ag
island film are possible. First, the more pronounced surface structures
of the thinner Ag film may provide better sites for chemisorption and
thereby increase the coverage (molecules/cm^2) of the adsorbed isonicotinic
acid within the laser focal area. Although an independent measurement on
the difference in adsorbed molecular coverage between 57 and 5 nm thick
films has not been performed, coverage effects are not expected to increase
the scattered intensity by more than a factor of ten [28]. In addition,
the submicron sized metal islands of the 5 nm thick film may enhance the
Raman polarizability by a charge transfer mechanism [29] made possible by
wave vector selection rule breakdown (uncertainty in wave vector $\Delta k \simeq 2\pi/a$
where a is the dimension of the islands) which greatly increases the number
of photon induced intraband transitions in the Ag sp-band. Finally, the submicron sized island structure may increase the local EM field intensity of
the incident excitation radiation near these structures as well as the reradiation efficiency of the inelastic emitters [9,30-33]. Both of these
mechanisms will be discussed further in conjunction with observed discrete
SERS peaks and continuum emission from molecules adsorbed on electrochemically roughened electrodes, as well as on metallic colloidal suspensions.

Metallic Colloidal Suspensions

SERS has been reported from pyridine in Ag and Au colloidal suspensions
which have the advantage that data are more readily compared with theory
since the metallic particles are immersed in a uniform medium [34]. However,
TEM photographs have shown chain-like aggregation of submicron sized particles upon the addition of pyridine [35], making the comparison between
the observed wavelength dependence of SERS intensity with calculations based
on Lorenz/Mie theory for isolated spherical particles difficult. Recently,
the wavelength dependence of SERS from Ag colloids for which citrate was

both the Raman-active adsorbate and the reducing agent in the colloid pre-paration was reported [36]. Although no TEM photographs were taken, isolated spherical Ag colloids of 21 nm particle radius were inferred from the measured extinction spectrum. The experimental wavelength determined in this study disagreed markedly with the calculated dependence based on the Lorenz/Mie theory for 21 nm particle radius [32,36].

We have observed SERS from $Au(CN)_2^-$ ions adsorbed on Au colloidal suspen-sions prepared by the incomplete reduction of aqueous $KAu(CN)_2$ (10 ml, 0.003 M) with aqueous $NaBH_4$ (40 ml, 0.01 M) under N_2 purging and at 90°C. When the resulting purple solution was stored cold, its physical appearance, extinction spectrum, SERS spectrum of $Au(CN)_2^-$ at 2138 cm^{-1} (see Fig.7), and continuum extending to beyond 4000 cm^{-1} were stable for months. TEM photo-graphs indicated that the individual particles were essentially spherical, having an average radius of 12 nm with a range of 3-30 nm. Random clusters (not chain-like) containing 2-15 particles and having radii up to 100 nm were more numerous than isolated single spheres.

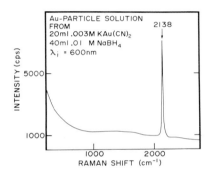

Fig.7 SERS from $Au(CN)_2^-$ adsorbed on Au colloidal suspensions. Be-low 400 cm^{-1}, the elastic scatter-ing masks out the lower frequency Raman modes which might exist

The wavelength dependences of the $Au(CN)_2^-$ vibrational intensity at 2138 cm^{-1} and the continuum arbitrarily selected to be near 2138 cm^{-1} (see Fig.7) have been measured relative to the depolarized peak of liquid benzene (1586 cm^{-1}). Within the wavelength range from 514.5 to 740 nm, the 2138 cm^{-1} peak intensity increased monotonically by 10^3x while the con-tinuum (near 2138 cm^{-1}) was virtually wavelength independent. No discrete SERS peak was detected at $\lambda_i \leq$ 514.5 nm. However, the strong inelas-tic continuum spectrum showed a maxi-mum near 530 nm regardless of the in-cident wavelength, even for $\lambda_i =$ 350.7 nm. The continuum emission from Au colloids was also detected without the presence of CN^- ions in solution. The observed continuum radiation spec-trum has characteristics similar to the photoluminescence observed from smooth Au surfaces and is ascribed to recombination of electrons from the Fermi level in the sp-band with holes in the d-band [37]. The inelastic continuum observed from Au colloids, when the incident photon energies $\hbar\omega_i$ are well below the energy separation of the Fermi level and the d-band ($\hbar\omega_i < E_{FL} - E_d \simeq 2.1$ eV), is not observed from smooth Au. We believe this continuum is from recombination radiation of intraband (sp-band) electron-hole pairs made possible by wave vector selection rule breakdown. No SERS was observed for the H_2O stretching and bending vibrations with or without the presence of Au colloids in the aqueous solution, while the $Au(CN)_2^-$ vib-rations were detectable only when Au colloids were present in the aqueous solution. We believe the lack of SERS for H_2O and the presence of SERS for $Au(CN)_2^-$ rule out the local EM field enhancement mechanism for the Au colloid case.

Possible Mechanisms

Any viable model or combination of models for SERS must explain the large enhancement and wavelength dependences for both the discrete peaks and the background continuum. A number of theories have been proposed for the enhancement of the discrete peaks, but few can account for the origin of the inelastic continuum radiation which can be detected even in the wavelength region where the discrete peaks are unobservable [8-11,22,29-33,38]. Although there is general agreement that some type of surface roughness or microstructure is essential to the observation of SERS, serious controversy remains over whether the role of the metallic microstructure is of a chemical nature (e.g., charge transfer between the metal and adsorbate [22,29]) and/or of an EM wave nature (i.e., enhancement of the local EM field intensity for the incident wave near the microstructure and enhancement of the reemission at the scattered wavelength [22,30-33,38]). Recent experiments have attempted to isolate the contribution of single mechanisms or a combination of several mechanisms to the observed SERS [28,39,40]. We shall briefly review both the charge transfer and local EM wave models and indicate their applicability to our experimental results.

(a)   Charge Transfer Mechanism

Mechanisms of charge transfer [29] between the metal and adsorbed molecule are shown in Fig.8 and can give rise to both discrete and continuum inelastic scattering. Arising from electron transitions within the sp-band of the noble metals, discrete SERS can be explained by the following processes: (1) photon induced electron-hole pair creation by intraband transitions from below to above the Fermi level; (2) charge transfer by electron tunneling to the adsorbed molecule; (3) formation of a temporary negative molecular complex, i.e., shape resonances [41]; (4) dissociation of the negative molecular complex with the ejection of a less energetic electron in conjunction with the excitation of vibrational modes of the adsorbed molecule; (5) charge transfer to the metal via tunneling of the ejected electron which has lost an amount of energy equal to the vibrational energy; and (6) electron-hole recombination to produce the SERS radiation.

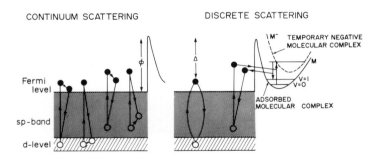

Fig.8   Schematic representation of the charge transfer mechanism for discrete SERS peaks and of electron-hole pair energy losses for coninuum radiation

Both processes (1) and (6) are greatly enhanced by the breakdown of the wave vector selection rule because the electrons are confined to metallic microstructures or surface roughness with dimension a. The charge transfer probabilities for processes (2) and (5) are sensitive to the difference (Δ) between the actual work function ϕ and the energy of the electron above the Fermi level (see Fig.8) and are difficult to estimate. The value of ϕ for a metal with an adsorbate and, in the case of an electrolyte with co-adsorbates, can be significantly lowered from its vacuum value [22]. For sufficiently low Δ, the electron should be able to tunnel through multilayers of molecules over distances up to 1000 Å [22]. Since strong vibrational mode overtones and combinations have rarely been observed [42], the lifetime of the negative molecular complex, process (3), must be assumed to be short compared to the vibrational period but long compared to the time duration associated with virtual electronic transitions to the higher electronic states in the nonresonant molecule adsorbed on the metal microstructure. For most molecules in the gaseous phase, the formation of negative molecular complexes occurs over only a limited range of incident electron energies, i.e., between 0.5 and 2.5 eV [41]. Consequently, within the context of the charge transfer model, the wavelength dependence of SERS is affected by a delicate balance between minimizing Δ for electron tunneling and optimizing the electron energy for formation of negative molecular complexes.

The charge transfer model can also explain the absence of discrete SERS when $\hbar\omega_i \geq (E_{FL} - E_d)$. Since the d-level is localized, surface roughness or microstructure related wave vector selection rule breakdown is not necessary for interband transitions between the d-level and states above the Fermi level. Thus, transitions from the d-level are more probable when $\hbar\omega_i$ is sufficiently large. However, electrons excited from the d-level are not able to tunnel into the adsorbed molecule because Δ is usually too large for readily available excitation sources having $\hbar\omega_i \geq (E_{FL} - E_d)$.

Continuum emission can result from sp-band electron-hole recombination after a series of nonradiative energy losses by the hole below the Fermi level and/or by the electron above the Fermi level without any charge transfer to the adsorbed molecule. Again, $\Delta k \simeq 2\pi/a$ can significantly increase intraband electron-hole pair creation and radiative recombinations. Continuum emission when the d-level electron-hole pairs are involved can be labeled as interband recombination after a series of nonradiative energy losses by the hole in the d-level and/or by the electron above the Fermi level without any charge transfer to the adsorbed molecule. The maximum of this interband luminescence is predicted to occur at $E_{FL} - E_d$ (e.g., at 530 nm for Au) regardless of $\hbar\omega_i$, assuming that $\hbar\omega_i > (E_{FL} - E_d)$ [37]. At present, no quantitative estimate of this charge transfer mechanism has been reported. We are, therefore, unable to critically test this model with the available experimental data.

(b) Local EM Field Mechanism

Both the excitation EM field intensity and the reradiation efficiency of the molecules near a metallic microstructure can be significantly enhanced [30-32]. Within the small particle Rayleigh limit, the near-field intensity at $\hbar\omega_i$ for an isolated spherical particle with dielectric constant $\varepsilon(\omega) = \varepsilon_1(\omega) + i\varepsilon_2(\omega)$ immersed in a medium with refractive index $n(\omega)$ is increased by $[\varepsilon(\omega_i) - n^2(\omega_i)]^2/[\varepsilon(\omega_i) + 2n^2(\omega_i)]^2$. The reradiation efficiency of the inelastic wave at $\hbar\omega_s$ is increased by a similar factor evaluated with $\varepsilon(\omega_s)$ and $n(\omega_s)$ [30-32]. The total enhancement at the surface plasmon resonances

$\varepsilon_1(\omega) + 2n^2(\omega) = 0$ for both ω_i and ω_s can reach 10^6 for Ag and somewhat less for Au and Cu. For larger particles, the Lorenz/Mie approach must be used [32]. While calculation of the integrated intensity cross section in the near field at a sphere $[Q_{nf}(\omega_i)]$ upon irradiation with a plane wave at $\hbar\omega_i$ is straightforward, calculation of the integrated intensity cross section at the far field resulting from molecular reradiation near the surface of a metallic sphere is difficult. Extension of the Lorenz/Mie approach to particles of different shapes (e.g., prolate or oblate spheroids) is possible [9,43,44]. However, the standard scattering formalisms are not applicable to aggregates of spherical particles or to spheres resting on a substrate (metallic or dielectric). The enhancement factor for SPP coupling of the incident and reradiated waves has been solved for a sinusoidal metallic grating [22].

Using bulk values of $\varepsilon(\omega)$ [45] and the Lorenz/Mie formalism, we have calculated $Q_{nf}(\omega_i)$ as a function of $\hbar\omega_i$ and metallic sphere radius for Ag and Au colloids immersed in water. For Ag, $Q_{nf}(\omega_i)$ reaches 2100x that of the incident intensity when the particle radius is around 25 nm and $\lambda_i \simeq$ 400 nm (see Fig.9). With increasing radius, the dominant peak in $Q_{nf}(\omega_i)$ decreases and moves to longer wavelengths. Furthermore, a second series of peaks in $Q_{nf}(\omega_i)$ which arise from surface plasmon resonances associated with higher order poles (e.g., magnetic dipoles and electric quadrupoles) become important for larger size particles (beyond the Rayleigh limit). For Au, the maximum in $Q_{nf}(\omega_i)$, $\sim80x$, is considerably smaller than that for Ag and occurs at larger radius (~50 nm) and longer wavelength ($\lambda_i \simeq 600$ nm; see Fig.9).

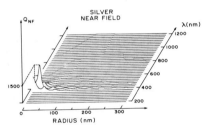

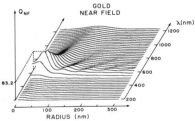

Fig.9 Integrated near-field EM intensity cross section Q_{nf} on the surface of Ag and Au spheres as a function of their radius and incident photon. The incident intensity has been subtracted and normalized in Q_{nf}. The maximum values of Q_{nf} are indicated

This is consistent with the dispersion of $\varepsilon(\omega)$ for Au since, at the surface plasmon resonance condition, $\varepsilon_1(\omega) + 2n^2 = 0$, $\varepsilon_2(\omega)$ for Au is much larger than that for Ag. Although the reradiation efficiency of the inelastic wave as a function of $\hbar\omega_s$ and particle radius is much more difficult to calculate [32], the total enhancement, including both $Q_{nf}(\omega_i)$ and the reradiation efficiency, can be approximated by $Q_{nf}(\omega_i)Q_{nf}(\omega_s)$ assuming a Raman shift of less than 200 cm^{-1}, which for Ag and Au particles yield a maximum of 4×10^6 and 6×10^3, respectively. Both the discrete and continuum inelastic radiation are enhanced by $Q_{nf}(\omega_i)$ and the increased reradiation efficiency, since the Maxwell wave equations and boundary conditions do not distinguish between the different types of physical processes responsible for the inelastic emission. Therefore, the local EM field mechanism does not explain the different wavelength dependences of the continuum and the discrete SERS peak observed for the Cu and Au electrodes, as well as for the Au colloids.

Our experimental data for the wavelength dependence of the SERS intensity for $Au(CN)_2^-$ adsorbed on random clusters of Au colloids cannot separate local EM field enhancements from the charge transfer enhancement of the Raman polarizability. However, our data are consistent with $Q_{nf}(\omega_i)$ calculations for a single Au sphere of radius 100 nm (TEM shows clusters of such size) with $Q_{nf}(\omega_i)$ peaking around $\lambda_i \simeq 700$ nm. This consistency may be coincidental since we have not yet measured the total enhancement factor as a self-consistency check. The EM field enhancement mechanism does not explain the ω_i dependence of the continuum emission or the lack of SERS from the surrounding H_2O molecules.

Conclusions

Our experimental results on SERS for molecules adsorbed on electrodes, films, and colloids have been summarized in the context of present models. Future experiments should emphasize isolating individual mechanisms. For example, it may be possible to separate the charge transfer model from the local EM field model by the determination of the SERS wavelength dependence in the interband region where $\hbar\omega_i > (E_{FL} - E_d)$. With increasing $\hbar\omega_i$, ε_2 increases, progressively lowering the local EM field contributions. In contrast, for the charge transfer mechanism, with increasing $\hbar\omega_i$, Δ decreases from $\Delta \simeq \phi$ to $\Delta \simeq 0$, causing the tunneling probability to increase and giving rise to SERS which may be less intense than that for $\hbar\omega_i < (E_{FL} - E_d)$ where both enhancement mechanisms can contribute to the overall SERS.

Acknowledgments

We thank Prof. Peter Barber and Ms. Barbara Messinger for making the Lorenz/Mie calculations on Ag and Au spheres. Partial support of this work by the Gas Research Institute (Basic Research Grant No. 5080-363-0319) and the Office of Naval Research (Contract No. N00014-76-C-0643) is also gratefully acknowledged.

References

1. S.P.S. Porto and D.L. Wood, J. Opt. Soc. Am. 52, 251 (1962).
2. J.M. Cherlow and S.P.S. Porto: "Laser Raman Spectroscopy of Gases", in Laser Spectroscopy of Atoms and Molecules , ed. by H. Walther, Topics in Applied Physics, Vol. 2 (Springer, Berlin, Heidelberg, New York 1976) p. 255
3. W.R. Fenner, H.A. Hyatt, J.M. Kellman, and S.P.S. Porto, J. Opt. Soc. Am. 63, 73 (1973).
4. W. Proffitt and S.P.S. Porto, J. Opt. Soc. Am. 63, 77 (1973).
5. M. Fleischmann, P.J. Hendra, and A.J. McQuillan, Chem. Phys. Lett. 26, 163 (1974).
6. D.L. Jeanmaire and R.P. Van Duyne, J. Electroanal. Chem. 84, 1 (1977).
7. M.G. Albrecht and J.A. Creighton, J. Am. Chem. Soc. 99, 5215 (1977).
8. E. Burstein, C.Y. Chen, and S. Lundqvist, in Light Scattering in Solids, edited by J.L. Birman, H.Z. Cummins, and K.K. Rebane (Plenum Press, New York, 1979), p. 479.
9. E. Burstein and C.Y. Chen, in Proceedings of the 7th International Conference on Raman Spectroscopy, edited by W.F. Murphy (North-Holland Publishing Co., Amsterdam, 1980), p. 346.
10. T.E. Furtak and J. Reyes, Surf. Sci. 93, 351 (1980).
11. R.P. Van Duyne, in Chemical and Biochemical Application of Lasers, Vol. 4, edited by C.B. Moore (Academic Press, New York, 1978), p. 101.
12. A. Otto, Proceedings of the 6th Solid-Vacuum Interface Conference,

Delft, The Netherlands, 1980 (to be published).

13. R. Dornhaus, M.B. Long, R.E. Benner, and R.K. Chang, Surf. Sci. 93, 240 (1980).
14. J. Billmann, G. Kovacs, and A. Otto, Surf. Sci. 92, 153 (1980).
15. A. Otto, J. Timper, J. Billmann, and I. Pockrand, Phys. Rev. Lett. 45, 46 (1980).
16. R.E. Benner, R. Dornhaus, R.K. Chang, and B.L. Laube, Surf. Sci. 101, 1980 (in press).
17. R.E. Benner, K.U. von Raben, R. Dornhaus, R.K. Chang, B.L. Laube, and F.A. Otter, Surf. Sci. (to be published).
18. J.P. Heritage and J.G. Bergman, in Light Scattering in Solids, edited by J.L. Birman, H.Z. Cummins, and K.K. Rebane (Plenum Press, New York, 1979), p. 167.
19. L.H. Jones and R.A. Penneman, J. Chem. Phys. 22, 965 (1954).
20. L.H. Jones, J. Chem. Phys. 43, 594 (1965).
21. J.C. Tsang and J.R. Kirtley, Phys. Rev. Lett. 43, 477 (1979).
22. S.S. Jha, J.R. Kirtley, and J.C. Tsang, Phys. Rev. B (to be published).
23. E. Kretschmann, Z. Phys. 241, 313 (1971).
24. Y.J. Chen, W.P. Chen, and E. Burstein, Phys. Rev. Lett. 36, 1207 (1976).
25. H.J. Simon, D.E. Mitchell, and J.G. Watson, Am. J. Phys. 43, 630 (1975).
26. R. Dornhaus, R.E. Benner, R.K. Chang, and I. Chabay, Surf. Sci. 101, 1980 (in press).
27. R.E. Benner, R. Dornhaus, and R.K. Chang, Opt. Commun. 30, 145 (1979).
28. J.E. Rowe, C.V. Shank, D.Z. Zwemer, and C.A. Murray, Phys. Rev. Lett. 44, 1770 (1980).
29. E. Burstein, Y.J. Chen, C.Y. Chen, S. Lundqvist, and E. Tosatti, Solid State Commun. 29, 565 (1979).
30. M. Moskovits, Solid State Commun. 32, 59 (1979).
31. S.L. McCall, P.M. Platzman, and P.A. Wolff, Phys. Lett. 77A, 381 (1980).
32. M. Kerker, D.-S. Wang, and H. Chew, Appl. Opt. (to be published).
33. M. Moskovitz, J. Chem. Phys. 69, 4159 (1978).
34. J.A. Creighton, C.G. Blatchford, and M.G. Albrecht, J. Chem. Soc., Faraday II 75, 790 (1979).
35. J.A. Creighton, C.G. Blatchford, and J.R. Campbell, in Proceedings of the 7th International Conference on Raman Spectroscopy, edited by W.F. Murphy (North-Holland, Amsterdam, 1980), p. 398.
36. M. Kerker, O. Siiman, L.A. Bumm, and D.-S. Wang, Appl. Opt. (to be published).
37. A. Mooradian, Phys. Rev. Lett. 22, 185 (1969).
38. J.I. Gersten, R.L. Birke, and J.R. Lombardi, Phys. Rev. Lett. 43, 147 (1979).
39. C.A. Murray, D.L. Allara, and M. Rhinewine, in Proceedings of the 7th International Conference on Raman Spectroscopy, edited by W.F. Murphy (North-Holland, Amsterdam, 1980), p. 406.
40. P.N. Sanda, J.M. Warlaumount, J.E. Demuth, J.C. Tsang, K. Christmann, and J.A. Bradley, Phys. Rev. Lett. (to be published).
41. I. Nenner and G.J. Schulz, J. Chem. Phys. 62, 1747 (1975); and G.J. Schulz, Rev. Mod. Phys. 45, 378 (1973).
42. B. Pettinger, in Proceedings of the 7th International Conference on Raman Spectroscopy, edited by W.F. Murphy (North-Holland Publishing Co., Amsterdam, 1980), p. 412.
43. J.I. Gersten and A. Nitzan, J. Chem. Phys. 73, 3023 (1980).
44. M. Kerker (private communication).
45. P.B. Johnson and R.W. Christy, Phys. Rev. B 6, 4370 (1972).

Inverse Raman Spectroscopy

A. Owyoung and P. Esherick

Sandia National Laboratories, Division 4216
Albuquerque, NM 87185, USA

1. Introduction

The development of high-power transform-limited tunable pulsed laser
systems has enabled major advances in the application of coherent Raman
techniques to high-resolution studies in gases. In particular quasi-cw
inverse Raman studies have demonstrated resolving powers which are
approximately a factor of 50 improvement over that which is generally
associated with conventional Raman techniques [1]. Even higher resolution is potentially possible as improved lasers are made available.

In this paper we will describe recent advances in inverse Raman
spectroscopy (IRS). We shall first briefly review the apparatus used in
an IRS study and describe improvements which have extended the accessible
IRS spectral region with greatly improved precision and accuracy
(~ 0.001 cm^{-1}). We then go on to describe two sets of experiments. The
first, a study of Q-branch spectra of the heavy spherical top molecule CF$_4$,
serves to illustrate the direct application of this apparatus to fundamental
studies in molecular spectroscopy. In the second, the apparatus is modified
to allow velocity-selective saturation of a Doppler broadened Raman line.
This has enabled the first studies of saturation dip type phenomena on
purely Raman-active transitions.

2. Experimental Apparatus

The apparatus used in the present IRS study closely parallels the
quasi-cw IRS apparatus described in Ref. 1. A pulsed pump laser and a
stable cw probe laser are overlapped in a focus in the gas sample. The
two sources are spectrally separated by the Raman shift of the mode under
study, with the probe laser on the shorter wavelength (anti-Stokes) side
of the pump laser. The probe thus experiences the "inverse Raman
absorption" induced by the pump and will display the Raman spectrum in the
form of an absorption signal as the pump source is spectrally scanned.

A schematic diagram of the IRS apparatus is shown in Fig. 1. The
pulsed pump laser source is an electronically scanned single-mode cw dye
oscillator (Coherent Model 599-021) which is pulse amplified by three dye
amplifiers pumped with a frequency-doubled Nd:YAG laser source, to produce
2MW pulses of 6 nsec duration. This system operates at a 10-pps repetition
rate and emits transform-limited pulses of 75 MHz spectral width (FWHM)
which now presents the primary limitation to the resolution of the system.

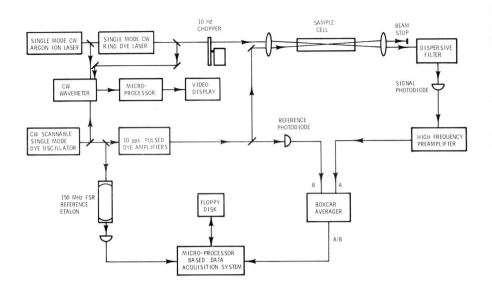

<u>Fig. 1</u>. Schematic diagram of the inverse raman spectroscopy apparatus.

    In order to extend the range of this IRS technique to lower Raman shifts, the cw argon-ion probe laser previously used is supplemented by a single-mode, cw ring dye laser (Spectra Physics Model 380A). Pumped with 2.5 watts from a single-mode argon ion laser, the Rh6G ring laser was able to provide over 300 mW output with a linewidth and frequency stability of $\pm$ 20 MHz. Amplitude modulation on the ring laser output was found to be excessive when the argon ion pump laser was operated multimode. However, under single-mode pumping conditions, the noise level was reduced to about twice the shot noise limit. By using dye lasers for both the pump and probe, one can in principle tune their frequency difference essentially to zero cm^{-1}. In fact, the minimum practical observable Raman shift will be limited by the dispersive filter used to separate the weak cw-probe signal from the high-intensity pulsed pump light at the detector. The Pellin-Broca prism, grating, and pinhole-pair system used here and in Ref. 1 was found to be perfectly adequate for a 700 cm^{-1} Raman shift. However, a double-grating system, in addition to a strongly crossed-beam configuration, may be required for studies of very small Stokes shifts e.g. several cm^{-1}. Nevertheless, discrimination is still expected to be orders of magnitude higher than that which is obtained in spontaneous Raman studies.

    Detection of the cw probe beam is accomplished using a Si photodiode (EGG Model FND-100) which is coupled through a high-frequency preamplifier to separate the transient 6 nsec absorption signal. A boxcar integrator (PAR Model 162/165) provides gated (5 nsec) signal averaging. By using the second channel of the boxcar to provide normalization of the signal to

the input pump power, the time-averaged boxcar output gives a direct
measure of the Raman spectrum. Also, a 10 Hz chopper synchronized to the
Nd:YAG laser serves as a 100 μsec optical gate to reduce the average power
of the probe source on the detector.

We have further updated the apparatus with a microprocessor-based data
acquisition system that can digitize and store data in real time. With
each laser shot, this system samples the output of the boxcar integrator
and averages an optional number of shots before saving the result. A
second input channel is used to monitor the transmission of unamplified
pump light through a 150 MHz confocal etalon, thus providing the data set
with a sequence of evenly spaced frequency markers as the pump source is
scanned. At the end of a scan the data is permanently stored on a floppy
disc for later retrieval and analysis. The latter is performed by transmitting the data to a CDC-6600 time-sharing system where it is wavenumber
scaled and plotted, and peak positions are identified, using a number of
interactive programs [2] written in APL.

Absolute wavenumber calibration of the spectra was accomplished by using
a cw wavemeter [3], with a microprocessor interface, to provide continuous
readout of the wavelengths of the pump and probe sources, and their relative
shift in cm^{-1}. The addition of a phase-locked-loop frequency multiplier to
the fringe counting electronics of this device increased its resolution to
about 5 parts in 10^8. The interactive microprocessor programs allows the
user to input a desired "target" Raman shift. The system then provides
the user not only with a continuous readout of the current Raman shift,
in cm^{-1}, of the pump and probe lasers, but also with the offset, in GHz,
from the target shift. The microprocessor can also average a specified
number of readings in real time. Typical standard deviations are somewhat
less than 0.001 cm^{-1}.

3. ν_1 Q-branch Studies of Heavy Spherical Top Molecules

Past IRS studies have been successfully applied to Q-branches in a variety
of light spherical top molecules [1, 4-5]. The lack of spectral accessi-
bility to smaller Raman shifts, however, had precluded the application of
high-resolution IRS techniques to heavier systems. The addition of the
ring dye laser as the cw probe gives the added versatility necessary to
access any new spectral region of interest in Raman studies.

In Fig. 2a we show a fully resolved spectrum of the ν_1 fundamental of
CF_4 taken at room temperature and 4 Torr total pressure. The spectrum is
most remarkable in its simplicity, and is readily fit by a rigid-rotor-
type two-parameter model with $\nu = \alpha + (\beta - \beta_o)J(J + 1)$. A linear least-
squares fit, using 42 experimentally determined peaks (J = 10 to 51) which
were measured in nine separate scans similar to that shown in Fig. 2a,
results in values of $\alpha = 909.0720 \pm 0.0001$ cm^{-1} and $\beta - \beta_o = (3.417 \pm$
$0.0006) \times 10^{-4}$ cm^{-1}. Here the uncertainties in α only reflect the
"goodness of fit", whereas the 0.001 cm^{-1} absolute accuracy of the band
origin is limited by our wavemeter accuracy. Assignment of the individual
rotational levels was achieved by trial and error, with an overall rms
deviation in the final fit of only 3.2×10^{-4} cm^{-1}. Varying our assignment of J by either +1 or -1 caused a fivefold increase in this deviation.

Using the determined constants, a simulation of the spectrum is obtained
by calculating the individual line intensities from the appropriate Boltz-

70

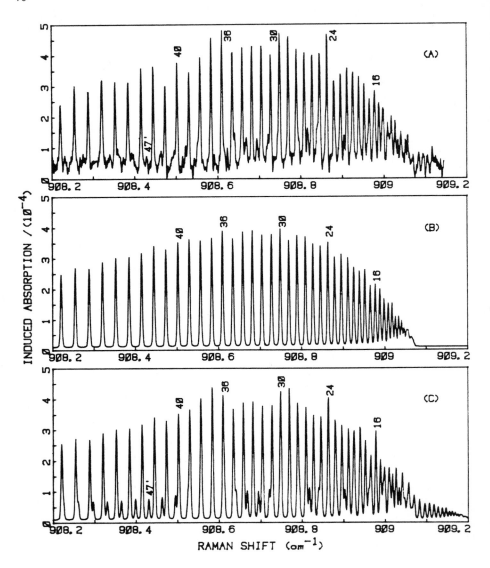

Fig. 2. Observed and calculated inverse Raman spectra in the region of the ν_1 fundamental of CF_4. The observed spectrum (a) was obtained at a pressure of 4 Torr CF_4 with pump and probe laser powers of 2 MW and 150 mW respectively. Calculated spectra are shown (b) for the ν_1 fundamental alone and (c) with the inclusion of the underlying $\nu_1 + \nu_2 \leftarrow \nu_2$ hotband.

mann and spin statistical formulas using a ground state value of 0.191 cm^{-1}. [6] Such a simulation is shown in Fig. 2b, where a Voight profile has been synthesized to include the Doppler and pressure widths of 0.0012 cm^{-1} and 0.0013 cm^{-1} respectively along with the 90 MHz instrumental linewidth.

Although the spectral fit of Fig. 2b is in excellent agreement with the experimental spectrum of Fig. 2a, a careful examination of the line intensities reflects some anomalies. In particular $J = 24, 29, 36,$ and 37 of the experimental spectrum are considerably more prominent than the calculations would bear out. Also, a careful examination of the region near the band origin using an expanded display of the digitized data reveals a remarkable periodicity in the features which at first appear to be "noise". These facts suggest the existence of an underlying band. In fact, further evidence of this band is seen in additional peaks which appear throughout the spectrum of Fig. 2a, particularly between $J = 30$ and 36 and between $J = 40$ and 45. A likely assignment for this band was the $\nu_1 + \nu_2 \leftarrow \nu_2$ hot band which would exhibit an intensity which is 25% of the ν_1 fundamental. Again a linear least squares fit was performed, assuming a rigid rotor model, this time using only 11 experimental peaks. The results are values of $\alpha' = 909.1997 \pm 0.0003$ cm^{-1} and $(\beta - \beta_o)' = (-3.405 \pm 0.004) \times 10^4$ cm^{-4} with a slightly larger standard deviation in the fit of 6.5×10^{-4} cm^{-1}. The completed simulation shown in Fig. 2c, which includes both bands, clearly improves the agreement with the experimental results.

4. Sub-Doppler Raman-Saturation-Dip Studies

Doppler broadening does not as yet impose a strong limitation on the resolving power of IRS techniques in the case of the heavier molecules. There are cases of lighter species however where Doppler broadening does limit the ability to obtain spectral information e.g. $^{13}CH_4$ [5]. Here techniques which reduce Doppler broadening by lowering the kinetic temperature of the gas, either in a static cell or in adiabatic free expansion, are certain to provide improvements in resolving power. It is interesting, however, to take an alternate approach and consider using a purely optical method of obtaining sub-Doppler spectra. In this context, the prospect of using a velocity-selective saturation technique is particularly appealing. Such a technique would not only offer sub-Doppler resolution, but might also allow the study of molecular dynamics in simple homonuclear diatomic systems (even those without low lying electronic resonances) where other optical techniques are not ordinarily applicable.

In Fig. 3 a schematic representation is given of the experimental configuration used in a "Raman saturation-dip" experiment. A pair of high-power pulsed saturating beams of frequency Ω and ω are used to saturate a Raman transition with resonant frequency $\nu_o \cong \omega - \Omega$ as they pass through the sample in the +Z direction. In analogy to the conventional inverse Raman experiment [1] we shall call these beams the "saturating pump" (Ω) and "saturating probe" (ω) beams. See Fig. 3. If the convolution of the linewidths of these two lasers is less than the Doppler width of the Raman transition, they will selectively saturate only a narrow velocity group out of the overall Boltzmann velocity distribution of the gas (Fig. 4a). It is then possible to use the inverse Raman technique described in previous sections of this work to observe this saturated spectrum using two additional "measurement lasers". These latter sources are temporally delayed from the saturating pair and inserted in the opposite, -Z, direction as shown in Fig. 3. This results in a display of the saturated velocity profile, where

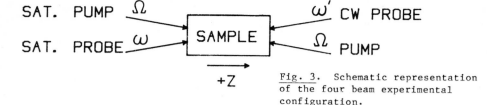

SAT. PUMP Ω ω' CW PROBE

SAT. PROBE ω SAMPLE Ω PUMP

+Z

Fig. 3. Schematic representation of the four beam experimental configuration.

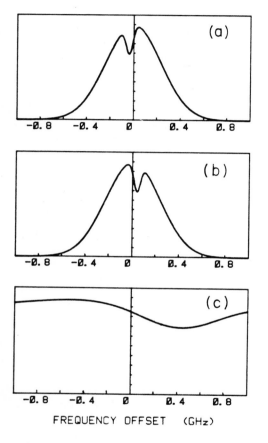

Fig. 4. Sub-Doppler saturated lineshapes plotted as a function of the frequency offset $\omega' - \Omega - \nu_o$ from line center for an independently tuned probe beam ω' which is (a) co-propagating with a pump beam Ω in the same (+Z) direction as the saturating beams, (b) co-propagating with a second measurement pump beam Ω in the $-Z$ direction, counter to the saturating beams, and (c) propagating in the $-Z$ direction in the absence of a second pump beam Ω.

the velocity group that has been saturated with a frequency offset $\Delta = |\omega - \Omega| - \nu_o$ now appears with the opposite Doppler shift $-\Delta$. This is illustrated in Fig. 4b for the case where the measurement lasers (Ω, ω' in Fig. 3) can be scanned over the lineshape independently from the saturating lasers. In our experiment, however, these frequencies are fixed equal, i.e. $\omega' \equiv \omega$. As the lasers are tuned towards line center, the saturation hole, at a frequency offset $-\Delta$, approaches the measurement frequency $+\Delta$ at twice the normal tuning rate until they overlap to produce the dip in signal at line center ($\Delta = 0$). Because both saturation hole and measurement frequency move simultaneously, the expected width of the observed saturation dip is one-half of the linewidth obtained by the spectral convolution of the four laser sources involved.

Preliminary experiments on the Raman-saturation-dip have been performed on the $Q_{01}(2)$ line of deuterium, D_2, at 2987.290 ± 0.002 cm^{-1} because of its large scattering cross section and sizeable Doppler width of 552 MHz. The cw single-mode argon-ion laser source operating at 514.5 nm is used as the cw probe with the additional saturating probe beam being provided by pulse amplifying a portion of this source using two coumarin-500 dye cuvettes transversely pumped by 40 mJ of energy from the frequency-tripled output of the Nd:YAG laser system. The addition of a Faraday isolator between these amplifiers and the argon laser is required to prevent light from the amplifiers from perturbing the stability of this laser. (A highly stable probe source is necessary for low-noise operation of the system.)

The saturating pump and probe beams are first focussed through the sample using a 40 cm focal length lens operating at f/150 and with a crossing angle at the focus of ~ 1.5 degrees. After re-collimation, a right-angle prism is used as a corner cube to return a time-delayed pump pulse back into the focal volume, thus inducing the inverse Raman absorp-

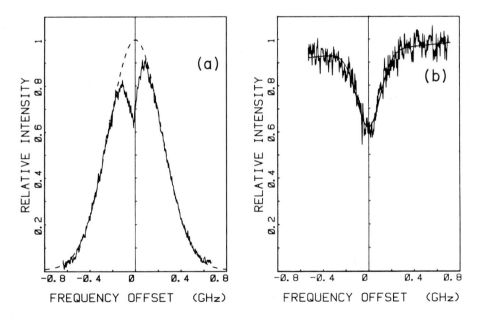

Fig. 5. Sub-Doppler saturation dip observed in deuterium using saturating beams at frequencies Ω and ω with powers of 2 MW and 100 kW respectively. (a) Forward (four-beam) configuration with 2 Torr of D_2 probing at 12 nsec after the peak of the saturating pulses and (b) backscattering (three-beam) configuration with 10 Torr of D_2 probing 2.5 nsec after the peak of the saturating pulses.

tion on the cw probe source. Experimentally, the pump frequency Ω is scanned rather than the probe frequency ω, but this does not pose any change in the conceptualization of the experiment.

In Fig. 5a a spectral scan of the Raman-saturation-dip on the $Q_{01}(2)$ transition in D_2 is shown. The saturation intensities are in good agreement with the measured IRS absorption cross section and the 552 MHz Gaussian Doppler lineshape is well fit to theory. The saturation dip is also well fit to a Gaussian lineshape, with a FWHM of 110 MHz ($\pm$ 10 MHz as averaged over several similar scans). Assuming saturating pump and probe sources of 75 MHz and 95 MHz Gaussian spectral widths respectively, the minimum dip-width observable in the experiment is calculated to be near 72 MHz by taking one half of the width of the convolution of the three pulsed sources. The natural width of the transition is expected to be much smaller from pressure broadening considerations. The contribution to the linewidth due to the slightly non-collinear geometry used in the study is also calculated to be negligible. The additional width observed could possibly be attributed to velocity relaxation during the 12 nsec delay between the saturating and probing steps. Although conclusive evidence will have to await further study, a preliminary scan of the saturation profile under the same conditions, but at 10 Torr, does indicate an appreciable filling in of the saturated hole, thus reflecting velocity-relaxation cross sections of the order $17A^2$. The small shift of the hole off of line center has not been adequately explained, but is likely to be the result of a frequency shift in the saturating probe beam in the amplification process.

Although the four-beam experiment described above is conceptually simpler, it is also possible to observe the Raman saturation-dip in a three-beam experiment which does not utilize the measurement pump beam of Fig. 3. As before, the saturating pump (Ω) and probe (ω) saturate a velocity group as shown in Fig. 4a. However in this case the saturated lineshape is observed by using the original pump beam (Ω) interacting with the counter-propagating weak cw probe beam ($\omega´$) in a backward Raman interaction. The Doppler shifts of the two frequencies Ω and $\omega´$ then add rather than nearly canceling, as occurs in the normal forward scattering experiment so that the back scattering Doppler width will be $R \equiv (\omega + \Omega)/(\omega - \Omega)$ times the forward scattering width, or 12 x 552 MHz = 6.63 GHz. Thus if we saturate, a 120 MHz wide velocity group, the saturation hole will appear in the back scattered profile with R times that width, or 1.44 GHz. Additionally, if the hole is created with a frequency offset Δ, it will appear on the opposite side of the profile with an offset $\Delta´ = -R\Delta$ as shown in Fig. 4c.

As the saturating frequency offset Δ is scanned towards line center, the hole at $\Delta´$ approaches from the opposite side at a rate R times faster. One can calculate the expected width of the saturation dip for the case $\omega = \omega´$ by convoluting the lineshape of the saturation hole (which is R times broader then the convolution of the lineshapes of the saturating lasers) with the lineshape factor of the measurement lasers, and then dividing the resultant linewidth by the relative scan rate (R + 1). For our case, this results in an instrumental lower limit for the dip-width of 112 MHz. [It should be noted here that if Raman gain rather than (inverse) Raman loss is used, i.e. if $\Omega > \omega$, the relative scan rate is reduced from R + 1 to (R − 1), which increases the width of the backscattering probed saturation dip to 132 MHz.]

In Fig. 5b we show a spectral scan of the sub-Doppler Raman-saturation dip which is observed using the backward-wave three-beam configuration. A 5 nsec boxcar gate was used to detect the IRS signal on the cw probe beam with a 2.5 nsec delay with respect to the peak of the temporal profile of the saturating beams. The saturation dip rests on a 6.65 GHz Gaussian profile as expected. A least-squares fit of the dip to a Gaussian yields a 245 MHz width (FWHM) and a sloping background contribution. Since this width is over 100 MHz broader than would be expected, this would infer that velocity relaxation is responsible for the added broadening. Indeed this is consistent with additional data taken at 19 Torr and with the results of the 4 beam saturation scans taken at 10 Torr, both of which exhibit broadening of the saturation dip with pressure.

Our preliminary observations of the Raman saturation dip reported here demonstrate that the phenomenon is clearly observable and does give sub-Doppler resolution. Noise levels are still an order of magnitude higher than one would expect in a direct IRS study. They arise in part from fluctuations in the intensities and temporal profiles of the interacting beams and in part from incomplete isolation of the cw probe source from the saturating probe amplifier chain. These problems should be solvable and the Raman saturation dip shows promise as a technique for obtaining sub-Doppler Raman data. Perhaps an even more important application will be the use of the technique as a means of studying molecular dynamics in homo-nuclear diatomic systems. Studies of the time evolution of the spectral shape of the saturation dip and its subsequent relaxation into a homogeneous saturation dip should provide a very effective means of studying velocity relaxation phenomena in such systems.

5. Conclusions

It is evident from the above illustrations that IRS can have wide application in the high-resolution Raman spectroscopy of gases. Even though the ν_1 mode of CF_4 is over an order of magnitude weaker than the ν_1 modes of the light spherical tops [1,2], further improvement in sensitivity will be required for the study of much weaker modes or overtones. Utilization of a higher power pump source and a multi-pass cell such as that used on cw stimulated Raman studies [4] should offer significant gains, thus making the IRS technique a powerful tool of general utility. Cooling of the sample, either in a cell or a molecular beam [7], should also provide a more favor-able partition function and thus a higher sensitivity for such studies. Increases in the pulsewidth of the pump source should also provide even higher resolution than is presently available.

The preliminary data gathered on the sub-Doppler Raman saturation dips illustrates the utilization of IRS for studying dynamic processes. We believe that IRS will be useful, not only for probing dynamic processes of this type, but also for probing photo-fragmentation products. The marriage of the IRS apparatus with a molecular beam will also provide a means of studying molecular dynamics on systems which have heretofore been inaccessible using conventional optical techniques.

We would like to gratefully acknowledge helpful discussions with Drs. L. A. Rahn and M. E. Riley and the expert technical assistance of R. E. Asbill.

76

References

1. A. Owyoung: "High Resolution Coherent Raman Spectroscopy of Gases", in Laser Spectroscopy IV, ed. by H. Walther, K. W. Rothe, Springer Series in Optical Sciences, Vol. 21 (Springer, Berlin, Heidelberg, New York, 1979) p. 175
2. J. A. Armstrong, J. J. Wynne and P. Esherick, J. Opt. Soc. Am. $\underline{69}$, 211 (1979).
3. F. V. Kowalski, R. T. Hawkins and A. L. Schawlow, J. Opt. Soc. Am. $\underline{66}$, 965 (1976).
   B. W. Petley and K. Morris, Opt. and Quant. Electron. $\underline{10}$, 277 (1978).
4. A. Owyoung, C. W. Patterson and R. S. McDowell, Chem. Phys. Lett. $\underline{59}$, 156 (1978) and $\underline{61}$, 636 (1979).
5. R. S. McDowell, C. W. Patterson and A. Owyoung, J. Chem. Phys. $\underline{72}$, 1071 (1980).
6. C. W. Patterson, R. S. McDowell, N. G. Nereson, R. F. Begley, H. W. Galbraith and B. J. Krohn, J. Mol. Spectrosc. $\underline{80}$, 71 (1980).
7. M. D. Duncan, and R. L. Byer, IEEE J. Quant. Electron. $\underline{QE-15}$, 63 (1979).

Surface Nonlinear Optics

Y.R. Shen, C.K. Chen, and A.R.B. de Castro

Materials and Molecular Research Division, Lawrence Berkeley Laboratory
Berkeley, CA 94720, USA, and

Department of Physics, University of California
Berkeley, CA 94720, USA

Surface electromagnetic waves are waves propagating along the interface of two media. Their existence was predicted by SOMMERFIELD in 1909 [1]. In recent years, they have found interesting applications in the study of overlayers and molecular adsorption on surfaces [2], in probing of phase transitions [3], and in measurements of refractive indices [4]. In our laboratory, we have been interested in the nonlinear interaction of surface electromagnetic waves. The motivation is two fold. First, while nonlinear optics in the bulk is a well developed field, surface nonlinear optics is still in its infant stage. Second, we would like to look into the possibility of using surface nonlinear optics for material studies. In this paper, we describe the preliminary results of our recent venture in this area.

Surface Plasmons

Surface plasmons (SP) are surface electromagnetic waves confined to the interface between a metal and a dielectric. Their propagation characteristics are governed by the dispersion relation

$$K_{\parallel} = K_{\parallel}' + i K_{\parallel}'' = \frac{\omega}{c} \left[\frac{\in_D \in_M}{\in_D + \in_M} \right]^{\frac{1}{2}}, \tag{1}$$

which can be readily derived from the boundary conditions at the interface. Here, $\in_D$ and $\in_M$ are the dielectric constants of the dielectric and the metal respectively, and $K_{\parallel}$ is the wavevector along the interface with $K_{\parallel}^2 > \omega^2 \in_D/c^2$, when $\in_M < 0$, $\in_D > 0$, and $|\in_M| > \in_D$. The wave is transverse magnetic and has the form

$$\vec{E} = (\mathscr{E}_{\parallel} \hat{\rho} + \mathscr{E}_z^D \hat{z}) e^{i K_{\parallel} \rho - \alpha_D z} \quad \text{for } z > 0$$

$$= (\mathscr{E}_{\parallel} \hat{\rho} + \mathscr{E}_z^M \hat{z}) e^{i K_{\parallel} \rho + \alpha_M z} \quad \text{for } z < 0, \tag{2}$$

where $z = 0$ is the interface between the dielectric medium in the upper half plane and the metal in the lower half plane. On either side of the interface, the wave amplitude drops off exponentially. Thus, in exciting the surface plasmon, the incoming laser power is squeezed into a layer of less than a wavelength thick $(\alpha_D^{-1} = \lambda |\in_D + \in_M|^{\frac{1}{2}}/2\pi |\in_M|, \ \alpha_M^{-1} = \lambda |\in_D + \in_M|^{\frac{1}{2}}/2\pi |\in_M|)$. The beam intensity is then greatly enhanced and the surface plasmon propagation characeristics appear to be rather sensitive to the interface structure.

78

Since small perturbation on the interface can be easily detected, surface plasmons can be used as a sensitive surface-specific probe. The enhanced beam intensity also facilitates the study of nonlinear optical effects on surfaces.

There are various methods one can use to excite a SP, either linearly or nonlinearly. The one most commonly used is the Kretschmann method, shown schematically in Fig. 1(a) [5]. The SP is excited when the angle of incidence of the exciting laser beam through the prism is properly adjusted so that its wavevector component along the surface is equal to $K_{\parallel}(\omega)$. This is seen by a corresponding reflectivity drop of the beam reflection from the prism. An example is shown in Fig. 1(b). In the experiments described below, the Kretschmann method is always used for SP excitation.

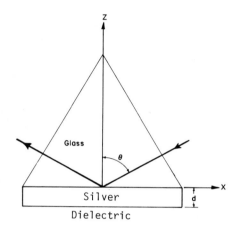

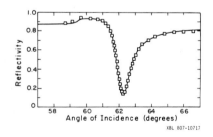

Fig.1(a) Kretschmann geometry for exciting surface plasmons

Fig.1(b) Reflectivity versus the angle of incidence θ showing the sharp dip resulting from surface plasmon excitation. The solid curve is a theoretical curve that fits the experimental data points

XBL 807-10717

Theory of Nonlinear Interaction of Surface Plasmons

We now describe briefly the theory of wave mixing of surface plasmons [6]. The output is governed by the wave equation

$$[\nabla \times (\nabla \times) - \omega^2 \epsilon/c^2]\vec{E}(\omega) = (4\pi\omega^2/c^2)\vec{P}^{(n)}(\omega)$$
$$\nabla \cdot (\epsilon\vec{E} + 4\pi\vec{P}^{(n)}) = 0, \tag{3}$$

where the source of mixing is the nonlinear polarization

$$\vec{P}^{(n)}(\omega) = \overset{\leftrightarrow}{\chi}^{(n)}(\omega = \omega_1 + \text{---} + \omega_n):\vec{E}_1(\omega_1)\text{---}\vec{E}_n(\omega_n)$$
$$\propto \exp(i\vec{k}_s \cdot \vec{r} - i\omega t), \tag{4}$$

with $\vec{k}_s = \vec{k}_1 + \text{---} + \vec{k}_n$. In our case, the fields $\vec{E}_1(\omega_1)$, ---, $\vec{E}_n(\omega_n)$ are assumed to be all SP. Equation (3) can be solved straightforwardly with the proper boundary conditions. Because of the limited space, we will not re- produce the results here. It is easy to see, however, that the output field is linearly proportional to $P_{||}^{(n)}(\omega)$ and $P_{z}^{(n)}(\omega)$. For $k_s > \omega \in_{D}^{\frac{1}{2}}/c$, the output can only appear as a coherent beam on the prism side with the beam direction determined by $\vec{k}_{s||}$, although surface roughness may couple part of the output out through the dielectric side. If $k_{s||} = K_{s||}' > \omega \in_{D}^{\frac{1}{2}}/c$, then $\vec{P}^{(n)}(\omega)$ drives the SP at ω resonantly. In other words, the SP at ω is now generated from optical mixing of the pump SP under the phase matching condition. For $k_s < \omega \in_{D}^{\frac{1}{2}}/c$, the output appears as coherent beams on both the dielectric and the prism sides with their directions determined by $\vec{k}_{s||}$.

Physically, the output arises from the collection of dipole radiation from a thin layer of oscillating dipoles induced by the mixing of pump SP at the interface. Thus, in the visible region, only a few hundred atomic or molecular layers effectively contribute to the nonlinear mixing. The output signal reduces by a factor of $10^4 - 10^5$ if only a monolayer of mater- ials is present.

Harmonic Generation by Surface Plasmons

The simplest nonlinear optical process is the optical second harmonic gener- ation. Second harmonic generation by SP is in fact readily observable even at the air-metal interface as first demonstrated by SIMON and coworkers [7]. In this case, metal is the nonlinear medium. Its second-order nonlinearity is, however, small because of the existence of inversion symmetry. The in- duced second-order nonlinear polarization arises from electric-quadrupole and magnetic-dipole contribution and can be written in the form [8]

$$\vec{P}^{(2)}(2\omega) = \alpha(\vec{E} \cdot \triangledown)\vec{E} + \beta(\triangledown \cdot \vec{E})\vec{E} + \gamma\vec{E} \times \vec{B}. \tag{5}$$

The first two terms are the electric-quadrupole terms. They are only non- vanishing within the Thomas-Fermi screening length (i.e., one or two atomic layers) near the surface. The last term is the magnetic-dipole term and is nonvanishing throughout the bulk. In practice, however, the total magnetic- dipole contribution is often negligible in comparison with the electric quadrupole contribution. This is because the nonlinearity from a single atomic layer without inversion symmetry is already appreciably larger than the nonlinearity from the magnetic-dipole contribution of a hundred atomic layers penetrated by the incoming pump field.

From the wavevector matching condition, it is easily seen that two coun- ter-propagating fundamental SP should generate a bulk second harmonic output propagating along the surface normal [9]. The corresponding nonlinear polar- ization obtained from (5) has however a vanishing component parallel to the surface. Consequently, no second harmonic generation along the surface nor- mal should be observed. We have verified experimentally that this is indeed the case. By varying the angle between the two fundamental SP, we should be able to determine the coefficients α, β, and γ in (5) separately. Such work

is presently still in progress. Harmonic generation by SP is actually a
viable method for studying optical nonlinearity of metals.

The second harmonic generation can be greatly enhanced if in the above
case, air is replaced by a nonlinear dielectric medium, for example, quartz
or KDP [7]. The process is now dominated by the nonlinearity of the nonlinear dielectric. Symmetry of the medium generally allows the existence of a
nonlinear polarization component parallel to the surface even for counter-
propagating SP. Thus, second harmonic generation along the surface normal
becomes easily observable [9]. That there is a one-to-one correspondence
between the propagation direction of the second harmonic beam in the three-
dimensional space and the interaction geometry of the two SP in the two-dimensional plane may find applications in information processing.

We have also observed third harmonic generation by SP at the interface
between metal and some organic liquid. The nonlinearity of the liquid appears to dominate, as is evidenced by the reduction of the third harmonic
signal beyond our detectability when the liquid is replaced by air. If the
SP dispersion curve $\omega(K_\parallel)$ concaves downward, then $2K_\parallel^i(\omega) < K_\parallel^i(2\omega)$ and it is
impossible to have phase-matched generation of a harmonic SP by fundamental
SP. Anomalous dispersion of the dielectric may be used to achieve phase
matching.

Surface CARS

Coherent anti-Stokes Raman scattering (CARS) has recently become a useful
spectroscopic technique [10]. The nonlinear polarization governing the anti-
Stokes output is

$$\vec{P}^{(3)}(\omega) = \overset{\leftrightarrow}{\chi}^{(3)}(\omega = 2\omega_1 - \omega_2):\vec{E}(\omega_1)\vec{E}(\omega_1)\vec{E}^*(\omega_2)$$

with the nonlinear susceptibility

$$\overset{\leftrightarrow}{\chi}^{(3)} = \overset{\leftrightarrow}{\chi}_{NR}^{(3)} + \frac{A}{[(\omega_1 - \omega_2) - \omega_v] + i\Gamma} \cdot$$

As $(\omega_1 - \omega_2)$ approaches the vibrational frequency ω_v of the nonlinear medium,
$\overset{\leftrightarrow}{\chi}^{(3)}$ is resonantly enhanced, and so is the anti-Stokes output. CARS can
therefore be used for probing Raman resonances. Clearly, the same process
can be extended to the surfaces using SP. Actually, with surface SP as pump
waves, the magnitude of $\vec{P}^{(3)}(\omega)$ can be greatly enhanced. So the anti-Stokes
output is still quite appreciable even though, as we discussed earlier,
$\vec{P}^{(3)}(\omega)$ is limited to a very thin layer near the interface.

We have demonstrated the feasibility of surface CARS using the setup
shown in Fig.2 [11]. The SP at ω_1 and ω_2 are efficiently excited by the input beams through the prism. In this case, phase-matched generation of anti-
Stokes SP is possible by properly adjusting the relative angle between the
pump SP in the surface plane. Then, the resonant spectrum of the anti-Stokes
output can be obtained by scanning $(\omega_1 - \omega_2)$. An example of pyridine on
silver is presented in Fig.3, where the experimental results are in good
agreement with the theoretical prediction. For an input of 2.8 mJ/cm^2 at ω_1
and 32 mJ/cm^2 at ω_2 with a pulsewidth of 30 ns and a beam cross section of
0.25 cm^2, the measured signal at the resonance 991 cm^{-1} peak is 1.5×10^4
photons/pulse, while the theoretical prediction is 3.3×10^4 photons/pulse.
Other characteristic features of the observed anti-Stokes output also agree
well with the surface CARS prediction.

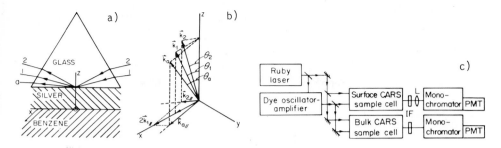

Fig. 2 (a) Prism-metal-liquid assembly for surface CARS measurement. Beam 1 is in the x-z plane, but beam 2 and the output are not. (b) Wavevectors in the glass prism with components in the x-y plane phase matched. (c) Diagram of the apparatus: IF is an interference filter and L is a lens

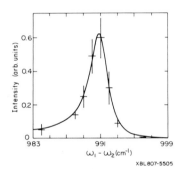

XBL807-5505

Fig.3 Anti-Stokes signal versus $\omega_1-\omega_2$ around the 991 cm^{-1} Raman resonance of pyridine

Surface plasmons actually have rather short attenuation lengths ℓ because of the large loss in the metal. Typically, $\ell = 1/K_\parallel \sim 10~\mu m$ on silver in the visible. This means that the interaction length in surface CARS is only of the order of 10 μm. Then, even if the dielectric medium is strongly absorbing, the anti-Stokes is not expected to be appreciably reduced by the absorption. We have tested this out with a 1 : 2 acetone-benzene mixture on silver. When 1.1 mM of oxazine 725 is dissolved in the solution so that it has an absorption coefficient of $\alpha \sim 400$ cm^{-1} at the anti-Stokes frequency, the anti-Stokes output remains essentially unchanged.

In summary, the surface CARS has the characteristics of large induced nonlinear polarization, small field penetration depth into the medium at the surface, short nonlinear interaction length, a highly directional coherent output, and the possibility of an effective reduction of luminescence background. It can therefore be used to study materials with strong absorption and luminescence, thin films, molecular overlayers, and adsorbed molecules, and other surface specific problems. The surface CARS output increases with the input laser intensities as $I^2(\omega_1)I(\omega_2)$. Its ultimate sensitivity is limited by optical damage on the surface. Since the damage usually has a fluence threshold rather than an intensity threshold, the ultimate sensitivity of surface CARS can be greatly improved by using picosecond pump pulses. From our experimental results with Q-switched pulses, we can estimate an output of $\sim 10^{11}$ photons/pulse for benzene on silver from an input of 10 μJ/pulse with a pulsewidth of 10 ps and a focal spot of 0.15 mm^2. As we mentioned earlier, the signal will reduce by a factor of $10^4 - 10^5$ if there is only a monolayer of benzene molecules on silver. Thus, with picosecond pulses, surface CARS should have the sensitivity of detecting sub-monolayer of adsorbed molecules.

Enhanced Second Harmonic Generation on a Rough Metal Surface

Recently, surface enhanced Raman scattering has attracted a great deal of attention [12]. The effective Raman cross section of some adsorbed molecules on a roughened silver surface seems to have increased by $10^5 - 10^6$ in comparison to that of the same molecules in solution. Various mechanisms have been suggested to explain the enhancement: some are purely electromagnetic in origin while others rely on the quantum-mechanical interaction between molecules and metal. A recent controlled experiment of ROWE et al. [13] indicates that local field enhancement in local regions of the rough surface is mainly responsible for the Raman enhancement. According to the simple theoretical model [14], the local field enhancement decays away in a distance of few molecular layers from the surface.

If the local field picture is correct, then the interaction of molecules with metal is only of secondary importance. Furthermore, the enhancement phenomenon should be fairly general. It should show up in all nonlinear optical effects involving metal surfaces. In this respect, second harmonic reflection from metal surfaces in air is most interesting because, as we mentioned earlier, only one or two atomic layers at the surface are supposed to contribute to the second harmonic generation, and the local field effect is particularly strong in the surface atomic layers.

We have carried out a preliminary study of second harmonic reflection on rough silver surfaces. We prepare the rough surfaces by cycling in electrolytic solution followed by dry cleaning with nitrogen gas. In comparison with a smooth surface, the roughened surface shows an enhancement of $> 10^3$ in the second harmonic output. Unlike the smooth surface, the second harmonic output from a rough surface is strongly diffused by scattering from surface roughness. Although theoretically the local field enhancement in second harmonic reflection may not be the same as that in the Raman case, our results do indicate that the local field effect is at least partially responsible for the surface Raman enhancement.

This work was supported by the Division of Materials Sciences, Office of Basic Energy Sciences, U.S. Department of Energy, under contract No. W-7405-ENG-48.

References

1. A. Sommerfeld, Ann. Physik 28, 665 (1909).
2. W. H. Weber, Phys. Rev. Lett. 39, 153 (1977); J. G. Gordon and J. D. Swalen, Optics Comm. 22, 374 (1977).
3. V. M. Agranovich, JETP Lett. 24, 558 (1976); K. C. Chu, C. K. Chen, and Y. R. Shen, Mol. Cryst. Liq. Cryst. 59, 97 (1980).

4. N. M. Chao, K. C. Chu, and Y. R. Shen, Mol. Cyrst. Liq. Cryst. (to be published).
5. E. Kretschmann, Z. Phys. $\underline{241}$, 313 (1971). The surface plasmon dispersion relation of (1) is slightly modified in the Kretschmann geometry.
6. F. DeMartini and Y. R. Shen, Phys. Rev. Lett. $\underline{36}$, 216 (1976).
7. H. J. Simon, D. E. Mitchell, and J. G. Watson, Phys. Rev. Lett. $\underline{33}$, 1531 (1974); H. J. Simon, R. E. Benner, and J. G. Watson, Opt. Commun. $\underline{23}$, 245 (1977).
8. N. Bloembergen, R. K. Chang, S. S. Jha, and C. H. Lee, Phys. Rev. $\underline{174}$, 813 (1968).
9. C. K. Chen, A. R. B. de Castro, and Y. R. Shen, Opt. Lett. $\underline{4}$. 393 (1979).
10. See, for example, M. D. Levenson, Phys. Today $\underline{30}$, 45 (1977).
11. C. K. Chen, A. R. B. de Castro, Y. R. Shen, and F. DeMartini, Phys. Rev. Lett. 43, 946 (1979).
12. See, for example, T. E. Furtak and J. Reyes, Surf. Sci $\underline{93}$, 351 (1980) and references therein.
13. J. E. Rowe, C. V. Shank, D. A. Zwemer, and C. A. Murray, Phys. Rev. Lett. $\underline{44}$, 1770 (1980).
14. S. M. McCall, P. M. Platzman, and P. A. Wolff (to be published).

A Quasi-Nonlinear Scattering Process: Probing of Short-Lived (μs to ps) Optically Pumped Excited States

J.A. Koningstein and M. Asano

Metal Ions Group, Department of Chemistry, Carleton University
Ottawa, Ontario K1S 5B6, Canada

Results and Discussion

The polarized Raman spectrum of corundum was recorded and assigned by PORTO and KRISHNAN [1] in 1967 and assignment of phonons of ruby is therefore straightforward. Contrary to corundum, ruby is characterized by electronic states which absorb light from the visible part of the electromagnetic spectrum while strong fluorescence occurs in the R_1 and R_2 lines. The emission is due to transitions between the $\bar{E}$ and $2\bar{A}$ levels - see Fig.1 - which are at 14418 cm^{-1} and 14447 cm^{-1} above the $^4A_2$ ground state.

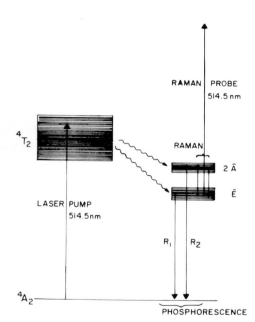

Fig.1 Part of the energy level diagram of Cr^{3+} in a field of trigonal symmetry. Also indicated is the optical pump which produces the population of the component levels of ^2E while part of the radiation not used for pumping excites the electronic Raman transition between these levels.

Population inversion of the $\bar{E}$ and $2\bar{A}$ levels and that of the ground state is rather easily obtained because the lifetime of this excited state is $\sim$ 3 ms.

The 514.5 nm radiation of an argon ion laser may be used to populate these states because the energy absorbed by ruby at that wavelength reaches the $\bar{E}$ and $2\bar{A}$ levels. If the laser beam is focussed down inside a 1 cm long ruby crystal rod then the trace inside this rod can be approximated by a 1 cm long cylinder with a diameter of 0.005 cm in which there are $\sim 10^{14}$ Cr^{3+} ions - for a ruby crystal which has 0.01% chromium by weight. If exposed to 10^{17} phonons/sec of an argon ion laser we find that per lifetime there are 3×10^{14} phonons for $\sim 10^{14}$ Cr^{3+} ions and consequently a large number of these ions are in the excited states. The Raman spectrum of $Cr^{3}:\alpha Al_2O_3$ shows a shift at 29 cm^{-1} - see Fig.2 - and the intensity of the band is linear in I^2_{laser} [2]. This is to be expected because upon a change

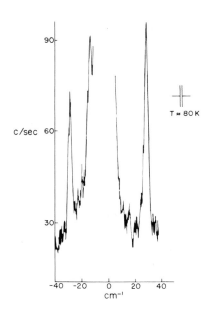

Fig.2 Part of the 80 K Raman spectrum of α-Al$_2$O$_3$:Cr^{3+}. The shift at 29 cm^{-1} is due to the electronic Raman transition between the components of ^{2}E.

of I_{laser} the population of the excited states changes linearly (assuming that saturation conditions are not met) and this is also the case for that part of the laser radiation at 514.5 nm which serves as the Raman probe. Other quasi-non-linear optical pumped excited state Raman transitions [3] were recorded for Cr^{3+} in $ZnAl_2O_4$, Cr^{3+} in YAlG and of Beryl. The Raman shift of Cr^{3+} in spinel is rather small (6.7 cm^{-1}) - see Fig.3, those for garnet and beryl are at 18.9 cm^{-1} and 62.8 cm^{-1}. In case the R_1 and R_2 levels are inhomogeneously broadened - low T - we predict that the width (ΔR) of the electronic Raman transition approaches the value $\Delta R = \Delta R_1 - \Delta R_2$ where ΔR_1 and ΔR_2 are the width of the R_1 and R_2 lines respectively. On the other hand for very high T, when all relevant phonons are populated, we expect the relation $\Delta R = \Delta R_1 + \Delta R_2$ to hold and this sum to be born out by experiments as shown in Figs.4 and 5 for ruby and spinel. Apparently for spinel the inhomogeneous broadening process plays a more important role than the homogeneous process in the temperature interval of 100 K - 200 K while for ruby homogeneous broadening becomes important for T > 150 K.

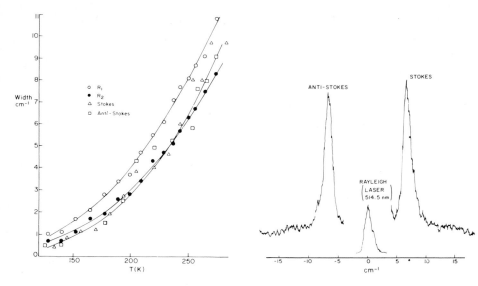

Fig.3 The optical pumped electronic Raman transition of $Cr^{3+}:ZnAl_2O_4$ at 77 K.

Fig.4

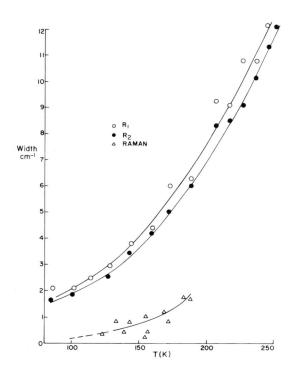

Fig.5 Plots of the temperature dependent line width of the R lines and electronic Raman transition of Cr^{3+} in yttrium aluminium garnet.

Before the availability of pulsed laser little attention has been given
to the fact that the PMT noise does not contribute to the S/N of pulsed
laser Raman spectra if they are detected with a PMT which rise time is
shorter than that of the pulse followed by gaited amplification of the PMT
signals. We have used a tunable dye laser giving 3.2 nm pulses with > 100 KW
peak power to induce Raman spectra and compared to the c.w. case we are in
a position to populated excited states having a lifetime of picoseconds and
induce light scattering thereof. If focussed down in a cell of 1 cm length
containing 10^{-4} molar solutions (or inside a rod like αAl_2O_3 containing
impurity ions) one calculates that there are $\sim 10^{10-11}$ molecules in the
trace of the beam. There are $\sim 10^{14}$ photons in each pulse, hence even if
the lifetime of an excited state of the solute or impurity is 1 ps does a
situation arise when the number of photons delivered per lifetime is larger
than the number of molecules in the trace of the beam. Hence a large number
of excited states are created as becomes evident from transmission studies.
A case in point is a 10^{-4} molar solution of the tris(2,2'-bipyridine)Cr III
complex in solution and the dependence of transmission or laser intensity is
shown in Fig.6 for λ_1 = 457.9 nm. This radiation is absorbed by the
$^4T_2 \leftarrow {}^4A_2$ band system, Fig.7, and for a defocussed laser beam the number of
photons is not large enough to populate any of the excited states. This is
quite different if the beam is focussed down and we observe a decrease in T
if I_ℓ increases. For 2.5 mW of average power ($\sim 10^{14}$ photons) we cannot
distinguish any absorptions from the ground state any more and the value of
T is dominated by that of $^4Y \leftarrow {}^4T_2$ transitions. If σ and σ^* are the cross
section for the transitions at λ_ℓ from respectively the ground and (populated)
excited state, then the number of photons (N_{ph}^{EST}) required to effectively
saturate (EST) the excited state is given by

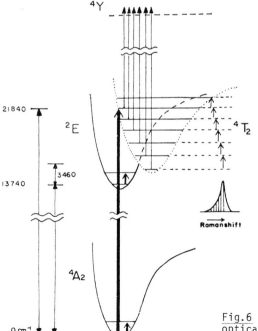

Fig.6 Level diagram of $Cr(bpy)_3{}^{3+}$ and
optical transitions for high laser power
(pulsed) at 458.0 nm.

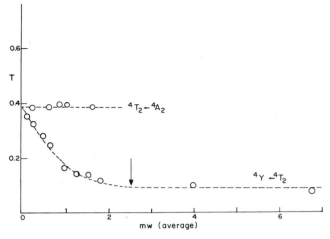

Fig.7 Transmission versus I_ℓ at 457.9 nm. Horizontal line, laser not focussed, lower curve, laser focussed.

$$N_{ph}^{EST} = N_v \cdot \frac{(\sigma + \sigma^*)}{\sigma} \cdot \frac{1}{\tau^*}$$

where N_v is the number of molecules in the path of the laser beam, σ^* and σ are obtained from respectively T at apparent saturation, and T for $I_\ell \rightarrow 0$. and τ^* is the lifetime of the excited state. We obtain the value $\tau^*_{^4T_2} = 10$ ps for $Cr(bpy)_3^{3+}$. A comparison of Raman bands of the same normal mode in $^4A_2$ and $^4T_2$ reveals (Fig.8) that the latter has a larger width and is asymmetric towards lower shift. This is due to the fact that successive vibrational levels of the mode in $^4T_2$ become populated. Also because $I_v = (v + 1)I_0$ where I_0 is the intrinsic intensity for the vibrational Raman transition from $v = 1 \leftrightarrow v = 0$, we find that during the rise time of the light pulse the time evolution of population and scattering of levels in the top of the surface is favoured over that of levels in the bottom of the electronic state, while the opposite holds during the decay part of the pulse and energy transfer via vibrational relaxation at the 1 - 10 ps time region takes place.

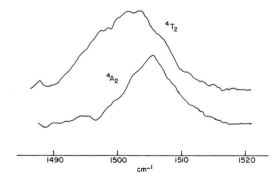

Fig.8 A comparison of band contours of Raman scattering for a normal mode in $^4A_2$ and the 10 ps lived $^4T_2$ state.

The results described above suggest that only a beginning has been made with the exploration of excitation of Raman transitions in optical pumped excited states and evidence point to the fact that the widths of the Raman bands of scattering from such pumped states are not much different from the widths of Raman bands of scattering from the ground state.

Compared to broad features associated with components of an electronic excited state we find that the excited state Raman spectra a yield much sharper spectra making it easier to study the differences of vibrational gaps of modes in ground and excited states of stable [3,5] as well as intermediate [4] species. The methods should also be capable to find scattering due to excitions but a complication may occur in that fluorescence or phosphorescence, originating from the excited state takes place. Quenching of this fluorescence can be achieved by a process where the excited gets populated to such a large degree that absorption from this state to an even higher lying and much shorter lived level occurs. This state in turn may release its energy via radiative or nonradiative paths which do not involve the excited state which is to be Raman probed. The chlorophyll molecule is a case in point. By pumping into the Soret band in the blue we suppress the fluorescence in the red which opens up the possibility to probe the lower lying singlet and triplet states and their lifetimes by applying (1). For chlorophyll in pyridine we obtained the following results. The first excited singlet state S_1 at 14780 cm^{-1} has a lifetime of $\tau_{S_1} = 45$ ns, S_3 at 19000 cm^{-1} has $\tau_{S_3} = 8 \pm 2$ ns, S_4 at 22625 has $\tau_{S_4} = 110$ ps. For the triplet system we found that the lifetime of T_0 (which is some 10150 cm^{-1} above the S_0 ground state) amounts to $\tau_{T_0} = .8 \times 10^{-4}$ s. Fluorescence due to $S_1 \rightarrow S_0$ of the dimer occurs at about the same frequency as that of the monomer but has a much longer lifetime ($\tau^0_{S_1} = 145$ ns in pyridine). Evidence has been found for the exciton splitting of Δ 110 cm^{-1} because these excited states too can be populated with pulses at $\lambda = 460.0$ nm which allowed us to record the (resonance enhanced) Raman exciton type electronic Raman effect between the dimer S_1 states in the wing of the Rayleigh line of Chl a in solution.

References

1   S.P.S. Porto and R.S. Krishnan, J. Chem. Phys. 47, 1009 (1967).

2   B. Halperin and J.A. Koningstein, J. Chem. Phys. 69, 3302 (1978).

3   B. Halperin, D. Nicollin and J.A. Koningstein, Chem. Phys. 42, 277 (1979).

4   G.H. Atkinson and L.R. Dossee, J. Chem. Phys. 72, 2195 (1980).

5   D. Nicollin and J.A. Koningstein, J. Chem. Phys. to be published.

6   ·M. Lutz, J. Raman Spectrosc. 2, 497 (1974)

Raman Scattering Study
of the Phase Transitions in $(NH_4)_2Cd_2(SO_4)_3$

J.C. Galzerani* and R.S. Katiyar

Instituto de Física "Gleb Wataghin", Universidade Estadual de Campinas
13100 - Campinas - SP - Brazil

1. Introduction

Di-ammonium di-cadmium sulfate (CAS) belongs to the langbeinite structure,
with space group T^4 containing four formula units in the primitive cell
[1,2]. JONA and PEPINSKY [3] were the first to discover the existence of a
phase transition in this crystal. The low-frequency dielectric studies by
OSHIMA and NAKAMURA [4] showed that $\varepsilon(0)$ remains practically constant from
room temperature to about 95 K, where it suddenly jumps to a higher value,
which decreases exponentially on lowering the temperature further. The sudden
jump in $\varepsilon(0)$ is accompanied by a paraelectric-ferroelectric phase transition
with the appearance of a constant spontaneous polarization [5]. For these
reasons, CAS is known as an improper ferroelectric.

From the free energy consideration, DVORAK [6] showed that in the langbeinite, four homogeneous structural phase transitions are possible. Three
of them are improper ferroelectrics and they result from the instability of
a M-point phonon; the ordered phase would belong to the space group C_2^2, C_3^4
or C_1^4. Additionally the calculations showed the possibility of a structural
phase transition from T^4 to D_2^4.

In the present work, we carried out systematic studies of polarized
temperature-dependent Raman scattering in CAS, in order to understand the
phonon behaviour in relation to the phase transition.

2. Vibrational Analysis in the Cubic Phase

Using the X-ray data of ZEMANN and ZEMANN [1], the total number of vibrational modes of CAS may be classified as $27A + 27E + 80F$. These modes can,
however, be grouped into two regions, namely, the internal modes which have

---

* Work supported partially by FAPESP and CNPq.

frequencies greater than ~ 300 cm^{-1}, and the external modes having frequencies less than this value. The total number of modes in the external frequency region should be classified as 10A + 10E + 29F neglecting the librational modes of NH$_4$ group. Comparing with the polarizability tensors for the point group T, the Raman spectra with the diagonal components should show peaks corresponding to A + E symmetries, whereas the F modes may be observed by studying the non-diagonal components.

3. Experimental

Single crystal with approximately 1 cm^3 was acquired from Minhorst Company of Germany. It was cut in small pieces and polished with diamond paste. The spectra were taken with a conventional Raman arrangement for 90° scattering. The source was an argon ion laser and the scattered light was analysed with the Spex 1401 double monochromator and detected by a photomultiplier FW 130. The resolution was about 1.5 cm^{-1}. The samples were mounted in a liquid nitrogen dewar for measurements from room temperature up to 82 K. For measurements below 82 K we used a close cycle helium system from Air Products, Inc. Two platinum resistances were used to monitor the temperature and using an Oxford temperature controller, we obtained stability better than 0.5 K.

4. Results and Discussion

4.1 External Modes and the Phase Transitions

In this section we discuss lattice modes of CAS and their variation with temperature. The Raman spectra for various polarizations and temperatures were studied and some of them with Y(ZZ)X orientation are shown in Fig.1.

It may be noted that the number of modes observed at room temperature is much less than predicted by group theory.

A careful study of low-frequency Raman spectra of CAS for the orientation Y(XX)Z at several temperatures is shown in Fig.2. The spectrum at 92 K shows a well-defined peak at 24.3 cm^{-1}, which decreases in frequency on increasing the temperature. At about 122 K, the frequency of this mode is so low that it appears as a wing to the Rayleigh line.

The temperature behaviour of the square of the frequency of this mode is plotted in Fig.3, and it appears to have linear temperature dependence,

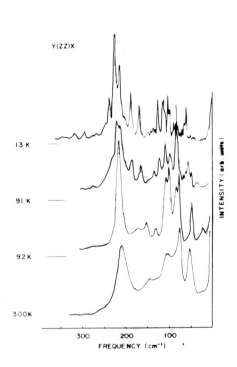

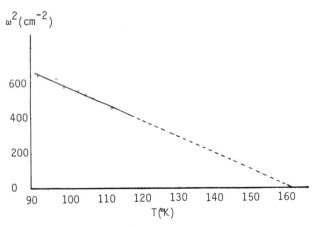

Fig.1. Raman spectra of CAS, with Y(ZZ)X orientation, at various temperatures

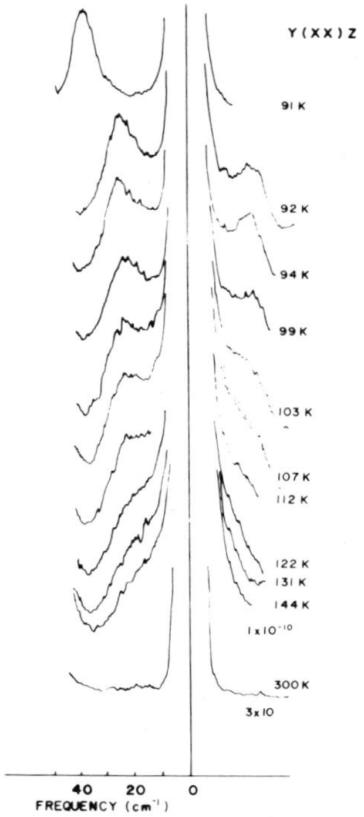

Fig.2. Temperature dependence of the low-frequency mode of CAS

Fig.3. Temperature dependence of the square of the low-frequency mode of CAS, above the ferroelectric phase transition

intercepting the temperature axis at ~ 162 K. This shows as if there had been some crystal instability around this temperature, similar to that observed in $SrTiO_3$ at 110 K [7]. In fact, if such a structural transition exists, it should be antidistortive, with a soft zone boundary mode in the cubic phase, which becomes Raman active in the lower phase, as the result of the doubling of the unit cell. Following DVORAK's analysis [6], this new intermediate phase (between 162 K and 91.5 K), may be assigned to D_2^4 symmetry, and from our spectra we have evidence of this. Other evidence in support of such a hypothesis may be drawn from the fact that the number of modes below 162 K in all the orientations studied increased approximately by a factor of two.

The Raman spectra reported in Fig.1 show that there are several changes in them, while lowering the temperature from 92 K to 91 K. Figure 2 shows that the mode at 24.2 cm^{-1} suffers a discontinuity and appears at 37 cm^{-1} in the 91 K spectrum and its frequency increases slightly on lowering the temperature further. These changes enabled us to determine the ferroelectric transition temperature to be 91.5 K. The spectra in the lower phase are in accordance with the C_2^2 group predicted [6]. This transition is known as improper, as it is triggered by a non-polar zone boundary phonon, which may couple (directly or indirectly) to a zone center polar mode, and, through condensation, induce a small polarization in an indirect way.

The transition at 91.5 K in CAS, may be compared with the transition in $Tb_2(MoO_4)_4$, studied by neutron inelastic scattering [8]. In the latter case, the actual transition takes place at T = 159° C, but the extrapolated value of T_0 for the doubly degenerate zone boundary soft mode, is 149° C, thereby classifying it as a first-order transition. Moreover, DORNER et al. [8] sug-

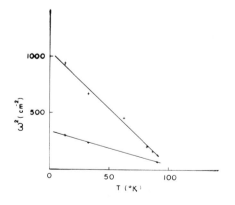

Fig.4. Square of the low-frequency modes in CAS (ferroelectric phase), as a function on temperature

gested that the above degenerate mode should give rise to two A_1 zone center modes in the ferroelectric phase. Our spectra of CAS, for temperature below 91 K, show two weak low-frequency modes, which increase in frequencies on lowering the temperature further. These modes are indeed predicted by DVORAK's theory [6] in the ferroelectric phase and the transition is triggered by the softening of a M-point doubly degenerate mode. The transition is of first order because the square of the frequencies of these modes (shown in Fig.4) do not go to zero at T_c. Moreover, their feeble appearance in the spectra is in accordance with the DVORAK's predictions [6]. Figure 4 should be compared to the scheme shown in [8] for a first-order improper transition.

4.2 The Rayleigh Line and the Critical Scattering

On heating the crystal through the ferroelectric transition, we noted that in the close vicinity of the transition temperature, the scattered light increases anomalously. This phenomenon could also be noticed, with lower intensity, while cooling the crystal slowly through the transition temperature. We repeated these spectra several times, and the results of a typical run are shown in Fig.5. The laser power used was 200 mW, attenuated by two neutral density filters, each with density 3. The effect is similar to the one observed in NH_4Cl by LAZAY et al. [9].

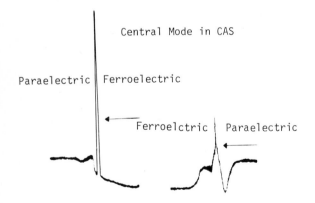

Central Mode in CAS

Paraelectric | Ferroelectric

Ferroelctric | Paraelectric

93 92 91 90 89 90 91 92
T(°K)   T(°K)

Fig.5. Temperature dependence of the intensity of the central mode in CAS, near the ferroelectric transition

Because of the instrumental limitations, we were unable to measure the width of this so called "central mode", in order to study its possible correlation with the divergence of the dielectric fluctuations, similar to that observed in KDP [10]. Another hypothesis is the existence of microdomains, located in a transitory manner [11-13]. OSHIMA and NAKAMURA [4] have indeed mentioned that the anomalous relationship between the dielectric constant and the temperature seems to show the existence of multidomains and that the motion of domain walls at the transition temperature could be responsible for it.

References

1. V.A. Zemann, J. Zemann: Cryst. 10, 409 (1957)
2. F. Jona, G. Shirane: "Ferroelectric Crystals", Pergamon Press, Oxford (1962)
3. F. Jona, R. Pepinsky: Phys. Rev. 103, 1126 (1956)
4. H. Oshima, E. Nakamura: J. Phys. Chem. Solids 27, 481 (1966)
5. M. Glogarová, J. Fousek: Phys. Status Solidi 15, 579 (1973)
6. V. Dvorak: Phys. Status Solidi (b) 52, 93 (1972)
7. P.A. Fleury, J.F. Scott, J.M. Worlock: Phys. Rev. Lett. 21, 16 (1968)
8. B. Dorner, J.D. Axe, G. Shirane: Phys. Rev. (b) 6, 1950 (1972)
9. P.D. Lazay, J.H. Lunacek, N.A. Clark, G.B. Benedek: in *Light Scattering Spectra of Solids*. Proc. Int. Conf. 1968, ed. by G.B. Wright (Springer, Berlin, Heidelberg, New York 1969)
10. M.D. Mermelstein, H.Z. Cummins: Phys. Rev. (b) 16, 2177 (1977)
11. T. Schneider, E. Stoll: Phys. Rev. Lett. 35, 295 (1975)
12. S. Aubry: J. Chem. Phys. 62, 3217 (1975)
13. J.A. Krumhansl, J.R. Schrieffer: Phys. Rev. (b) 11, 3535 (1975)

Part II

Laser Spectroscopy

Magic Angle Line Narrowing in Optical Spectroscopy[*]

S.C. Rand, A. Wokaun, R.G. Devoe, and R.G. Brewer

IBM Research Laboratory, San Jose, CA 95193, USA

ABSTRACT: Spin decoupling and line narrowing are observed for the first time in an optical transition, $^3H_4 \leftrightarrow {}^1D_2$ of Pr^{3+} in LaF_3 at 2°K, using optical FID. The ^{19}F nuclei, when irradiated by an appropriate rf field, undergo forced precession about an effective field at the magic angle in the rotating frame. The fluctuating $^{19}F-^{19}F$ dipolar interaction is thereby quenched and the optical linewidth drops from ~10 to ~2 kHz, as predicted in a theory of spin diffusion.

In the field of nuclear magnetic resonance (NMR), there exist several ways of reducing the time-dependent magnetic dipolar interaction between spins. Examples are motional narrowing [1,2], macroscopic sample rotation [3], spontaneous spin flip-flop processes [4], and forced spin precession [5-10]. In this article, we report the first observation of this kind in an optical transition of a low temperature solid, $LaF_3:Pr^{3+}$, where the dilute nuclear spin (I) is praseodymium and the abundant spin (S) is fluorine. The Pr^{3+} ions are coherently prepared by a laser field and thereafter exhibit nonlinear optical free induction decay (FID). The Pr^{3+} dephasing time, as suggested in earlier work [11], is limited by spin diffusion among the ^{19}F nuclei which undergo resonant flip-flops and impress weak fluctuating fields on the ^{141}Pr nuclei. This action adiabatically modulates the Pr^{3+} optical transition frequency through the dipolar I-S interaction and broadens the line. We now show that the half-width half-maximum (HWHM) optical linewidth is reduced from ~10 to ~2 kHz when the fluorine spin diffusion process is quenched by application of suitable magnetic static and rf fields, causing the ^{19}F nuclei to precess about an effective magnetic field at the *magic angle* in the rotating frame. This observation enables us to identify in an unambiguous way that the $^{19}F-^{19}F$ dipolar interaction is the dominant optical line broadening mechanism and provides the first test of spin diffusion theory in an optical transition. As we shall see, the behavior at optical and rf frequencies is different.

Imagine that the Pr^{3+} ions are coherently prepared by a laser field in the optical transition $1 \leftrightarrow 2$ and then experience nonlinear FID when the laser frequency is switched outside the Pr^{3+} homogeneous linewidth. The novelty of this technique [11] is that a single homogeneous packet ($\leq$10 kHz width) can be selected from the much broader inhomogeneous lineshape (5 GHz width). The FID signal, expressed in terms of the induced polarization, is of the form

$$\langle p(t) \rangle = \left\langle p(0) e^{i(\omega_{12}t + \int_0^t \delta\omega_{12}(t')dt')} \right\rangle \tag{1}$$

[*] Work supported in part by the U.S. Office of Naval Research.
Published in Physical Review Letters 43, 1868 (1979).

where $\delta\omega_{12}(t')$ represents the fluctuation in the optical transition frequency ω_{12} due to the nuclear dipolar S-S and I-S interactions. The bracket denotes an average over the optical inhomogeneous lineshape, the geometric variables of the I-S interaction, and the S-S spin fluctuations. With the assumption of Markoffian spin statistics, we apply the spin diffusion theory of Klauder and Anderson [12] and find that (1) predicts a Lorentzian homogeneous lineshape having a HWHM linewidth

$$\Delta\nu = \frac{4\pi\hbar}{9\sqrt{3}} \left| \left(\gamma_I'' I_z'' - \gamma_I' I_z' \right) \gamma_S S_z \right| n_S \frac{r}{R} .$$ (2)

Here, γ_I ($\sim$23 kHz/G) denotes the enhanced gyromagnetic ratio of $^{141}Pr^{3+}$ for the lower (double prime) or upper (single prime) electronic state. The fluorine spin S has a gyromagnetic ratio γ_S (4 kHz/G), number density n_S, flips at the intrinsic rate r and has a macroscopic rate parameter R, introduced by Klauder and Anderson [12] to assure stationarity.

Now consider the application of a static magnetic field B_0 and a radio frequency field $B(t)=2B_x\cos\omega t$ which is detuned from the fluorine Larmor frequency $\gamma_S B_0$ by $\Delta_S = \gamma_S B_0 - \omega$. In a frame rotating at the frequency ω, the effective field $B_e = \left(B_x^2 + (\Delta_S/\gamma_S)^2 \right)^{\frac{1}{2}}$ is stationary and makes an angle $\beta = \tan^{-1}(\gamma_S B_x/\Delta_S)$ with the static field B_0. When the ^{19}F precession frequency $\gamma_S B_e$ exceeds the square root of the S-S second moment, the ^{19}F precessional motion at the angle β will tend to be uninterrupted, and it is then advantageous to perform a transformation to a second rotating frame where the axis of quantization is parallel to $\vec{B}_e$ [8-10]. In this tilted rotating frame, the secular part of the dipolar Hamiltonian $\mathcal{H}=\mathcal{H}_{SS}+\mathcal{H}_{IS}$ contains the time-independent terms

$$\mathcal{H}_{IS}(\beta) = \cos\beta\mathcal{H}_{IS}^{(0)}$$ (3a)

$$\mathcal{H}_{SS}(\beta) = \frac{1}{2}(3\cos^2\beta-1)\mathcal{H}_{SS}^{(0)}$$ (3b)

where $\mathcal{H}_{IS}^{(0)} = \sum_j 2(b_j''-b_j')S_{jz}I_z$ and $\mathcal{H}_{SS}^{(0)} = \sum_{k<j} a_{kj}(3S_{kz}S_{jz}-\vec{S}_k\cdot\vec{S}_j)$
are the corresponding dipolar terms in the absence of an rf field, a_{kj} and b_j being the usual geometric factors [8-10]. We are now able to modify Eq. (2) by including the effect of an rf field on the Pr^{3+} optical linewidth. First, (3a) implies that $S_z(\beta)=\cos\beta S_z(0)$ [5], which replaces the heteronuclear term S_z in (2). Second, we associate the S spin flipping rate r in (2) with an inverse correlation time, $r(\beta)\equiv 1/\tau_c(\beta)=|\frac{1}{2}(3\cos^2\beta-1)|/\tau_c(0)$, derived previously by Mehring et al. for NMR using the result (3b). Equation (2) now takes the form

$$\Delta\nu(\beta) = \Delta\nu(0)\left|\cos\beta \cdot \frac{1}{2}(3\cos^2\beta-1)\right| ,$$ (4)

and gives the Pr^{3+} optical linewidth as a function of the off-resonance angle β. Equation (4), which is shown in Fig. 3, predicts that the linewidth vanishes at the magic angle $\beta=\cos^{-1}\sqrt{1/3}$ or 54.7° and also at $\beta=\pi/2$, the fluorine resonance condition.

The Pr^{3+} transition [11] examined, $^1D_2 \leftrightarrow ^3H_4$ at 5925Å, involves the
lowest crystal field components of each state. These are electronic
singlet states that couple to the Pr^{3+} nuclear spin I=5/2 in second order,
producing quadrupolar splittings of order 10 MHz and an enhanced nuclear
gyromagnetic ratio γ_I. Three equally intense optical transitions occur,
$I''_z \leftrightarrow I'_z = \pm 5/2 \leftrightarrow \pm 5/2$, $\pm 3/2 \leftrightarrow \pm 3/2$, and $\pm 1/2 \leftrightarrow \pm 1/2$, and overlap due to the large
inhomogeneous strain broadening of ~5 GHz. We investigated the $\pm 1/2 \leftrightarrow \pm 1/2$
transition, which is easily identified because it appears as a long decay
following the faster 5/2 and 3/2 components, consistent with Eq. (2) and a
computer simulation of a triexponential decay.

Optical FID is monitored by laser frequency switching [11] using the
pulse sequence of Fig. 1. An acousto-optic Bragg modulator, driven by a
110 MHz rf pulse generator, deviates the beam of a cw ring dye laser
through a 1.5 mm aperture while imparting a 110 MHz laser frequency switch.
Before and after the pulse the undeviated beam is blocked and since the
pulse repetition rate is 1 Hz, complications due to optical pumping [11]
are reduced. The transmitted beam propagates along the c axis (laboratory
y axis) of a $5 \times 6 \times 7$ mm^3 crystal of $LaF_3:Pr^{3+}$(0.1% Pr^{3+}) before striking a
p-i-n photodiode. The laser field is linearly polarized along the
laboratory x axis and has a 1.0 mm diameter in the crystal at a power of
10 mW. In Fig. 1, we see that the Pr^{3+} ions are coherently prepared by the
optical field during the initial 200 µsec of the pulse, and then FID occurs
when the laser frequency shift is suddenly reduced from 110 to 108 MHz, the
laser field remaining constant. Most important, broadening due to laser
frequency jitter is reduced by frequency-locking the ring laser to a
passive external reference cavity, the laser stability being ~1 kHz over
the duration of the measurement, ~50 µsec.

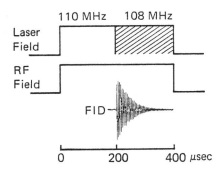

Fig.1. Pulse sequence showing the laser
field and its frequency shift and the
spin decoupling radio frequency with
time. The Pr^{3+} ions are coherently prepared by the laser field in the initial
200 µsec interval and then exhibit optical FID when the laser frequency is
suddenly switched 2 MHz at t = 200 µsec.

For the spin decoupling experiment, an rf coil oriented along the x
axis is in close contact with the crystal and provides a pulse with a
rotating component B_x=25 G which is variable over the frequency range 0 to
600 kHz and is time-coincident with the optical pulse of 400 µsec duration
(Fig. 1). Both coil and crystal are immersed in liquid helium at 1.8°K. A
static magnetic field B_0=130 Gauss is oriented either along the z or y
laboratory axes ($\perp$ or $\parallel$ to the crystal c axis) and exceeds the local field
so that the nonsecular dipolar terms are small. Individual FID signals,
which appear at a 2.003 MHz beat frequency because of heterodyne detection
with the laser field, are captured with a Biomation 8100 Transient Recorder
for reproduction on an X-Y chart recorder (Fig. 2). In the absence of
power broadening, the observed dephasing time is $T_2/2$.

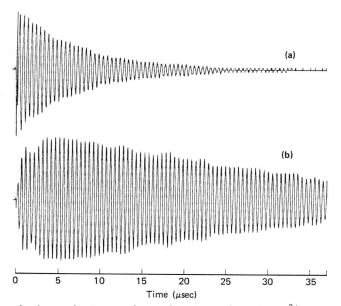

Fig.2. Optical free induction decay is LaF$_3$:Pr^{3+} at 1.8°K in the presence of a static magnetic field B$_o$ = 130 G $\perp$ c axis and (a) with no rf field where T$_2$ = 15.6 μsec (10.2 kHz) and (b) with an rf field B$_x$ = 25 G where T$_2$ = 66 μsec (2.4 kHz).

A clear demonstration of optical line narrowing by spin decoupling is indicated in Fig. 2 for the case B$_o \perp$'c axis. In the absence of rf, trace (a), the linewidth at HWHM is 10.2 kHz (T$_2$=15.6 μsec). In the presence of rf, trace (b), the value is 2.4 kHz (T$_2$=66 μsec) where the rf frequency ω/2π=450 kHz corresponds to the magic angle 54.7°. Additional confirmation is obtained by varying the rf frequency over the range from off-resonance, β=0, to on-resonance, ω/2π=250 kHz or β=π/2. Figure 3 shows that the observed optical linewidth for the case B$_o \parallel$c axis follows the theoretical expectation Eq. (4), where a frequency offset of 3 kHz is added to account for residual broadening. Furthermore, at frequencies above resonance (β>π/2), the experimental pattern repeats with the mirror symmetry predicted by (4). For the case B$_o \perp$c axis, the behavior is similar but not identical to Fig. 3 because of the Pr^{3+} hyperfine anisotropy. Our theoretical model for spin decoupling therefore is confirmed in some detail and clearly exposes the magnetic origin of the optical line broadening mechanism.

On the other hand, spin decoupling in NMR has revealed different characteristics, partly because the FID observed in the rf region is a first order process and thus both dynamic and static interactions contribute to the linewidth. In systems such as AgF, the F spin diffusion process can motionally narrow the NMR ^{109}Ag resonance [4] and when the spin diffusion process is suppressd as it is at the magic angle, the linewidth *broadens* rather than narrows [9,10]. Line narrowing [9,10] is observed, however, at the F resonance condition β=π/2.

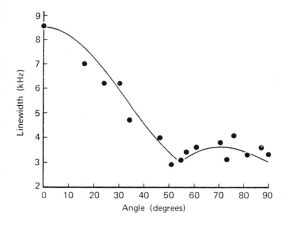

Fig.3. Pr^{3+} optical linewidth versus the angle $\beta = \tan^{-1}(\gamma_S B_X / \Delta_S)$ expressed in degrees. Solid circles ●: experimental points for the case $B_O = 130$ G ‖ c axis and $B_X = 25$ G. Solid curve—: Eq. (4) with frequency offset of 3 kHz included for residual broadening.

At the magic angle, the optical linewidth of ~2 kHz appears to be limited by residual laser frequency jitter. Of course, a fundamental limit of 160 Hz is set by a $^1D_2$ radiative lifetime of 0.5 msec. We estimate that at 1.8°K the phonon broadening linewidth is only 7 Hz, and in the preparative FID stage, laser power broadening and the effect of a finite optical pulse width of 200 μsec are negligible. As the laser frequency stability is improved further, it will be possible to examine weaker interactions which otherwise would be obscured in the absence of spin decoupling, an example being the ^{141}Pr-^{141}Pr interaction. Thus, a new family of spin decoupling experiments can be carried out for the first time at optical frequencies, allowing the manipulation and study of the basic dynamic processes.

The technical assistance of D. Horne and K. L. Foster are acknowledged.

REFERENCES

1. N. Bloembergen, E. M. Purcell, and R. V. Pound, Phys. Rev. **73**, 679 (1948).

2. A. Abragam, Principles of Nuclear Magnetism, (Oxford, London, 1961).

3. E. R. Andrew, A. Bradbury, and R. G. Eades, Nature **182**, 1659 (1958).

4. A. Abragam and J. Winter, C. R. Acad.-Sci. (Paris) **249**, 1633 (1959).

5. F. Bloch, Phys. Rev. **111**, 841 (1958).

6. M. Lee and W. I. Goldburg, Phys. Rev. **140**, A1261 (1965).

7. D. A. McArthur, E. L. Hahn, and R. E. Waldstedt, Phys. Rev. **188**, 609 (1969).

8. J. S. Waugh, L. M. Huber, and U. Haeberlen, Phys. Rev. Lett. $\underline{20}$, 180 (1968); W-K. Rhim, A. Pines, and J. S. Waugh, Phys. Rev. $\underline{B3}$, $\overline{684}$ (1971).

9. M. Mehring, G. Sinning, and A. Pines, Z. Phys. $\underline{B24}$, 73 (1976).

10. M. Mehring and G. Sinning, Phys. Rev. $\underline{B15}$, 2519 (1977)1

11. R. G. DeVoe, A. Szabo, S. C. Rand, and R. G. Brewer, Phys. Rev. Lett. $\underline{42}$, 1560 (1979).

12. J. R. Klauder and P. W. Anderson, Phys. Rev. $\underline{125}$, 912 (1962).

Superhigh-Resolution Spectroscopy

V.P. Chebotayev

Institute of Thermophysics, Siberian Branch of Academy of Sciences of USSR
630090 Novosibirsk-90, USSR

1. Introduction

For a long time a resolution of the optical methods was worse
compared to that of the methods used in the other wavelength
ranges. The last years saw a considerable progress in optical
spectroscopy in improving a resolution. This is due to the
development of new physical concepts underlying the attainment of nonlinear supernarrow optical resonances with no
Doppler broadening.

At present three methods are used to obtain narrow optical
resonances: the method of saturated absorption, two-photon
resonances in a standing-wave field, and resonances in separated optical fields. The three methods involve the elimination
of the influence of Doppler broadening, so the narrowness of
resonances is mostly determined by the time of coherent particles-field interaction. The first two methods are well known
and developed (see /1/). Just these methods have permitted
receiving the principal results on obtaining and using supernarrow resonances. The saturation resonances with a relative
width of $< 10^{-11}$ have been obtained by using telescopic beam
expanders to increase the time of particles-field interaction.
The development of the third method has been started in 1976
(see /2/). This method has showed interesting potentialities.
The use of forbidden transitions and tunable lasers will permit to obtain resonances with a relative width of $< 10^{-13}$.

Narrow resonances enabled one to observe in optics some
phenomena that could not be earlier studied. Some results in
this field received in the Institute of Thermophysics of the
Siberian Branch of the USSR Academy of Sciences are discussed
in this report.

2. Resonance Width

Let us consider very briefly the principal factors that
influence the shape of narrow optical resonances, as we cannot
analyze all the factors in detail.

2.1 Collisional Broadening and Shift

Collisional broadening is about 10 mHz/Torr. So to obtain
resonances with a width of about 100 Hz requires a pressure of
about 10^{-5} to 10^{-6} Torr. As has been experimentally shown, the
broadening of narrow resonances has some qualitative peculiarities. Refs. /3/ and /4/ reported on the observation of nonlinear dependences of collisional width and shift on particle
density.

Figure 1 shows the experimental dependence of a resonance
width in methane on pressure. In the pressure range of 1 to 5
mTorr the slope of the dependence is 30 ± 3 MHz/Torr. With an
increase of pressure the slope becomes less and at a pressure
of about 20 mTorr amounts to about 10 MHz/Torr. In this range
the slope of the curve coincides with Doppler contour broadening. The difference in the slopes is indicative of the great
role of elastic collisions with no phase mismatch. In the range
of low pressures the Doppler frequency shift $kv\theta$ at scattering
at a typical angle θ is more than a resonance width 2Γ.

So almost all atoms contribute to broadening. With an increase of pressure a resonance width increases, and some part of
atoms in collision does not go out of the region of interaction with field, the slope of the curve of broadening is decreased. The nonlinear dependence of a resonance width on pressure permits the determination of elastic and inelastic scattering cross sections and a typical scattering angle.

Of great importance in finding the mechanism of collisions
in low-pressure gases and in determining the limiting values
of frequency reproducibility in gas lasers are experiments on

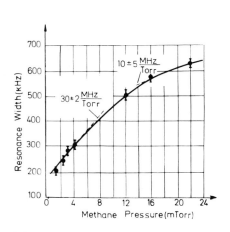

Fig.1   Fig.2

observation of a nonlinear dependence of resonance shift on gas pressure. These first experiments were carried out on the $F_2^{(2)}$ line in methane in 1972. However due to the influence of a magnetic hyperfine structure on an operating transition of methane an unambiguous interpretation of experimental results proved to be impossible. Recently we have carried out addition-al investigations of maximum resonance shifts on the $F_2^{(2)}$ line in methane on pressure.

Figure 2 shows the experimental dependence of maximum res-onance shift in methane on pressure. The value of frequency for each pressure is determined by extrapolating to zero inten-sities of the field in a cavity. The nonlinear shift on pres-sure is observed. In a pressure range of about 1 mTorr the shift is small and amounts to 10 to 20 Hz/mTorr. With an incr-ease of pressure the shift is increased. In a pressure range of 3 to 4 mTorr the shift slope amounts to about 400 Hz/mTorr. The nonlinear shift of a maximum resonance is due to the in-fluence of a magnetic hyperfine structure (MHS) and to collis-ional shift in methane. The dotted line shows the calculated dependence of resonance shift due to the MHS. The difference between experimental and calculated curves indicates the in-fluence of collisional shift in methane. An analysis has showed that the collisional shift in methane is nonlinearly dependent on pressure. In the range of high pressures of 3-4 mTorr the shift slope is about ten times as large as that in the range of low pressures (about 1 mTorr). A decrease of the slope of the collisional shift in the range of low pressures promotes the achievement of high-frequency reproducibility of optical standards. The theory of resonance broadening in a low-pressure gas has been suggested in /5, 6/.

2.2 Transit Effects

At a very low gas pressure the time of coherent interaction with field is determined by the time of transit of particles. As usual we deal with an ensemble of particles with Maxwell velocity distribution. Therefore at a nonlinear interaction the contribution of particle into a resonance depends on velo-city. The contribution of particles with smaller velocities is increased and an effective resonance width is decreased as compared to a width that would be provided by the particle with an average thermal velocity. The influence of transit effects on a saturation resonance shape was theoretically studied in/7-9/.

Figure 3 shows the results of the first direct observation of narrowing of a nonlinear resonance in methane due to the influence of slow atoms /10/. The experiments have been carried out in a He-Ne laser with a methane cell. The light beam dia-meter in the cell was 4 mm. The resonance broadening due to the finite time of atom-field interaction was about 20 kHz. At a methane pressure of 360 μTorr a resonance of Lorentzian shape is observed whose width is mostly determined by transit broadening. With a decrease of pressure down to 60 μTorr the

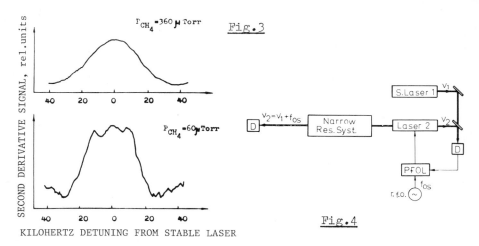

Fig.3

Fig.4

SECOND DERIVATIVE SIGNAL, rel.units

KILOHERTZ DETUNING FROM STABLE LASER

resonance shape is largely changed and three maxima are observed which correspond to the hyperfine components. The spacing between these components was about 11 kHz. The homogeneous width of an individual hyperfine component corresponding to the observed resolution was about 9 kHz, which is much less than the transit resonance width for an atom with an average thermal velocity (20 kHz). The observed effect of resonance narrowing may be due to the influence of slow atoms only.

At transit effects the resonance shape is sensitive to the field geometry. For example, in /9/ it has been shown that with divergence a resonance shift appears at different intensities of oppositely traveling waves. In separated optical fields slow atoms also give rise to the resonance narrowing. In a three-beam system a resonance width depends on both the transit time of a particle with an average thermal velocity and on a homogeneous linewidth. The resonance width is equal to $\Gamma_r \sim \sqrt{\Gamma/T_o}$, where Γ is a homogeneous width, T_o is the transit time of a particle with an average thermal velocity between fields /10/.

A somewhat another situation is observed for two-photon resonances where the influence of slow atoms is less. Here the resonance width is determined by the transit time of a particle with an average thermal velocity /11/.

2.3 Other Effects

There is a lot of other factors that influence a resonance shape. If a homogeneous linewidth becomes comparable with the shift that is due to a second-order Doppler effect, the latter can largely influence the resonance shape. There appears one more mechanism of inhomogeneous line broadening depending on the square of particle velocity. The influence of this effect on the lineshape of two-photon resonance has been discussed in /12/. Note that the allowance for this effect is important in experiments on the observation of two-photon resonance on the 1s-2s transition of hydrogen. Residual electric and magnetic fields can shift and broaden a resonance. The influence of these factors depends on the type of transition. The earth gravitational field of some geometry may also result in resonance shift.

3. Applications of Narrow Resonances

In this section we shall consider some applications of narrow resonances that have been performed in our laboratory.

3.1 Spectroscopic Applications

The attainment of very narrow resonances requires low gas pressures. A decrease in absorption and saturation parameter results in a rapid decrease of a resonance intensity. The accumulation of a signal must be therefore used to record resonances. In addition, the recording of narrow resonances requires lasers with narrow lines. It is therefore necessary to use special spectrometers in studying supernarrow resonances. The scheme of such a spectrometer is given in Fig.4 /13/. It consists of a high-stable laser, a tunable laser, and a system for obtaining supernarrow resonances.

3.1.1 Line Structure

Supernarrow optical resonances allowed studies into a magnetic hyperfine structure (MHS) on vibrational-rotational molecular transitions. Very careful studies of the MHS of the $F_2^{(2)}$ line in methane (λ = 3.39 µm) have been carried out by HALL /14/ using a telescopic beam expander outside a cavity. We used a telescopic beam expander inside a cavity of a He-Ne laser. This enabled us to obtain intense resonances with a relative width of about 1 kHz and to use them for frequency stabilization /15/. Fig.5 shows the record of the MHS of a line in methane. With stable lasers the distance between the hyperfine components have been measured with an absolute accuracy of about 10 Hz. The results of these investigations are summarized in Table. Recently the spectrometer with a telescopic beam expander of 80 cm diameter and about 20 m length has been made to study the MHS of molecules in the 10 µm range on the basis of CO_2 /16/.

Transition $F' \rightarrow F''$	Splitting frequency [kHz]		Intensity of lines $([d_f^{f}]^2$, theory)	Intensity of nonlinear resonances	
	Hall and Borde	Our experiment		Theory	Our experiment
8→7	11.4	11.4 ± 0.3	1.157	1.168	1.20 ± 0.10
7→6	0	0	1	1	1
6→7	− 11.1	− 10.8 ± 0.3	0.864	0.874	0.90 ± 0.05

110

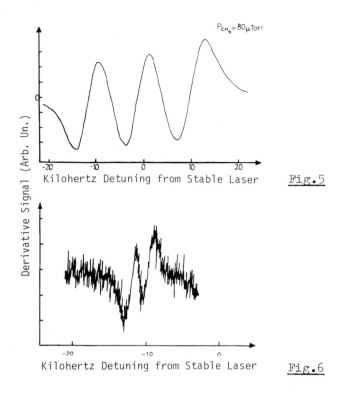

$P_{CH_4} = 80 \mu Torr$

Kilohertz Detuning from Stable Laser   Fig.5

Derivative Signal (Arb. Un.)

Kilohertz Detuning from Stable Laser   Fig.6

3.1.2 Recoil Effect

The resonances of about 1 kHz wide permitted a direct observation of recoil effect in optical spectra /17,18/. As has been theoretically shown, a resonance is split due to a recoil effect /19-21/. Fig.6 shows the field resolution of the doublet, the component spacing equal to about 2 kHz being in agreement with theory. Note that an investigation of the recoil doublet permits studies of relaxation processes on each level separately.

3.1.3 Temperature Shift of Resonance

In /22/ we have investigated the temperature shift of the frequency of a narrow resonance on the $F_2^{(2)}$ line in methane that is due to a second-order Doppler effect. The temperature shift was observed on the frequency displacement of a laser stabilized to the resonance maximum in methane with heating an absorption cell. Within the limits of a measurement error the dependence of the shift was linear with the slope of 0.5 ± 0.05 Hz/degree, which is in good agreement with a calculated value 0.52 Hz/degree (dashed line).

3.1.4 Elastic Scattering of Excited Particles

In /23/ we have performed a direct observation of elastic scattering of particles at small angles by using resonances. Fig.7 shows the resonance shape at various helium pressures in the methane cell. At a methane pressure of about 10^{-3} Torr the resonance has a Lorentzian shape and width of about 70 kHz. With an addition of helium the resonance is broadened and the pedestal of 2-3 MHz wide appears. The formation of the pedestal is due to elastic scattering of particles. A little shift of the pedestal is observed with respect to the resonance center, which may be due to their separated collisional shifts. At low pressures (about 10 mTorr) when the frequency of collisions is small, the form of the pedestal is directly connected with the characteristics of a differential scattering cross section. The experimentally found value of a typical scattering angle is about 1°. With an increase of pressure the width and amplitude of the pedestal is increased. With $P_{He}=0.1$ Torr the amplitude of the pedestal is compared with that of a sharp part of the resonance. With a further increase of pressure the resonance shape is determined by the pedestal whose width may be considerably more than a collisional one. In this region of pressures the resonance shape is determined by diffusion of particles at repeated events of scattering.

3.1.5 An Anomalous Zeeman Effect on Vibrational-Rotational Molecular Transitions

The attainment of supernarrow optical resonances of about 1 kHz wide has opened up a possibility to study an anomalous

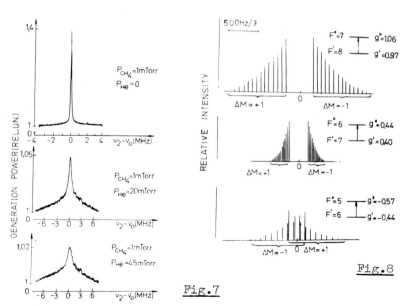

Fig.7

Fig.8

Zeeman effect in weak magnetic fields of about 1-10 Gauss, where the magnitude of Zeeman level splitting is less than spacings of a magnetic hyperfine structure.

Figure 8 shows the splitting of the main hyperfine component of the $F_2^{(2)}$ line in methane in a magnetic field. For a linearly polarized light and axial field the transitions are allowed with a variation of the projection M of the total momentum $\Delta M = \pm 1$. Each transition of a hyperfine structure is split into 2F-1 components. The frequency spacing between M components is far less than the experimentally obtained halfwidth of a resonance in methane of about 2 kHz. These are therefore not resolved in experiment.

Figure 9 shows the records of the MHS of the $F_2^{(2)}$ line in methane with no magnetic field and in a longitudinal magnetic field of 5 Gauss. The application of a magnetic field to the absorption cell results in variation of the shape of hyperfine components. A high-frequency component (F=8 → 7) is split by a value of about 3.5 kHz, a central one (F=7 → 6) is broadened by 1.5 kHz and becomes asymmetric. No considerable changes of the resonance shape at the F=6 → 5 transition are observed.

The obtained experimental results are qualitatively explained by line splitting on the MHS transitions into $\sigma_\pm$ components. In a weak magnetic field of about 0-5 Gauss an anomalous Zeeman effect is observed. The MHS lines are split in a different way, as g factors of levels on the MHS transitions are different and depend on the total angular momentum F_1. It is seen in this figure that a theoretical splitting into σ_+ and σ_- components for lines on the MHS transitions are in good agreement with that observed in experiment.

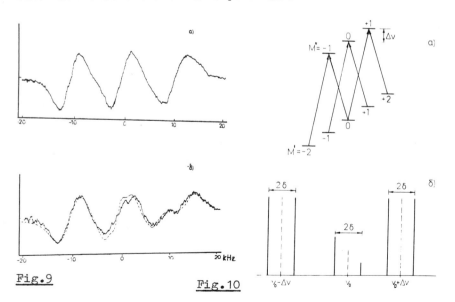

Fig.9   Fig.10

A new physical effect observed in experiment is the line asymmetry on the F=7 → 6 transition. This phenomenon is due to the influence of a recoil effect in the magnetic field and takes place both for a normal and for an anomalous Zeeman effect. A qualitative picture of the occurrence of the asymmetry is as follows (see Fig.10). Along with the resonances on 6_+ and 6_- components ($\Delta M = \pm 1$), additional resonances arise on the transitions with a common upper and lower levels (crossing resonances) at the resonant nonlinear interaction of a standing electromagnetic wave of linear polarization with a gas in the longitudinal magnetic field. Due to a recoil effect each resonance of the components is split by a value 2δ ($\delta = \hbar k^2/2m$, k is a wave number, m is the mass of a molecule). The frequencies of crossing resonances with common lower levels are shifted by $+\delta$ to a blue region with respect to the transition frequency, for the resonances with common upper levels by $-\delta$. Since the number of transitions with common upper and lower levels is different, the total intensities of the above crossing resonances differ. The recoil effect results in the occurrence of asymmetry of nonlinear resonances in a magnetic field.

4. Other Applications

4.1 Frequency Stabilization of Lasers

Most important application of resonances is the frequency stabilization of lasers. Fig.11 shows the progress in this field (for details see /24/). The short-term stability of a He-Ne laser with a methane absorber is much better than the stability of masers. Our direct measurements have showed that the laser line width is about 0.2 Hz. A laser is the best monochromatic source of coherent electromagnetic waves.

4.2 Measurement of Small Displacements

Now let us dwell on the possibility of using narrow optical resonances to record small displacements /25/. The used method is based on the recording of small variations of phase and frequency of a laser that are converted into oscillations of a radiation intensity with the aid of narrow resonances. Small periodic displacements on the large base of about 5 m with a relative sensitivity of 10^{16} have been obtained in experiment. The technique of measurement was as follows. The laser frequency was tuned to the region of maximum steepness of an optical resonance, and a periodic modulation of the cavity length was performed. With the aid of resonance the variation of laser frequency was converted into the variation of radiation intensity that was recorded.

Figure 12 shows a typical record of the signal of first harmonic in the radiation intensity at different quantities of a periodic modulation signal. The noise level corresponded to the resolution of 10^{-16}, which agreed with an absolute change of length of 10^{-5} Å.

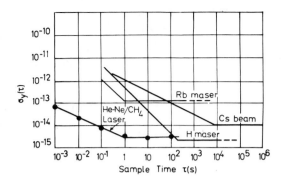

Fig.11

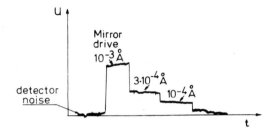

Fig.12

Note that promising may be the use of two-photon resonances. The following parameters of the system have been estimated:

$\lambda = 0.5$ μm, $P = 1$ W, $\ell = 1$ cm, $\Gamma \simeq 1$ kHz. This estimation yields $\Delta\ell/\ell \sim 5 \cdot 10^{-21}$. This permits the use of the method to produce a laser detector of gravitational waves.

5. Linear Doppler Free Absorption Resonances of Trapped Particles

Until very recently the attainment of supernarrow optical resonances was connected with the nonlinear interaction of optical radiation with particles. In this section we shall discuss new ideas the realization of which should result in obtaining linear Doppler free resonances with a homogeneous width. Consideration will be given to linear resonances of absorption (emission) of trapped particles. For definiteness we shall consider ions in a potential trap. In a radio-frequency range this problem was discussed in a good many of works (see/26/). When an oscillation amplitude of an ion in the trap A is much less than a wavelength λ, an absorption line undergoes no Doppler broadening. With an increase of the oscillation amplitude the resonance intensity rapidly decreases, when $A \sim \lambda$. This method was therefore assumed to be unapplicable in the optical band[1].

---

[1] We do not consider the case of copious particle cooling /28/, when a free path length may be much less than a wavelength. The Doppler effect does not manifest itself here.

In /27/ we discussed the possibility to obtain a linear absorption resonance of particles in traps free of Doppler broadening. As particles may be confined in traps at a vacuum of about 10^{-10} Torr for a long time /29/, this permits one to obtain lines of about 1-10 Hz wide. In this section we shall consider the principal physical peculiarities of the new method.

5.1 The Nature of Resonance

In optical and shorter wavelength ranges we are interested in an amplitude of ion oscillations in the pit that is much more than a wavelength, which should results in line broadening due to the Doppler effect. Two cases may be distinguished here. If an ion oscillation frequency Ω is far less than a homogeneous linewidth $2\Gamma_{21}$, the shape of an absorption line is the same as that for untrapped particles. We are interested in the case where the frequency of ion oscillations is much more than a homogeneous width ($\Gamma = \Gamma_{21}/\Omega \ll 1$). Then for the ion trapped in an oscillatory potential the absorption line shape is an equidistant array of resonances with a homogeneous width that arise on the background of a Doppler contour due to the periodicity of atomic motion. If a particle motion in the oscillatory potential is considered in a quantum-mechanical way, its energy on an upper and lower levels is equal to

$$E_n^{(2)} = E_2 + h\Omega\,(n + 1/2)$$

$$E_m^{(1)} = E_1 + h\Omega\,(m + 1/2), \tag{1}$$

where n and m are quantum numbers of the oscillators (see Fig. 13). The resonances in absorption arise when

$$\omega = (E_n^{(2)} - E_m^{(1)})/\hbar = \omega_{21} + \Omega\,(n - m). \tag{2}$$

There is a great **number** of transitions having the same resonance frequency $\omega_{21} + \Omega k$, where $k = n - m$ is the number of a resonance in the array.

In a classical treatment of particle motion it is more convenient to use the system of reference related to the particle itself. In this case an immovable particle is acted upon by a frequency-modulated field rather than a monochromatic one. The modulation frequency is equal to the particle oscillation frequency Ω. The spectrum of such a field contains the set of frequencies $\omega_n = \omega + n\Omega$.

When harmonic frequencies coincide with a transition frequency, absorption resonances arise. Since in the oscillatory potential the frequency Ω is independent of the particle velocity, averaging over velocities does not destroy the resonance array. Finally, the line shape of absorption of a particle in the oscillatory potential is of the form:

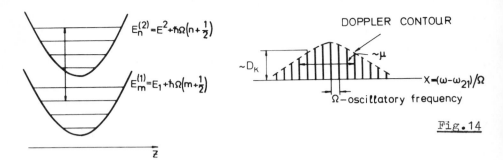

Fig.14

Fig.13

$$L(x) = \sum_{k=-\infty}^{\infty} D_k \delta (k - x), \qquad (3)$$

$D_k = 1/\sqrt{\pi} \mu \exp(-k^2/\mu^2)$, $\mu = \omega_d/\Omega$ is a Doppler linewidth
in units of Ω , $x = (\omega - \omega_{21})/\Omega$ is the field frequency detuning relative to the transition frequency in units of Ω . The
formula for L(x) shows that an absorption is non-zero only at
frequencies of an incident field $\omega = \omega_{21} + k\Omega$, $k = 0, \pm 1, \pm 2, \ldots$
The absorption probability at these frequencies is proportional
to the function D_k that describes a "discrete" Doppler contour
(Fig.14).

5.2   Radiation and Stark Line Broadening

With allowing for radiation decay the line shape of absorption
will take the form

$$L(x) = \sum_{k=-\infty}^{\infty} D_k \frac{1}{\pi} \frac{\Gamma}{\Gamma^2 + (x - k)^2} \quad , \qquad (4)$$

where $\Gamma = \Gamma_{21}/\Omega$ is a homogeneous linewidth in units of Ω ,
$\Gamma_{21} = (\gamma_1 + \gamma_2)/2$, γ_1 and γ_2 are inverse lifetimes of levels 1 and 2. When $\mu \gg \Gamma$, D_k may be put before the sign of the
sum in the point k = x, i.e.,

$$L(x) = D_x f(x). \qquad (5)$$

After summation over k we have

$$f(x) = \text{sh } 2\pi\Gamma/(\text{ch } 2\pi\Gamma - \cos 2\pi x). \qquad (6)$$

With $\Gamma \ll 1$ f(x) is a sum of nonoverlapping Lorentzian contours,
with $\Gamma \gg 1$ f(x) = $1 + 2e^{-2\pi\Gamma} \cos 2\pi x$, i.e., f(x) $\longrightarrow$ 1.

Another mechanism of broadening is related to the Stark

effect, as a confining potential is usually produced by an electric field. A moving atom in various points of the pit is acted upon by various electric fields, which results in broadening of line components. Thus, an atom in the oscillatory pit is acted upon by the electric field that gives rise to the Stark level shift. The interaction associated with this effect may be taken into account in the form of a small addition to the atom energy

$$V_\nu = - \frac{\alpha^{(v)} E_o^2}{2a^2} (x^2 + y^2 + z^2), \qquad (7)$$

where $\alpha^{(v)}$ is the atom polarizability on levels $V = 1, 2$, E_o is a typical intensity of the electric field in the trap, a is a typical dimension of the trap. If the atom polarizability on upper and lower levels is different, the shift and broadening of an individual component of the array is determined by a value $\gamma = \Delta\omega(L^2/a^2)$, where $\Delta\omega = (\alpha^{(1)} - \alpha^{(2)}) E_o^2/2h$ is the maximum Stark shift, L is an amplitude of thermal oscillations. The factor $L^2/a^2 \sim T$ (T is an absolute temperature) allows for a decrease of the Stark shift in cooling, as the typical dimension connected with the amplitude of thermal oscillations becomes less than the trap dimension. Let us perform a numerical estimation. For the trap with a linear dimension of about 250 μm we have $E_o \sim 10^4$ V/cm, $\Omega \sim 2.5\,mHz/30/$. If the trap confines the ions for which $L \sim a$, $(\alpha^{(1)} - \alpha^{(2)})$ 10^{-24} cm^3, then $\gamma \sim 10^5$ Hz. This means that individual components of the line may be resolved either for the transitions which have $(\alpha^{(1)} - \alpha^{(2)})/ \ll 10^{-25}$ cm^3 or by cooling ions.

5.3 The Influence of Anharmonism

In a real trap the potential with an ion differs from an oscillatory one. This means that Taylor's series expansion of the potential near the pit bottom contains corrections that are due to anharmonism. The difference of this potential from the oscillatory one results in smearing of the periodic array in the line shape, except for the resonance in the line center. Note that in a quantum consideration the energy of a particle excited on an upper and lower levels for an arbitrary potential is written in the form:

$$E_{n_1 n_2 n_3}^2 = E_2 + E_{n_1 n_2 n_3} ,$$

$$E_{m_1 m_2 m_3}^{(1)} = E_1 + E_{m_1 m_2 m_3} , \qquad (8)$$

where n_1, n_2, n_3 are the quantum numbers characterizing the state of the center of gravity of an atom in the potential. There is a lot of transitions (of an order of $\bar{n} = kT/\hbar\Omega$) with $n_1 = m_1$, $n_2 = m_2$, $n_3 = m_3$ that have the same frequency $\omega_{21} =$

$= (E_2 - E_1)/\hbar$. This sequence of transitions gives a resonance in the line center. For different quantum numbers of an upper and lower levels it is impossible in a general case to find a sequence of transitions with the same frequencies, i.e., no resonances arise at the other frequencies.

Thus, for an arbitrary potential there is a resonance with homogeneous width in the absorption line shape on the background of a Doppler contour (Fig.15).

5.4 Resonance Amplitude

We have studied the resonance in the one-dimension oscillatory pit. Its amplitude is $D_o/\pi\Gamma$. In a three-dimensional case we have three oscillators with different frequencies in three mutually perpendicular directions. A wave that propagates at an arbitrary angle interacts with the three oscillators and the line shape is a superposition of three periodic arrays of resonances. An increase of the number of resonances is accompanied by a decrease of their amplitudes. The resonance amplitude in the center of an absorption line is of an order of $D_o^3/\pi\Gamma$. For $D_o \sim 10^{-3}$ the amplitude A decreases by 10^{-6} times as compared with a one-dimension case. Thus, the resonance intensity may be considerably increased if a light beam is oriented along a symmetry axis of the pit.

If a wave propagates along the z axis $(\Theta = 0)$, we have a one-dimension case

$$A = D_o/\pi\Gamma. \tag{9}$$

A further analysis will be made for small angles $\Theta \ll 1$ so that $\mu\Theta \gg 1$. When $\mathcal{Y} \ll 1$, i.e., a wave propagates almost in the xz plane, we can write formulas for two cases

1. $\mu\Theta\mathcal{Y} \ll 1$

$$A = \mathcal{V}_1/\pi^2\Gamma\mu^2\Theta \; ;$$

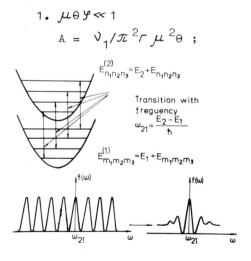

$E_{n_1 n_2 n_3}^{(2)} = E_2 + E_{n_1 n_2 n_3}$

Transition with frequency
$\omega_{21} = \dfrac{E_2 - E_1}{\hbar}$

$E_{m_1 m_2 m_3}^{(1)} = E_1 + E_{m_1 m_2 m_3}$

$f(\omega)$ $f(\omega)$

ω_{21} ω ω_{21} ω

Fig.15

2. $\mu\theta\psi \gg 1$

$$A = \nu_1 \nu_2 / \pi^{5/2} \Gamma \mu^3 \theta^2 \psi \ . \tag{10}$$

These formulas show a sharp dependence of the resonance on angles θ and ψ. $\nu_1 = \Omega_1/\Omega$, $\nu_2 = \Omega_2/\Omega$, $\Omega_3 = \Omega$; Ω_1, Ω_2, Ω_3 are frequencies of oscillators along the axes.

Orienting a beam accurately along the symmetry axis of the trap we can obtain a resonance amplitude $D_o/\pi\Gamma$ only in the case of an ideal three-dimension oscillator. However the an-harmonism results in the fact that a wave interacts not only with an oscillator in the direction of its propagation but also with transverse oscillators. As we have found out, the interaction of a wave with additional oscillators results in a decrease of the resonance amplitude in the line center. We shall show the way the anharmonism influences the quantity of a resonance. Calculations have been made for the correction to a potential

$$V = - \frac{2M\Omega^2}{s} (x^2 + y^2)z^2, \tag{11}$$

where s having a dimension of area determines the quantity of anharmonism. For a wave propagating along the z axis the quan-tity of a resonance turns out to be equal to

$$A = \frac{2.8}{\pi^2 \Gamma} \ \frac{1}{\mu^2 \gamma} \ , \tag{12}$$

where $\gamma = L^2/s$ is broadening due to anharmonism in units of , L is an amplitude of thermal oscillations. From a comparison of A in eq. (9) with $D_o/\pi\Gamma$ it is seen that the anharmonism results in decreasing the resonance amplitude by $2.8/\pi^{1/2}\mu\gamma$ times. The influence of anharmonism is appreciable only for very small angles.

Thus, our studies have showed that an observation of a linear resonance of absorption of trapped particles with a homogeneous width in the line center in the optical region can be realized. This requires the potentials close to oscil-latory ones and orientation of a trap with a desired accuracy relative to an incident wave.

The author is indebted to Drs. S.N.Bagayev, E.V.Baklanov, and E.A.Titov for valueable discussions and help in preparing the paper.

References

1. V.S.Letokhov, V.P.Chebotayev: Nonlinear Laser Spectroscopy (Springer, Berlin, Heidelberg, New York 1977)

2. V.P.Chebotayev: In Coherence in Spectroscopy and Modern Physics, ed. by F.T.Arecchi, R.Bonifacio, and M.O.Scully (Plenum Press, New York, London 1978) p. 173

3. S.N.Bagayev, E.V.Baklanov, V.P.Chebotayev: Zh.Eksp.Teor. Fiz. Pis'ma 16, 15 (1972)

4. ibid. 16, 344 (1972)

5. V.A.Alekseyev, T.L.Andreyeva, I.I.Sobelman: Zh.Eksp.Teor. Fiz. 64, 833 (1973)

6. E.V.Baklanov: Opt. i Spektr. 38, 24 (1975) (see preprint No. 22, Institute of Semiconductor Physics, Novosibirsk 1972)

7. S.G.Rautian, A.M.Shalagin: Zh.Eksp.Teor.Fiz. 58, 962 (1970)

8. E.V.Baklanov, B.Ya.Dubetsky, E.A.Titov, V.M.Semibalamut: Kvantovaya Elektronika 2, 11 (1975)

9. C.J.Borde, J.L.Hall, C.V.Kunasz, D.G.Hummer: Phys.Rev. A 14, 236 (1976)

10. S.N.Bagayev, L.S.Vasilenko, A.K.Dmitriyev, M.N.Skvortsov, V.P.Chebotayev: Zh.Eksp.Teor.Fiz.Pis'ma 23, 399 (1976)

11. E.V.Baklanov, B.Ya.Dubetsky, V.P.Chebotayev: Appl.Phys. 11, 201 (1976)

12. E.V.Baklanov, V.P.Chebotayev: Kvantovaya Elektronika 2, 606 (1975)

13. S.N.Bagayev et al.: Appl.Phys. 13, 291 (1977)

14. C.Borde, J.L.Hall: Phys.Rev.Lett. 30, 1101 (1973)

15. V.P.Chebotayev: In Proceedings of 2nd International Conference on Laser Spectroscopy (Megeve, France, June 1975) ed. by S.Haroche, J.C.Pebay-Peyroula, T.W.Hänsch, and S.E. Harris (Springer, Berlin, Heidelberg, New York 1975)

16. C.J.Borde, M.Ouhayoun, A. van Lerberghe, C.Salomon, S. Avrillier, C.D.Cantreall, J.Borde: In Laser Spectroscopy IV ed. by H.Walther (Springer, Berlin, Heidelberg, New York 1979) p. 142

17. V.P.Chebotayev: In Proceedings of the International School on Physics "Enrico Fermi" (Varenna, Italy, July 1976), Metrology and Fundamental Constants (North-Holland Publishing Company, Amsterdam, New York, Oxford 1980)

18. J.L.Hall, C.J.Borde, K.Uehara: Phys.Rev.Lett. 37, 1339(1976)

19. A.P.Kolchenko, S.G.Rautian, R.I.Sokolovsky: Zh.Eksp.Teor. Fiz. 55, 1864 (1967)

20. E.V.Baklanov: Opt.Commun. 13, 54 (1975)

21. C.G.Aminoff, S.Stenholm: J.Phys.B $\underline{9}$, 1039 (1976)

22. S.N.Bagayev, V.P.Chebotayev: Zh.Eksp.Teor.Fiz. Pis'ma $\underline{16}$, 614 (1972)

23. S.N.Bagayev, A.S.Dychkov, V.P.Chebotayev: Zh.Eksp.Teor. Fiz. Pis'ma $\underline{29}$, 570 (1979)

24. V.P.Chebotayev: Report at the XIX General URSI Assembly (Helsinki, Finland, August 1978)

25. S.N.Bagayev, A.S.Dychkov, V.P.Chebotayev: Report at VI Vavilov Nonlinear Optics Conference (Novosibirsk, USSR, June 1979)

26. H.G.Dehmelt: Adv.Atom. and Molec.Phys. $\underline{3}$, 53 (1967)

27. E.V.Baklanov, V.P.Chebotayev, E.A.Titov: Appl.Phys. $\underline{20}$, 361 (1979)

28. D.J.Wineland, R.E.Drullinger, F.L.Walls: Phys.Rev.Lett. $\underline{40}$, 1939 (1978)

29. Proceedings of 2nd Frequency Standards and Metrology Symposium (Copper Mountain, USA, July 1976)

30. W.Neuhaser, M.Hohenstoff, P.E.Toschek, H.G.Dehmelt: Appl.Phys. $\underline{17}$, 123 (1978).

Opto-Acoustic Spectroscopy of Condensed Matter

C.K.N. Patel and E.T. Nelson

Bell Laboratories, Murray Hill, NJ 07974, USA, and

A.C. Tam

IBM Research Laboratory, San Jose, CA 95193, USA

1. INTRODUCTION

The availability of tunable lasers in the visible and infrared regions of electromagnetic spectrum has led to a resurgence of a variety of new application of opto-acoustic detection technique first described[1] a hundred years ago. The combination of a reasonably high power tunable spin-flip Raman laser and opto-acoustic cells using very sensitive acoustic microphones have resulted[2] in an ability to measure gaseous absorption coefficients as small as $\sim 10^{-10}$ cm^{-1} using optical cell lengths of only 10 cm and a total gas volume of ~ 1 cm^3. Extensions of this remarkable ability to measure small absorption coefficients in gases to measurements of small absorption coefficients in condensed media has been slow in coming. Straight-forward adaptation of the gas-phase acoustic microphone technique, as discussed in Ref. 3, to the study of condensed media has not been very successful for measuring small absorption losses. During the last two years we have developed a different opto-acoustic technique which is ideally suited for measuring weak optical absorption spectra of condensed phase materials including liquids and solids. The technique differs from the earlier gas phase studies in respect of both the sources of radiation used in the study as well as the transducer used for detection of the acoustic signals. Instead of using CW tunable lasers, we have used pulsed laser sources and instead of sensitive microphones we used piezo-electric transducers for the conversion of the transient acoustic pulses (generated by the absorption of the optical radiation and the subsequent nonradiative decay of the absorbed fraction). The transducers are either immersed in the liquid or contacted to solids. Using a flash-lamp pumped tunable dye laser, capable of generating ~ 1 mJ pulses of ~ 2 μsec duration and using lead-zirconate-titanate transducers, we have demonstrated an ability to measure $\alpha l \lesssim 10^{-6}$ cm^{-1} where α is the absorption coefficient and l is the length of the sample.

In the present review, Section 2 will describe the principles of acoustic pulse generation and we will give a brief description of the dependence of the acoustic signal on the relevant material parameters. In Section 3 we will describe the experimental techniques for the opto-acoustic absorption measurements and techniques for calibration for obtaining absorption coefficients. The application of opto-acoustic absorption measurements to the study of a variety of linear and nonlinear absorption phenomena will be described in Section 4. In the case of linear absorption measurements we have measured 1) the vibrational overtone absorption spectra of a number of aromatic hydrocarbons in the visible region, 2) absorption spectra of water and heavy water, 3) absorption spectra of thin liquid films where we have demonstrated an ability to measure spectra of films as thin as 10—100 Å when the absorption coefficients are in the $\sim$100 cm^{-1} range, 4) high resolution absorption spectra of powders where a novel experimental geometry allows the measurements of small absorption coefficients in the presence of large amounts of scattered light, and 5) absorption spectra of condensed molecular gases ,at cryogenic temperatures. The nonlinear opto-acoustic spectroscopic studies of condensed matter include 1) the measurements of weak two photon absorption spectra, and 2) the measurements of weak Raman gain spectra of liquids. In the last section, in addition to giving our conclusions, we will include some future applications of the opto-acoustic spectroscopy described here.

2. OPTO-ACOUSTIC SIGNAL GENERATION

The opto-acoustic signal generation relies on the conversion of the absorbed optical radiation into heat through nonradiative relaxation processes. Detailed derivation of the acoustic pulse generation following the absorption of a pulsed of optical radiation is given in Ref. 4 and we will not reproduce it here. However, we will go through a phenomenological description which contains most of the physics underlying the opto-acoustic signal generation. The key parameter is the rapid conversion of the absorbed energy into heat by nonradiative relaxation processes. Figure 1 shows schematically a laser beam of radius R traversing a cell

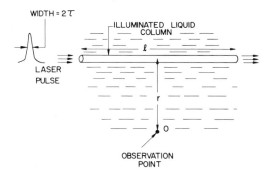

Figure 1. Schematic of opto-acoustic signal generation.

containing a liquid and a piezoelectric transducer placed within. The laser illumination is provided by a flash-lamp pumped dye laser producing pulses of duration τ_p and a repetition rate of f. For the purposes of the present discussion, we assume that the laser pulse length is much longer than the nonradiative relaxation time, τ_{NR}, and the time, τ_a, for an acoustic pulse to travel across the laser illuminated region in the liquid, i.e.,

$$\tau_p \gg \tau_{NR} \tag{1}$$

$$\tau_p \gg \frac{2R}{v_a} \equiv \tau_a \tag{2}$$

where v_a is the acoustic velocity in the medium. Further, we assume that the laser pulse is much longer than the response time of the piezoelectric transducer, τ_{pzt}, i.e.,

$$\tau_p \gg \tau_{pzt} \ . \tag{3}$$

Under these situations, achievable when using a flash-lamp pumped dye laser, with an input pulse energy of E per pulse, the energy absorbed by the medium having an absorption coefficient of α and length l is given by

$$E_{abs} = E \ (1 - e^{-\alpha l}) \ . \tag{4}$$

For $\alpha l \ll 1$, a situation which we want to explore,

$$E_{abs} \approx E\alpha l \ . \tag{4}$$

Since we are assuming that nonradiative relaxation predominates in the medium, the thermal energy E_{th} is given by

$$E_{th} = E_{abs} = E\alpha l \ . \tag{6}$$

Knowing the specific heat at constant pressure, C_p, we can evaluate the rise in the temperature, ΔT, of the illuminated volume (neglecting thermal conduction) from

$$E_{th} = C_p V \Delta T \rho \tag{7}$$

where V is the illuminated volume and ρ is the density of the condensed material. With

$$V = \pi R^2 l \tag{8}$$

we obtain

$$\Delta T = \frac{E\alpha}{\pi R^2 C_p \rho} \quad .$$
(9)

Now if we assume adiabatic, isobaric expansion, we can calculate the new volume of the illuminated region. With ΔR being the increase in the radius of the illuminated volume,

$$\pi R^2 l - \pi (R+\Delta R)^2 l = \beta V \Delta T$$
(10)

where β is the volumetric expansion coefficient. By manipulation of Eq. (10), we obtain for $\Delta R << R$

$$\Delta R \approx \frac{R}{2} \beta \Delta T \quad .$$
(11)

Inserting ΔT from Eq. (9)

$$\Delta R = \frac{E\alpha\beta}{2\pi R C_p \rho} \quad .$$
(12)

This expansion creates a pressure wave which travels radially outwards from the illuminated cylinder at sound velocity. The change in pressure, Δp, created at a point is related to the frequency of the sound wave f_a and the displacement Δx through

$$\Delta p = 2\pi f_a v_a \Delta x \rho \quad .$$
(13)

But Δx is proportional to ΔR evaluated in Eq. (12). Thus we have $\Delta x = \text{const} \times \Delta R$, and

$$\Delta p = \text{const.} \frac{f_a}{R} \frac{\beta v_a}{C_p} E \alpha \quad .$$
(14)

For a fixed geometry of laser illumination and laser properties, f_a and R can be lumped into the constant. Thus we have

$$\Delta p = \text{const.} \ \frac{\beta v_a}{C_p} \ E\alpha \ . \tag{15}$$

The condition described in Eq. (2) is just satisfied for $R \approx 1mm$, $v_a \approx 2 \times 10^5$ cm sec^{-1} and $\tau_p \approx 1-2$ μsec for a flash lamp pumped dye laser. Further, we see that the results obtained in Eq. (15) have also been obtained somewhat more rigorously by Naugol'nykh.[5]

When the illuminated cylinder is "thick" compared to the acoustic pulse propagation distance during the excitation pulse, i.e.

$$\tau_a \gg \tau_p \tag{16}$$

we can show[6] that the transient pressure change caused by absorption of the pulse optical radiation is given by

$$\Delta p = \text{const.} \ \frac{\gamma \beta v_a^2}{C_p} \ E \ \alpha \tag{17}$$

where $\gamma = \dfrac{C_p}{C_v}$ with C_v being the specific heat at constant volume. Thus, some care has to be exercised in using the functional dependence desdribed in Eq. (15) or Eq. (17) for comparing opto-acoustic signals from different materials.

Since the electrical signal output from the piezoelectric transducer is given by

$$V_{oa} = \text{const} \times \Delta p \tag{18}$$

we obtain

$$V_{oa} = K' \ \frac{\beta v_a}{C_p} \ E\alpha \tag{19}$$

where K is a constant that include the geometrical parameters as well as the response properties of the piezoelectric transducer. Now we see that the normalized opto-acoustic signal, S, is given by (for $\tau_p \gg \tau_a$)

$$S = \frac{V_{oa}}{E} = K' \frac{\beta v_a}{C_p} \alpha \qquad (20)$$

$$= K\alpha \quad . \qquad (21)$$

Thus a measurement of the normalized opto-acoustic signal for a given substance gives directly the information regarding the absorption coefficient α. Notice that while the phenomenological description obtained in Eq. (20) contains most of the important parameters necessary for the practical use of opto-acoustic spectroscopy for quantitative measurements of small absorption coefficients in condensed matter, it does not provide any information regarding the shape of the opto-acoustic output pulse. Detailed derivation[4] of the acoustic pulse generation shows that the pressure pulse contains a positive going pulse followed by a negative going pulse separated by a time of the order of the laser pulse duration. A further point not seen from Eq. (20) is the time delay between the laser pulse and the arrival of the pressure pulse at the piezoelectric transducer placed at a distance r from the laser illuminated cylinder as shown in Fig. 1. The time delay, τ_d, is given by

$$\tau_d = \frac{r}{v_a} \quad . \qquad (22)$$

This time delay is a crucial parameter in implementation of the opto-acoustic techniques for absorption spectra measurements in that the delay allows the use of proper time gating of the output signal, using a boxcar averager to discriminate against undesirable transients.

3. EXPERIMENTAL DETAILS

a. *Techniques*
Figure 2 shows a typical experimental setup for pulsed laser immersed or attached transducer gated detection opto-acoustic spectroscopy.[7] In most of the experimental results reported in this review, the pulsed laser radiation was obtained from a flash lamp pumped tunable dye laser (Chromatix). Using a variety of dyes we have obtained tunable laser radiation from 4500 Å to 7500 Å having a pulse width of ~ 2 μsec, a peak pulse energy of approximately 1 mJ, and a repetition rate of 10-20 pulses per second. The tunable dye laser radiation subsequently traverses through an opto-acoustic cell which is filled with the liquid to be investigated and, in this particular case, has an immersed piezo-electric transducer. The transmitted radiation is collected by a broad band pulse energy detecting system which is used to monitor the reference laser pulse energy on a pulse-to-pulse basis. The piezo-electric transducer assembly is shown in detail in Fig. 3 where we see that the active element for conversion

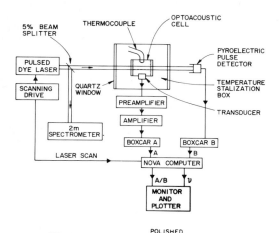

Figure 2. Schematic experimental setup for opto-acoustic spectroscopy.

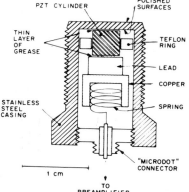

Figure 3. Cut-away schematic of the tranducer assembly.

of pressure to electrical signal is a lead zirconate-titanate cylinder approximately 4 mm long and 4 mm in diameter. The piezo-electric transducer is housed inside an all stainless steel assembly which is in contact with the liquid through a thin, polished stainless steel diaphragm to avoid the problems associated with absorption of scattered radiation from a bulk liquid which may be intercepted by the transducer surface and give rise to false signals. Further, since we wish to avoid the contamination of the liquids used in the studies by any noninert materials, the entire opto-acoustic cell is constructed out of stainless steel and only inert materials are in contact with the liquid housed inside the cell. The output from the piezo-electric transducer is amplified using a fast preamplifier designed by G. L. Miller of Bell Labs and is further amplified using a commercial amplifier. The opto-acoustic pulse which occurs in the form of a prolonged ringing signal because of the natural resonance frequency of the piezo-electric transducer and because of the reflection of the acoustic signals from the back surface of the opto-acoustic cell is detected at a specific time using a boxcar averager. The transmitted laser pulses, to a very large extent, are unattenuated since we are dealing with very transparent materials. The transmitted pulses are detected using a pyroelectric detector, the output from which is fed into a second boxcar averager for obtaining a reference signal. The averaged signals from the opto-acoustic boxcar, as well as the reference boxcar, are digitized and fed into a Nova minicomputer for further averaging and ratioing for obtaining the normalized opto-acoustic signal. The information regarding the dye laser frequency is also monitored by the minicomputer and the output from the computer is in the form of a plot which gives the normalized opto-acoustic signal as a function of the dye laser frequency. Since the peak pulse energy is of the order of 1 mJ and the time duration of the laser pulses is the order of $1-2$ μsec, the peak powers are typically of the order of 1 kW. The diameter of the unfocused laser beam traversing the opto-acoustic cell is of the order of 4 mm and thus the laser intensity within the cell is typically of the order of 10 kW cm^{-2}. A small fraction of the dye laser output is monitored using a 2 meter Bausch and Lomb spectrometer and the dye laser frequency is accurately calibrated against simultaneously observed spectra of noble gas lamps. The dye laser frequency is thus determined to an accuracy of the order of $\pm$ 1 cm^{-1}. The typical linewidth of the dye laser output is of the order of 1 cm^{-1} which is more than sufficient for the investigations of both liquids and solids.

b. *Calibration*
The experimental setup described above allows us to obtain the relative values of normalized opto-acoustic signal from the absorption of the materials. As mentioned in Section 2, if all the parameters regarding the geometry as well as the properties of the PZT transducer together with the coupling efficiency are accurately known, we can obtain an absolute value of the absorption coefficient from the measured values of normalized opto-acoustic signal. However, this has been thought to be a difficult, if not impossible, problem and heretofore we have relied upon an experimental technique for obtaining absolute values of absorption coefficients from measured normalized opto-acoustic signal. The technique[8] we use involves comparison of the measured signal for a neat liquid, for example, with that of the liquid doped with an appropriate dye which has an absorption coefficient large enough so that it can be accurately measured using conventional spectrometric techniques. We also assume that only a very small concentration of the dye is used for such calibration which implies that the overall mechanical and thermal properties of the liquid remain unchanged from the neat situation to the doped situation. Once the measurement of absorption coefficient with the dye as well as its opto-acoustic signal are obtained we can very easily dilute the dye concentrations to values appropriate for the opto-acoustic signals used for neat signals because we want to avoid the problems of electronic nonlinearity of the amplifiers. For a normalized opto-acoustic signal of S for an absorption coefficient of α for the neat liquid and S_D and α_D as the values of the normalized opto-acoustic signal and the absorption coefficients for the liquid doped with a dye, the calibration factor K in Eq. (21) is obtained through the following arguments. (The absorption coefficient of the doped

liquid includes the absorption of the dye as well as the absorption of the neat liquid.)

$$S = K\alpha \tag{22}$$

and

$$S_D = K(\alpha + \alpha_D) \; . \tag{24}$$

Hence

$$K = \frac{S_D - S}{\alpha_D} \; . \tag{25}$$

Thus Eqs. (22) through (25) finally give us a method by which we obtain the value of the constant K in Eq. (21) for obtaining absorption coefficient from the normalized opto-acoustic signals. In general we have used this technique for obtaining absorption coefficients and it works exceedingly well in the case where one can find dyes which are highly absorbing and can be dissolved easily in the case of a variety of liquids. For example, with benzene we have used iodine as a dye, with water we have used potassium permanganate as a dye, and so on. In the cases where appropriate dyes are not available we have relied upon the scaling dependance on the expansion coefficient, specific heat ,and the velocity of sound in different liquids for calibrating one liquid against another liquid. (Note, however, that the use of Eq. (15) or Eq. (17) will depend upon the relative magnitude of τ_a and τ_p as described in Section 2.) In general we find that calibrations when carried out against a dye give us an absorption coefficient which is accurate to $\pm$ 10% while calibrations against liquid through transfer by scaling of the expansion coefficient, specific heat and sound velocity give us values which are accurate to about $\pm$ 30%.

c. *Sensitivity*
The sensitivity of the opto-acoustic technique described above was determined by measuring the weak 6th harmonic of vibrational absorption[7,8] in C_6H_6 (see Section 4 for spectroscopy) with increasing dilution in CCl_4 which has very small absorption in the wavelength region of interest. Without going through details, we have shown[8] that absorption coefficients as small as $\sim 10^{-6} - 10^{-7}$ cm^{-1} can be measured (with a signal-to-noise ratio of l, input pulse of energy of ~ 1 mJ/pulse, and a pulse repetition frequency of 10 sec^{-1}, and an integration time of 1 sec).

The factors affecting and limiting the sensitivity of the measurement technique include, a) optical absorption signals from windows, b) light scattering in the bulk of the liquid, and c) electrostriction. The signals arising from the first two can be minimized by 1) choice of low loss windows, 2) by reducing scattering impurities in the liquid, 3) by appropriate time gating of the acoustic signals observed from the PZT transducer. The time gating is very important since scattered light travels at the velocity of light in the medium, $v_c = \dfrac{c}{n}$, where n is the refractive index of the material, while the acoustic signal generated in the bulk will travel at the acoustic velocity v_a in the medium. Thus the signals arising from the scattered light being absorbed by the transducer will occur essentially promptly after the laser pulse while the bulk acoustic signal is delayed by τ_d defined in Eq. (22). For the case of acoustic signals arising from optical absorption of windows, the acoustic pulse is delayed from that originating from the bulk. The acoustic pulse generated due to electrostriction processes in the laser irradiated region is, however, of the same form as that arising from the bulk absorption and hence there is no reasonable way in which the two can be distinguished. The electrostriction signal pulse is estimated[6] to be of the same size as that arising from optical absorption for $\alpha \approx 10^{-6}$ cm^{-1}. However, there is no (or only weak) dependence of the electrostriction pulse on the laser wavelength and hence the 10^{-6} cm^{-1} level does not represent a limitation to the smallest α that can be measured using the opto-acoustic spectroscopy of the material being studied having a wavelength dependent absorption. In such a case, the minimum absorption measurement capability can be improved by increasing the input pulse energy. All that is necessary now is the ability to subtract out the constant background signal.

4. APPLICATIONS

The ability to measure $\alpha l \approx 10^{-6}$ using path length < 1 cm has clear application to a variety of areas. We briefly summarize in this section some of these applications.

a. Linear Absorption Studies

i. Benzene and benzene derivatives

The first application of the opto-acoustic absorption measurement technique was the study of weak harmonic absorption in aromatic hydrocarbons for the 6th, 7th, and 8th harmonic of the vibrational overtones[7-9] of the fundamental vibrational C-H stretch frequency. In Fig. 4, we see these overtone absorptions which are centered at 6071, 5315, and 4752 Å, respectively. The absorption constant is plotted in absolute values. We see that the peak absorption decreases as we go to higher harmonics together with increasing linewidth. The information on the linewidths, and the peak absorption coefficients, can now be used for obtaining a better model of harmonic absorption. Because of the resolution possible (~ 1 cm^{-1}) even when measuring the very small absorption coefficients, we can make definitive remarks regarding the shapes and asymmetrics of the absorption structures seen in Fig. 4. The overtone spectra of gaseous benzene measured in the visible show none of the increasing asymmetry with increasing n seen for liquid benzene. These differences are

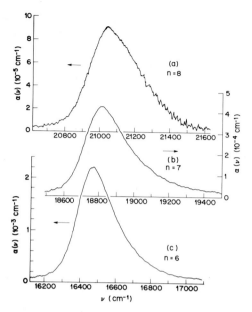

Figure 4. Overtone absorption spectra of benzene for n =6, 7, and 8.

thought to arise from the local environment effects on the benzene molecules in the liquid state and are discussed in detail in Refs. 8 and 9.

Using a recently developed simplified scheme of liquid opto-acoustic spectroscopy,[10] it is now possible to rapidly measure spectra of a variety of liquids. Fig. 5 shows a preliminary study of the sixth harmonic of benzene, bromobenzene, chlorobenzene, fluorobenzene, and toluene. The strengths of the peaks are comparable but there is a well-defined shift in the peak absorption frequency which can be correlated with the electronegativity of Br, Cl, F and CH_3, respectively.

ii. Water and D_2O

Because of the industrial and technological importance, the optical absorption spectra of water have been extensively measured over the last hundred years. A look at earlier measured spectra reveals a lack of consistency in the reported absolute values of the absorption coefficient of water in the 4000 Å to 7000 Å region where the smallest absorption coefficients are between 10^{-4} cm^{-1} and 2×10^{-3} cm^{-1}. The opto-acoustic spectroscopy method, because of the ease with which small absorption can be rapidly and accurately determined, was applied to the measurement of liquid water absorption spectrum in the visible region. Fig. 6 shows our measured absorption spectra[11] together with earlier data. The opto-acoustic measurements were calibrated using $KMnO_4$ as the dye for water, and $K_2Cr_2O_7$ as the dye for heavy water and are accurate to $\pm 10\%$ over the entire region. We believe that these data now represent the most accurate measurements of the visible absorption spectrum of water to date.

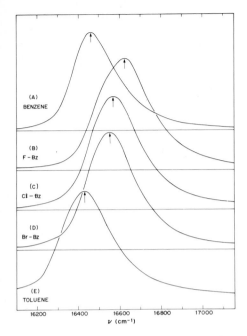

Figure 5. Sixth harmonic absorption spectra of benzene, bromobenzene, chlorobenzene, fluorobenzene, and toluene.

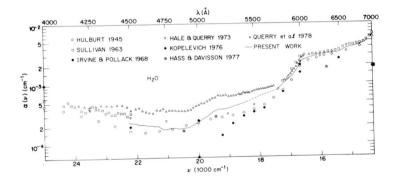

Figure 6. Absorption spectrum of water measured by opto-acoustic technique, and compared with earlier data.

Similar measurements were also carried out for D_2O. Details of the measurements as well as discussions can be found in Ref. 12.

iii. Liquid Films and Powders

Since we are able to accurately measure the absorption spectra of materials in the range where $\alpha l \le 10^{-6}$, it is clear that for strong absorption features with $\alpha \ge 10^2$ cm^{-1}, we should be able to measure absorption spectra of thin liquid films of thickness ~ 1 Å. The technique for opto-acoustic spectroscopy measurements of thin films is slightly different from that shown above for bulk liquids and is shown in Fig. 7

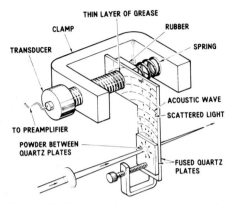

Figure 7. Schematic of experimental setup for opto-acoustic absorption studies of liquids and powders.

where the liquid film is contained between the two quartz plates. The same setup is also useful for the study of powders where there is considerable amount of scattered light which under normal circumstances would be intercepted by the opto-acoustic transducer and will give false signals. The bend in the quartz substrate, as shown, minimizes the scattered light problems. We have shown that by increasing the number of bends, it is possible to about completely eliminate the effects of scattered light.[4] Figs. 8 and 9 show the liquid film and powder absorption spectra which are described and

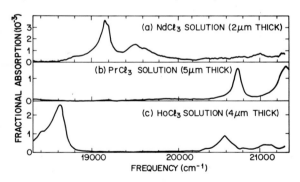

Figure 8. Opto-acoustic absorption spectra of thin liquid films of NdCl$_3$, PrCl$_3$, and HoCl$_3$.

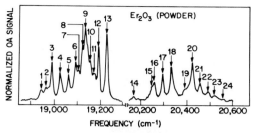

Figure 9. Opto-acoustic absorption spectrum of Er_2O_3 powder.

discussed in detail in Refs. 13 and 14. For the present purposes, it suffices to say that from the observed signal-to-noise ratio seen in Fig. 8, we have the capability of measuring submicron thick liquid films when the liquid exhibits strong absorption features.

iv. Low Temperature Liquified Gases
The technique of measurement of low absorption coefficient optical spectra using opto-acoustic spectroscopy is easily extendable to lower temperatures since the solid state piezoelectric transducer materials continue to function without any problems at any temperature below the Curie temperature. For example, the Curie temperature for the transducer used in most of the present experiments (LZT-5) is 500 K. Thus by providing a proper cooling scheme, opto-acoustic spectra at T < 300 K can be easily obtained.

We have recently obtained the optical spectra of liquid methane[15] (94K) and liquid ethylene[16] (110K) in the visible region. These gases, and especially methane, are important minor constituents of the atmospheres of the outer giant planets such as Jupiter, Saturn, Neptune, etc., where the pressure and temperature conditions are such that these gases could exist as liquid droplets in the atmospheres of these planets. Figs. 10 amd 11 show the spectra of liquid methane and ethylene where

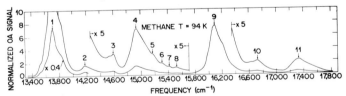

Figure 10. Opto-acoustic absorption spectrum of liquid methane (94K).

we see that unlike the spectra of aromatic hydrocarbons such as benzene, the spectra of methane and ethylene are very rich and complicated. The peaks denoted by the arrows have been identified and these arise from a combination of local modes and normal modes of various vibrations of the respective molecules. (See Refs. 15 and 16 for details.) It is clear that we have unlocked a new region of study of low absorption spectra of cryogenic liquids.

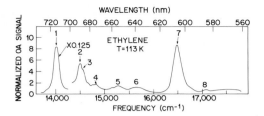

Figure 11. Opto-acoustic absorption spectrum of liquid ethylene (113K).

b. *Nonlinear Absorption Studies*

i. Two Photon Absorption

A careful look at the earlier discussion is sufficient to convince that the opto-acoustic spectroscopy technique developed here is able to measure small absorption coefficient spectra regardless of the origin of the absorption mechanism. Since two photon (and multiphoton) absorption mechanisms generally give rise to only weak absorptions, the opto-acoustic absorption measurement technique should be ideally suited for these measurements also. The experimental scheme is that used above in the linear absorption studies with the only exception that a focusing lens is incorporated into the path of the pulsed dye laser beam to give higher peak power inside the opto-acoustic cell. As opposed to primarily the high vibrational harmonic absorption studies

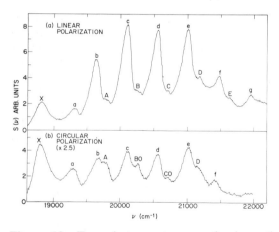

Figure 12. Two photon opto-acoustic absorption spectrum of benzene.

described for example in Section 4.a.i, the excitation by two photon absorption is into an electronic state. Generally, most neat liquids when excited to the first electronic state lose the excitation by nonradiative processes involving rapid transfer to a nearby triplet state and subsequent decay back to ground state. The predomince of nonradiation decay channels is ideally suited for opto-acoustic spectroscopy. We have measured[17] the two photon opto-acoustic spectra of benzene in the 5200 Å − 4500 Å region (corresponding to the A_{1g} to B_{2u} excitation) as shown in Fig. 12. The various peaks marked are identified as arising from the vibronic transitions in the electronic systems except the peak marked X which arises from the 6th overtone absorption described in Section 4.a.i. Since the absorption coefficient of the 7th overtone of C-H stretch in benzene has been accurately measured,[9] we now are able to provide for the first time the absolute values for the two photon absorption cross-section for the $A_{1g} − B_{2u}$ band. Further, from the identifications we are also able to provide the first data on the electronic origin of the $A_{1g} − B_{2u}$ band which differs significantly from that in the gas phase benzene. (See Ref. 17 for details.)

ii. Raman Gain

Raman scattering process leaves a molecule in its excited state when the scattered radiation is Stokes shifted from the incident radiation. In case of liquid, the excited state is generally the first vibrational level which decays via nonradiative mechanisms. Thus Raman scattering process can be observed using the opto-acoustic detection technique. However, spontaneous Raman scattering is weak, the typical Raman scattering efficiences being $\sim 10^{-8} − 10^{-10}$ cm^{-1}. Moreover the spontaneous Raman scattering detection using opto-acoustic technique can not give any spectral information. But by providing, in addition to the pump radiation, a signal radiation at the Stokes frequency, we can cause the Raman scattering process to be stimulated, thereby increasing the vibrational excitation rate and providing spectral information by tuning either the pump on the signal radiation frequency, keeping the other one fixed. For typical pump radiation intensities of $\sim 10^6$ W/cm^2, Raman gain of $\sim 10^{-3}$ cm^{-1} is possible. Thus for every photon stimulated at the signal frequency, a pump photon is lost and a quantum of vibrational excitation created in the medium. The decay of this "stimulated" vibrational excitation can be measured with the opto-acoustic technique and it gives a measure of the Raman gain.

The measurement technique involves incorporation of a second laser in the basic opto-acoustic absorption measurement scheme as shown[18] in Fig. 13. We have measured Raman gain spectra of benzene, acetone, 1,1,1 trichloroethane, toluene, and n-hexane. Fig. 14 shows sample spectra of benzene. The peak measured Raman gain is $\sim 10^{-3}$ cm^{-1}. From the observed signal-to-noise ratio, we have a capability of measuring Raman gain as small as $\sim 10^{-5}$ cm^{-1} for the 1 mJ, 2 μS duration pump and signal laser pulses used in the studies so far. Reference 18 should be consulted fro additional details.

5. CONCLUSIONS

We have shown in this brief review that the pulsed laser, immersed or contacted transducer, gated detection opto-acoustic absorption method is ideally suited for rapid study of weak absorption spectra of condensed matter. By no

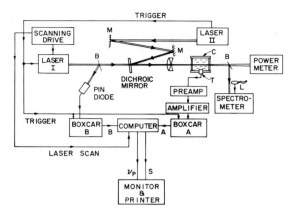

Figure 13. Experimental setup for opto-acoustic Raman gain studies. .

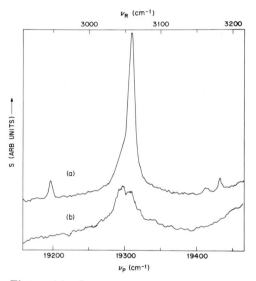

Figure 14. Opto-acoustic Raman gain spectra of benzene for pump and signal polarizations (a) parallel, and (b) orthogonal.

means have we exhausted the areas of optical spectroscopy of condensed matter to which the opto-acoustic scheme will make significant contributions. We have barely scratched the surface. Thus it is worthwhile ennumerating some of the future directions of research using this technique. (Extensions to longer wavelengths appear to be of immediate importance. We have already made significant advances here[19] by using a high pressure Raman cell for shifting the pump dye laser frequencies to higher and to lower frequencies. We have started using the tunable pulsed laser radiation in the 8000 Å − 2.5 μm region for a variety of studies, some of which are included in the list below.)

1. Materials testing such as measurements of absorption losses of preforms used for making low loss optical fibers, semiconductors, etc.

2. Trace impurity detection in liquids and solids.

3. High pressure gas studies at gas densities approaching liquid densities.

4. Higher order Raman processes such as hyper-Raman scattering, etc.

5. Forbidden transitions in condensed media being made partially allowed by external perturbations such as electric and magnetic fields, pressure, etc.

6. Excited state and chemical kinetics.

7. Monolayers.

8. OA microscopy.

9. Use of other excitation sources such as electron beams and X-rays for study of electron loss and X-ray absorption, respectively.

REFERENCES

1. A. G. Bell, *Proc. Am. Assoc. Adv. Sci.* **29,** 115 (1880).

2. C. K. N. Patel and R. J. Kerl, *Appl. Phys. Letters* **30,** 578 (1977).

3. W. R. Harshbarger and M. B. Robin, *Acct. Chem. Res.* **6,** 329 (1973); A. Rosencwaig, *Anal. Chem.* **47,** 592A (1975).

4. C. K. N. Patel and A. C. Tam, *Revs. Mod. Physics,* (to be published).

5. K. A. Naugol'nykh, *Sov. Phys. Acoust.* **23,** 98 (1977).

6. E. T. Nelson and C. K. N. Patel, (in preparation).

7. C. K. N. Patel and A. C. Tam, *Appl. Phys. Lett.* **34,** 767 (1979).

8. A. C. Tam, C. K. N. Patel, and R. J. Kerl, *Opt. Letters* **4,** 81 (1979).

9. C. K. N. Patel, A. C. Tam, and R. J. Kerl, *J. Chem. Phys.* **71,** 1470 (1979).

10. A. C. Tam and C. K. N. Patel, *Opt. Letters,* **5,** 27 (1980).

11. C. K. N. Patel and A. C. Tam, *Nature,* **280,** 302 (1979).

12. A. C. Tam and C. K. N. Patel, *Appl. Optics,* **18,** 3348 (1979).

13. C. K. N. Patel and A. C. Tam, *Appl. Phys. Lett.* **36,** 7 (1980).

14. A. C. Tam and C. K. N. Patel, *Appl. Phys. Lett.* **35,** 843 (1979).

15. C. K. N. Patel, E. T. Nelson, and R. J. Kerl, *Nature,* **286,** 368 (1980).

16. E. T. Nelson and C. K. N. Patel, (to be published).

17. A. C. Tam and C. K. N. Patel, *Nature,* **280,** 304 (1979).

18. C. K. N. Patel and A. C. Tam, *Appl. Phys. Lett.* **34,** 760 (1979).

19. E. T. Nelson and C. K. N. Patel, (in preparation).

IR Laser Absorption Spectroscopy of Local Modes of the H⁻ Ion in Pure and Rare-Earth-Doped CaF_2

E.C.C. Vasconcellos, S.P.S. Porto
Instituto de Física G. Wataghin, Universidade de Campinas
13100 - Campinas, S. Paulo - Brazil

C.A.S. Lima
Departamento de Física, Universidade de Brasília
70910 - Brasília - D.F.,Brazil

1. Introduction

IR spectroscopy of hydrogenated calcium fluoride has attracted some attention in the past few years [1-7]. As has long been known [1] negative hydrogen ions can be incorporated into pure calcium fluoride in high concentrations. The H⁻ ion substitutes for the fluorine ion in a regular site of tetrahedral symmetry. The crystal shows strong IR absorption due to localized modes of the hydride ion. In CaF_2 crystals doped with trivalent rare-earth ions the H⁻ can serve as a charge compensator for the impurity ions. In this role it replaces the corresponding fluorine ions in interstitial lattice sites [5].

2. Laser Powered IR Spectroscopy of Local Modes

We give below a brief account of studies of IR absorption carried on crystalline samples of CaF_2 either pure or doped with Dy^{3+} or Er^{3+} ions due to H⁻ ions sitting both on C_{4v} and T_d symmetry sites. Experimental conditions differed from those reported in previous work [5] in that our source of excitation was a CO_2 laser rather than a conventional low intensity incoherent source. This enabled us to attain beam intensities unavailable with commom IR spectrometers. However, it introduced the limitation that the available frequencies were discrete rather than continuous. To bypass this difficulty we devised an spectroscopic exploration technique which exploits the thermal shift of local mode frequencies. We were able to make accurate determinations of the resonance frequencies as a function of temperature and also to study thermally induced changes in absorption of a given laser line. By combining the results for different lines, chosen around the position of a given local mode, at a given temperature, it is possible to obtain the mode profile as a function of temperature. Thus, with this technique we obtain all the information conveyed by conventional IR spectroscopy. It has, however, the advantage of making it possible to excite samples with intensities extending from low to very high values, by use of adequately focussed high power IR laser beams. Details on sample preparation, and experimental setup and procedures are given elsewhere [8].

Making use of the forementioned exploration technique we studied the behavior of the sample transmittance under irradiation with a CO_2 laser beam tuned to a frequency around that of a given H^- local mode. Changing the sample temperature between $20\ ^\circ K$ and room temperature we varied the mode frequency to match it with that of the appropriate laser line. In this way we determined the functional dependence of the mode frequency on temperature.

Once tuned to a local mode, we held fixed the laser frequency and the sample temperature while varying the beam intensity at the sample. From the observed variation of the transmittance we studied the intensity dependence of the resonant absorption by the corresponding local mode.

3. Experimental Results

In Fig. 1(a) we present our experimental data for the transmittance (I_{tr}/I_{in}) as a function of temperature for the $CaF_2:Er^{3+}$: H^- sample irradiated with CO_2 laser lines having frequencies around that of the C_{4v} local mode. Similar measurements made on the $CaF_2:Dy^{3+}:H^-$ sample are presented in Fig. 1(b). The irradiation frequencies are indicated in the figures.

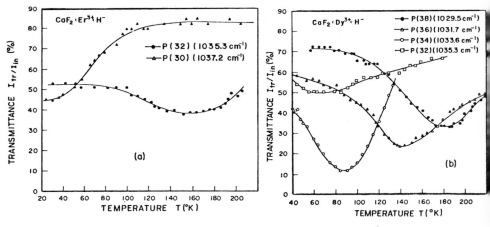

Fig.1 Transmittance vs. temperature for C_{4v} modes in $CaF_2:Er^{3+}:H^-$ and $CaF_2:Dy^{3+}:H^-$ samples.

For all the samples examined one also expected strong absorption by local modes associated with the H^- ion occupying tetrahedral sites in the host matrix. In the $CaF_2:H^-$ crystal this corresponds to a situation where the H^- substitutes for a regular fluorine ion. The same is true of the $CaF_2:RE^{3+}:H^-$ samples provided the substitution occurs in a lattice site well separated from any RE^{3+} impurity. Our results concerning these T_d symmetry local modes are presented in Figs. 2(a) to 2(c).

As to the frequencies of both the C_{4v} and T_d local modes, as obtained from the observed minima in the transmittance vs. temperature curves in Figs. 1 and 2, a summary is given in Table 1. Also given in the Table are the corresponding values, as determined by other authors using conventional IR spectroscopy.

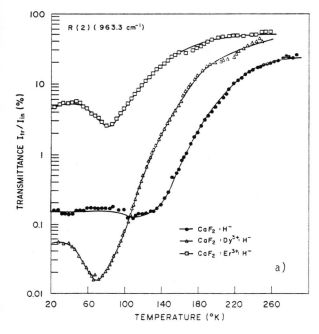

Fig.2 Transmittance vs. temperature for T_d modes in $CaF_2:H^-$, $CaF_2:Dy^{3+}:H^-$ and $CaF_2:Er^{3+}:H^-$ samples.

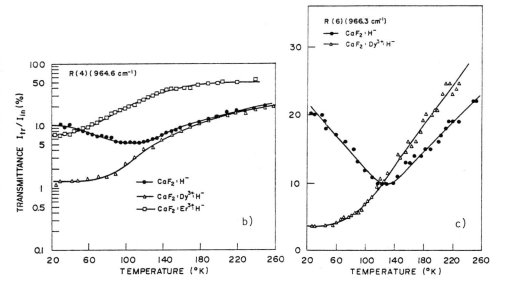

Table 1 H⁻ ion local modes. The "laser lines" are the CO_2 laser lines whose frequencies matched those of the modes, at the specified temperature. Observations: (a) mode not seen; frequency outside our working range; (b) mode not seen; Ref. given reports it being anomalous, paired with that at 1081.1 cm⁻¹ (800°K); (c) we observed a signal enhancement but could not be totally conclusive as to the location of the resonance.

Sample	I (this work)			Symmetry	II (others)			OBS.
	Laser line	Mode frequency (cm⁻¹)	Temperature (°K)		Mode frequency (cm⁻¹)	Temperature (°K)	Ref.	
CaF_2:Er³⁺:H⁻ (0.5 molar %)	P(32)	1035.3	150	C_{4v}	1035.8	77	5	
	P(30)	1037.2	20					
	.	.	.	C_{4v}	1085.6	77	5	(a)
	.	.	.	C_{4v}	1037.0	77	5	(b)
	R(24)	1081.1	80	C_{4v}	1081.1	77	5	(c)
	R(2)	963.3	80	T_d	965	77	5	
	R(4)	964.6	~ 20	T_d				
CaF_2:Dy³⁺:H⁻	P(32)	1035.3	65	C_{4v}	1033.1	20	5	
	P(34)	1033.6	90					
	P(36)	1031.7	140	C_{4v}	1102.4	77	5	(a)
	P(33)	1029.5	185					
(0.05 molar %)	.	.	.	T_d	965	77	5	
	R(2)	963.3	70	T_d				
	R(4)	964.6	~ 20					
CaF_2:H⁻	.	.	.		965.6	20	1	
	R(2)	963.3	110	T_d	965.1	77		
	R(4)	964.6	110		957.8	290		
	R(6)	966.3	130					

Measurements have also been carried out for the transmittance vs. beam intensity on both the CaF_2:Dy³⁺:H⁻ and CaF_2 samples, at fixed temperatures. The corresponding data are presented in Figs. 3(a) to 3(d) for irradiation with the indicated CO_2 laser lines.

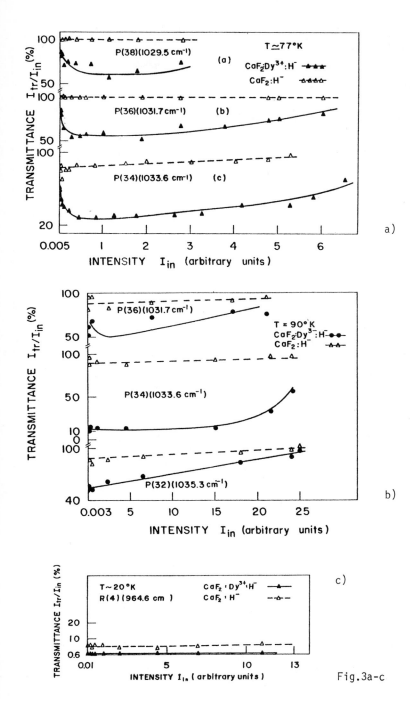

Fig. 3a-c

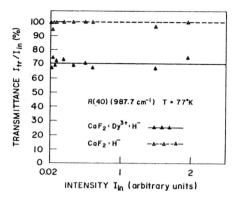

d)

<u>Fig.3</u> Transmittance vs. incident intensity for the CaF_2:H^- and CaF_2:Dy^{3+}:H^- samples.

4. Remarks

The investigation being presented here is, in many aspects, a natural extension of the work presented in Ref. [2]. There, CO_2 laser illuminated calcium fluoride was first studied. Such experiments are attractive in that high-intensity resonance effects can be studied in a resonant system which is free of inhomogeneous broadening, and which is very simply described as a spherical mechanical vibrator in lowest order but with degeneracies lifted by higher orders so that distinct level intervals are isolated in frequency. In the present work, we have set about firstly to alleviate the difficulties caused by the lack of continuous tunability of the CO_2 laser and, secondly, to extend the method to rare-earth-doped crystals which afford a more generous set of resonance and which have sites of lower symmetry.

However, due to space limitations, no attempt will be made in this short presentation, to bring in an interpretation of our data. This will be done in a forthcoming full article.

We would like to close our remarks by stating that this contribution, presented at this Symposium held in memory of Professor Sergio Porto represented his last scientific accomplishment, shortly before his untimely death.

5. References

1. R. J. Elliot, W. Hayes, G. D. Jones, H. F. MacDonald, Proc. Roy. Soc. A 289, 1 (1965).
2. L. C. Lee and W. L. Faust, Phys. Rev. Lett. 26, 648 (1971)
3. J. A. Harrington, Ph. D. Dissertation, Northwestern University, Illinois (1971)
4. J. A. Harrington and R. Weber, Phys. Stat. Sol. (b) 56, 541 (1971)
5. G. D. Jones, S. Peled, R. Rosenwaks and S. Yatsiv, Phys. Rev. 183, 353 (1969).
6. R. C. Newman, Advances in Physics 18, 545 (1969)
7. J. W. J. Timans, H. W. den Hartog, Phys. Stat. Sol. (b) 73, 283 (1976)
8. E. C. C. Vasconcellos, Ph. D. Dissertation, Universidade de Campinas (1978).

Nonlinear Optics of Cryogenic Liquids*

S.R.J. Brueck and H. Kildal

Lincoln Laboratory, Massachusetts Institute of Technology
Lexington, MA 02173, USA

I. Introduction

Cryogenic liquids are being used as the active media for an increasing
number of nonlinear optical devices. Results have been reported on stimulated Raman oscillators and amplifiers[1-4], third-harmonic generators[5],
four-wave difference frequency generators[6,7] and infrared Kerr
switches.[8] Photo-acoustic and photo-refractive measurements of weak
absorptions in these liquids are also a subject of current interest [9,10].
In this paper, we report third-harmonic generation (THG) and ac Kerr
effect measurements which allow us to obtain the electronic nonlinearity
(hyperpolarizability), the vibrational two-photon resonance nonlinearity,
and the molecular reorientation contributions to the third-order
susceptibilities, $\chi^{(3)}$, of the cryogenic liquids CO, O_2, N_2 and Ar.
Measurements of the electrostrictive and absorptive coupling of a laser
radiation field to the liquid hydrodynamic modes are presented. These
measurements are of interest because of their implications for the efficiencies and limitations of the nonlinear optical devices. Further, the
relative simplicity of these media allow many of these quantities to be
calculated or extrapolated from low density gas-phase measurements for comparison with these results. This should lead to an increased understanding
of liquid-state properties.

Several mechanisms with varying response times, dispersion characteristics and tensor symmetry relationships contribute to this third-order
susceptibility. At frequencies which are far from any resonances, isolated
atoms and molecules exhibit a nonlinear optical response which may
heuristically be thought of as arising from an intense field driving the
oscillations of the electron cloud relative to the nuclei beyond the harmonic regime. This hyperpolarizability has a response time of
10^{-15}-10^{-16} s corresponding to electronic frequencies and therefore contributes to both THG and Kerr experiments.

In addition to the purely electronic hyperpolarizability, molecular
species also have a vibrational contribution which exhibits strong dispersion effects for infrared frequencies near vibrational resonances of the
medium. Far from any vibrational resonance, the electronic hyperpolarizability dominates the response. Of particular interest to us will be the
case of a vibrational two-photon resonance between the incident CO_2 laser
frequencies and the Raman-active vibrational mode of CO. Due to the two-
photon resonance requirement, the vibrational resonance contributes to
THG but not to the two-frequency Kerr effect measurements. This process

*This work was sponsored by the Department of the Air Force and the
Department of Energy.

may be characterized by a susceptibility for pulse durations longer than vibrational dephasing times of 10-100 ps for simple diatomic liquids.

A torque is exerted on anisotropically polarizable molecules in a radiation field which tends to orient the molecules with their axis of maximum polarizability along the field polarization direction and thereby induces a birefringence in the medium. The time scale of this process, set by the liquid orientational relaxation time, is typically 0.3-1 ps. Thus, molecular orientation contributes only to the Kerr measurements.

Symmetric molecular and atomic species also exhibit a smaller effective anisotropic polarizability due to translational correlations. For low pressure gases this is evidenced by a quadratic contribution to the density dependence of the Kerr coefficient. Similarly to the case for anisotropic molecules this contribution to $\chi^{(3)}$ responds on a molecular (atomic) translational time scale of approximately 0.1-1 ps.

Absorption and electrostriction couple the electromagnetic field to the liquid acoustic waves and thermal diffusion modes; this also contributes to $\chi^{(3)}$. These processes, which do not induce any birefringence, respond at approximately an acoustic transit time across the laser focal diameter and have been monitored using photo-acoustic and photorefractive techniques. These methods have also been used to study weak absorptions and vibrational kinetics in liquids.

2. THG Experiments

THG experiments using the CO_2 R(20) laser line at 944.2 cm^{-1} as the pump have been carried out to determine the relative hyperpolarizabilities of liquid O_2, N_2, and Ar. These liquids all have a positive dispersion between the pump and the third-harmonic frequency. A few percent of liquid CO was therefore added to the liquids to give a negative dispersion which allowed optimization of the phase-mismatch integral for a tight-focusing geometry (10-cm cell length, 1.5-cm confocal parameter). Figure 1 shows the dependence of the third-harmonic output energy on the CO relative peak absorbance for a fixed input power into the cell. The CO peak absorbance refers to the second overtone transition at 1.58 μm and is measured relative to the absorbance in undiluted CO. The variation in the THG output with CO concentration is due to changes in the phase-mismatch integral. The square root of the ratio between the maximum third-harmonic powers gives the ratio between the hyperpolarizabilities. In reducing the data, we have not corrected for the $\chi^{(3)}$ contribution due to the few percent CO concentration in the liquids. This is permissible since liquid CO should have about the same nonresonant hyperpolarizability as the other liquids. As we did not have an appropriate phase-matching scheme for high concentration CO we were not able to measure its hyperpolarizability.

Liquid CO has a vibrational two-photon resonance in the region of the CO_2 laser R(6) line at 1069 cm^{-1}. This gives rise to a much larger $\chi^{(3)}$ in this spectral region and leads to the possibility of efficient THG.[5] Detailed measurements of the two-photon resonance parameters have been carried out and are reported elsewhere.[5,11] The measurements of $\chi^{(3)}$ are in excellent agreement with values extrapolated from gas-phase measurements of the Raman cross section. A detailed experimental and theoretical investigation of the efficiency limitations of this THG process including the

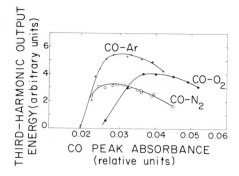

<u>Fig. 1.</u> Relative third-harmonic output power vs CO peak absorbance for liquid $CO-O_2$, $CO-N_2$ and $CO-Ar$ mixtures at a CO_2 laser frequency of 944.2 cm^{-1}.

effects of pump depletion, two-photon absorption and Raman scattering processes has also been carried out.[5] Energy conversion efficiencies of 4% to the third-harmonic of the CO_2 pump laser frequency were observed.

In order to calibrate the measured hyperpolarizabilities on an absolute scale, we have monitored the interference between the hyperpolarizability of liquid O_2 and this two-photon resonant third-order susceptibility. As the CO_2 pump laser is tuned through the vibrational two-photon resonance in liquid $CO-O_2$, the variation in the THG signal due to this interference for a $CO-O_2$ mixture at 77 K with a 0.077 relative CO peak absorbance at constant input power is shown in Fig 2. These measurements were carried out using a tight-focusing geometry with a 0.65-cm confocal parameter in a 9.6-cm path-length cell. The changes in the phasematching condition and the confocal parameter were negligible for the frequency range of 1035 to 1084 cm^{-1} over which the CO_2 laser was stepwise tuned. The signal varies

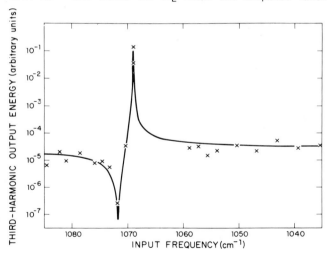

<u>Fig. 2.</u> Interference between resonant and nonresonant contributions to $\chi_{1111}(-3\omega,\omega,\omega,\omega)$ for a $CO-O_2$ mixture with 0.077 CO relative peak absorbance.

by more than six orders of magnitude. A destructive interference between the resonant and nonresonant susceptibilities is evident on the high-frequency side of the two-photon resonance at a frequency mismatch of about -6 cm^{-1}.

In order to determine accurately the signal level and frequency position for maximum destructive interference we have tuned through this region by changing the CO concentration while keeping the pump frequency constant at 1071.88 cm^{-1} [CO$_2$ R(10) line]. Figure 3 shows the measured concentration dependence of the third-harmonic signal. A minimum is observed at a CO relative peak absorbance of 0.081. Without the interference the third-harmonic signal would have followed a curve similar to the curves in Figure 1 which are characteristic of a tight-focusing geometry. The solid curves are fits of the data using the hyperpolarizability and the signal normalization as the only adjustable parameters.

Analysis of this data clearly shows the presence of a broad linewidth contribution to the two-photon resonant $\chi^{(3)}$ in addition to the much more intense narrow linewidth component. These are associated with respectively the anisotropic and isotropic components of the Raman polarizability tensor. This is the first time that the anisotropic scattering component has been observed in a polarized geometry in a liquid. The solid curves in the figures are the least squares fits to the data points with only the O$_2$ hyperpolarizability and a normalization constant as adjustable parameters. Details of this analysis are given in Ref. [12].

3. AC Kerr Measurements

The ac Kerr coefficients of the cryogenic liquids CO, O$_2$, N$_2$ and Ar have been measured by monitoring the induced polarization rotation of a low-power visible probe laser beam due to an intense copropagating CO$_2$ TEA laser pulse. The results for the power p transmitted through crossed polarizers, normalized to the incident probe laser power p_0, is given in Fig. 4 for liquid O$_2$. In this figure α represents a weak component of the probe laser beam polarized in the analyser direction. In the dashed region of the curve the Kerr measurement is dominated by a homodyne signal between

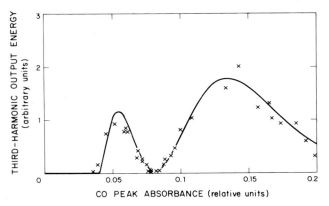

Fig. 3. Third-harmonic signal near cancellation of resonant and nonresonant contributions to $\chi_{1111}(-3\omega,\omega,\omega,\omega)$ in liquid CO-O$_2$ (laser frequency = 1071.88 cm^{-1}, confocal parameter = 0.65 cm).

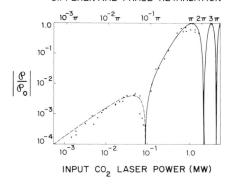

DIFFERENTIAL PHASE RETARDATION

INPUT CO_2 LASER POWER (MW)

Fig. 4. Relative He-Cd laser power transmitted through analyser due to CO_2-laser-field-induced Kerr rotatio in liquid O_2. (CO_2 laser frequency 1039.4 cm^{-1}, T = 77.4 K, confocal parameters $\sim$ 23 cm).

this component and the component rotated from the polarizer direction. At higher CO_2 input powers the more familiar quadratic Kerr effect dominates. The curve in the figure is a least squares fit to the data points with the Kerr nonlinearity as the only adjustable parameter (cf. [12]).

4. Coupling to Hydrodynamic Modes

The characteristic times for the liquid hydrodynamic modes generated by the laser radiation field are the acoustic propagation and the thermal diffusion times across the laser spot size. For laser pulse durations shorter than these times, the laser excitation sets up initial conditions for the hydrodynamic modes which then evolve after the laser pulse. Both absorptive and electostrictive coupling mechanisms result in the generation of propagating acoustic waves which may be detected with a piezoelectric transducer located outside the interaction region (photo-acoustic technique) and a nonpropagating diffusive (entropy) mode which may be detected by monitoring the refractive index variations associated with its density and temperature perturbations (photo-refractive technique). The acoustic wave has a time scale of 100 ns for a 100-μm spot size whereas diffusion times are much longer ($\sim$ 100 ms). PATEL and TAM [9] have recently demonstrated the detection of weak absorptions in liquids by measuring the acoustic waves generated by absorptive coupling. An analysis[10] of the hydrodynamic equations describing this system shows that the electrostrictive coupling sets a lower bound on the absorption constants that may be measured by both photo-acoustic and photo-refractive techniques. This limitation is more severe for the photo-acoustic tech- niques than for the photo-refractive technique in accord with our measurements. An interesting example of another measurement which may be carried out using the photo-refractive technique is shown in Fig. 5 which shows the temporal development of the thermal lensing of a He-Ne laser beam collinear with a CO_2 pump laser in pure liquid CO (top trace) and CO with 100 ppm of CH_4 added (bottom trace). The CO_2 laser is tuned to the R(6) line at 1069.14 cm^{-1} which is in two-photon resonance with the CO vibrational mode. The rise-time of 3.5 ms for the pure CO is determined by the slow decay of the CO vibrational excitation into thermal excitation of the liquid; the long time decay of the signal is due to thermal diffusion. With the addition of CH_4 the vibrational decay is much more rapid due to CH_4-CO collisions.[13] After this measurement was made more extensive

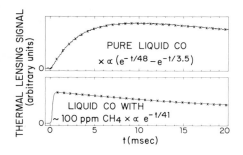

Fig. 5. Vibrational energy relaxation in liquid Co monitored by the thermal lensing induced on a visible laser probe beam. Excitation is by twophoton absorption of the CO_2 R(6) line at 1069.14 cm^{-1}.

purification of the CO gave a measurement for the CO lifetime of 5.5 ms in good agreement with the results of LEGAY-SOMMAIRE et al.[14] It is interesting to note that no acoustic waves were generated during this process since the coupling of energy into the thermal modes occurs too slowly to generate any acoustic modes.

5. Discussion

No previous measurements of the hyperpolarizabilities of cryogenic liquids have been reported. Gas phase measurements have been performed, however, for all of the atomic and molecular species considered here using a variety of techniques. We can use these results to test the scaling characteristics of the hyperpolarizabilities to liquid densities according to the simple extrapolation

$$\chi_{el}^{(3)} = \left(\frac{n^2 + 2}{3}\right)^4 N \chi_{atomic}^{(3)} \tag{1}$$

where N is the liquid density, χ_{atomic} the susceptibility per atom or molecule and a local field correction has been included. The extrapolated values together with our experimental results are listed in Table I. The good agreement, on an absolute as well as a relative scale, suggests that the hyperpolarizabilities of the condensed media are determined by adding up the contributions from individual molecules ignoring any correlation effects.

The measured total Kerr susceptibilities are listed in Table II along with the χ_{or} determined by subtracting the electronic contribution $2\chi_{el}/3$ obtained from THG measurements (cf. Table I). The major source of error in this evaluation of χ_{or} is in the relative calibration of the two sets of data. A comparison can be made with the previous measurement of MCWANE and SEALER[15] who measured χ_{1221} $(-\omega,\omega,\omega,-\omega)$ = χ_{or} + $2\chi_{el}/3$ of liquid N_2 relative to CS_2 using a Q-switched ruby laser at 694 nm. Also shown in the table is a comparison with measurements[16] of the total integrated anisotropic Rayleigh scattering cross section $d\sigma^{Ray}/d\Omega$ = $(d\sigma^{Ray}/d\Omega)_{VH}$ + $(d\sigma^{Ray}/d\Omega)_{VV}$ for liquid CO, O_2, N_2 and Ar in a 90° scattering geometry. The values for the diatomic liquids are in reasonable agreement with our data and show that the largest contribution to the Kerr susceptibility arises from molecular orientation.

Table 1.

ELECTRONIC CONTRIBUTIONS TO $\chi^{(3)}$ FOR SIMPLE LIQUIDS

LIQUID	T (K)	EXTRAPOLATED $10^{34} \chi_{el}^{(3)}$ (Asm/V^3) FOUR-WAVE MIXING (a)	THG (b)	KERR (c)	DC FIELD INDUCED SECOND HARMONIC (d)	MEASURED $10^{34} \chi_{el}^{(3)}$ (Asm/V^3)
CO	77.4	10 ± 1	—	—	5.5 ± 0.3	—
O$_2$	77.4	10 ± 1.	—	—	5.0 ± 0.1	3.9 ± 1.2
N$_2$	77.4	7.4 ± 0.7	4.0 ± 0.6	4.5 ± 0.5	3.2 ± 0.1	3.5 ± 1.0
Ar	87.3	11 ± 1	6.2 ± 1.1	4.8 ± 0.3	5.8 ± 0.2	4.5 ± 1.3

(a) W.G. RADO, APPL. PHYS. LETT. 11, 123 (1967).

(b) J.F. WARD AND G.H.C. NEW, PHYS. REV. 185, 57 (1979).

(c) A.D. BUCKINGHAM et. al., TRANS. FARADAY SOC. 64, 1776 (1968); 66, 1548 (1970).

(d) R.S. FINN AND J.F. WARD, PHYS. REV. LETT. 26, 285 (1971) J.F. WARD AND C.K. MILLER, PHYS. REV. A19, 826 (1979).

Table 2.

ORIENTATIONAL CONTRIBUTIONS TO $\chi^{(3)}$ FOR SIMPLE LIQUIDS

LIQUID	KERR MEASUREMENTS $\chi_{or} + 2\chi_{el}^{(3)}/3$	THG MEASUREMENTS $2\chi_{el}^{(3)}/3$	ORIENTATIONAL SUSCEPTIBILITY $\chi_{or}^{(3)}$ PRESENT RESULTS	POLARIZATION ROTATION (a)	ANISOTROPIC RAYLEIGH SCATTERING (b)	EXTRAPOLATED FROM ANISOTROPIC POLARIZABILITIES (c)
CO	8.3 ± 1.7	(2.4 ± 0.8)	5.9 ± 2.5	—	7.3 ± 1.9	22.4
O$_2$	59 ± 9	2.6 ± 0.8	56.4 ± 9.8	—	62 ± 14	132.0
N$_2$	13 ± 3	2.3 ± 0.7	10.7 ± 3.7	13 ± 1.3	17 ± 3.6	37.3
Ar	1.9 ± 0.6	3.0 ± 0.9	−1.1 ± 1.5	—	0.37 ± 0.06	

(a) P.D. McWANE AND D.A. SEALER, APPL. PHYS. LETT. 8, 278 (1966).

(b) J. BRUINING AND J.H.R. CLARKE, CHEM. PHYS. LETT. 31, 355 (1975).

(c) N.J. BRIDGE AND A.D. BUCKINGHAM, PROC. R. SOC. (LONDON) A295, 334 (1966).

It is again interesting to contrast the liquid measurements with extrapolations based on gas-phase measurements[17] of the polarizability anisotropy, $\alpha_\parallel - \alpha_\perp$. The extrapolated χ_{or}, obtained using an expression similar to Eq. (1) are approximately a factor of three larger than the measured values and are given in Table II. This discrepancy is due to orientational and translational correlations. The relative values, however, are in better agreement with the experiment confirming that polarizability anisotropy is the dominant source of the Kerr susceptibility for the diatomic liquids.

It is clear from Table II that the orientational susceptibility for liquid Ar is much smaller than for the diatomic liquids and is not accurately determined by the present measurements due to uncertainties in the calibrations of the two independent experimental techniques. The Kerr measurements, combined with the Rayleigh measurements, show that the electronic contribution dominates the Ar Kerr susceptibility.

References

1. J. B. Grun, A. K. McQuillan and B. P. Stoicheff, Phys. Rev. 180, 61 (1969).
2. A. Z. Graziuk and J. G. Zubarev, Appl. Phys. 17, 211 (1978).
3. R. Frey, F. Pradere, J. Lukasik and J. Ducuing, Optics Commun. 22, 335 (1977).
4. E. Wild and M. Maier, J. Appl. Phys. 51, 3078 (1980).
5. S. R. J. Brueck and H. Kildal, Opt. Lett. 2, 33 (1978); H. Kildal and S. R. J. Brueck, IEEE J. Quantum Electron. QE-16, 566 (1980).
6. R. D. McNair and M. L. Klein, Appl. Phys. Lett. 31, 750 (1977); 32, 346 (E) (1978).
7. H. Kildal and S. R. J. Brueck, Appl. Phys. Lett. 32, 173 (1978).
8. S. R. J. Brueck and H. Kildal, Appl. Phys. Lett. 35, 665 (1979).
9. C. K. N. Patel and A. C. Tam, Appl. Phys. Lett. 34, 467 (1979); 34, 760 (1979).
10. S. R. J. Brueck, H. Kildal and L. J. Belanger, Optics Commun. 34, 199 (1980).
11. S. R. J. Brueck, Chem. Phys. Lett. 53, 273 (1978).
12. Helge Kildal and S. R. J. Brueck, J. Chem. Phys. (to be published, November 1980).
13. W. F. Calaway and G. E. Ewing, J. Chem. Phys. 63, 2842 (1975).
14. N. Legay-Sommaire and F. Legay, Chem. Phys. Lett. 52, 213 (1977).
15. P. D. McWane and D. A. Sealer, Appl. Phys. Lett. 8, 278 (1966).
16. J. Bruining and J. H. R. Clarke, Chem. Phys. Lett. 31, 355 (1975).
17. N. J. Bridge and A. D. Buckingham, Proc. Roy. Soc. London, Ser. A 295, 334 (1966).

Part III

Laser Photochemistry

Bond Selective Excitation of Molecules

J.S. Wong and C.B. Moore

Department of Chemistry, University of California and Materials and
Molecular Research Division of the Lawrence Berkeley Laboratory
Berkeley, CA 94720, USA

Bond selective photochemical reactions have provided an elusive target for
laser photochemists. Early claims of mode selectivity in multiphoton dissociation have largely been discredited [1]. Although the laser pumps a
single vibrational mode, energy is transferred to other modes more rapidly
than the molecule is excited above dissociation thresholds by sequential
absorption of photons. Single photon excitation appears more promising for
mode selective vibrational photochemistry [2]. All of the energy is absorbed at once. The strongest high overtone absorptions come from the most
anharmonic excitations, local modes (LM), in which a single bond is excited
[3 - 5]. Small differences in C-H stretching potential functions are resolved. In the example presented here the chemically equivalent but spatially inequivalent in-plane and out-of-plane C-H oscillators are resolved in

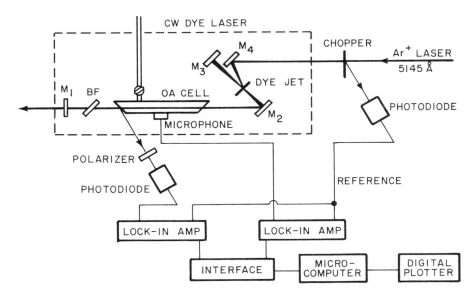

Fig.1 Experimental apparatus for optoacoustic (OA) high overtone spectroscopy. Not shown is the stepping motor that rotates the birefringent
filter (BF) of the dye laser

the high overtone spectra for the methyl groups of gaseous propane. The ability to excite a single C-H oscillator in a particular steric position suggests a highly selective form of vibrational photochemistry [2, 6, 7].

High Overtone Excitation of Propane and CHCl₂F

In the infrared spectrum of $CD_3CD_2CHD_2$, it was possible to resolve two C-H stretching fundamental bands from the two spatially inequivalent C-H bonds [8]. Selective deuteration decouples the remaining C-H bond from the rest of the molecule resulting in LM character even in the fundamentals. Chemically inequivalent (aryl vs alkyl, methylene vs methyl) C-H oscillators have been resolved in liquid phase overtone spectra [9, 10]. The integrated absorptions of peaks corresponded roughly to the number of LM's of each type [9]. Unfortunately, intermolecular interactions in the liquid phase broaden the individual peaks and obscure the dynamical information obtainable from resolved spectra.

The experimental apparatus for gas phase, high overtone spectroscopy is illustrated schematically in Fig. 1. A nonresonant optoacoustic cell is

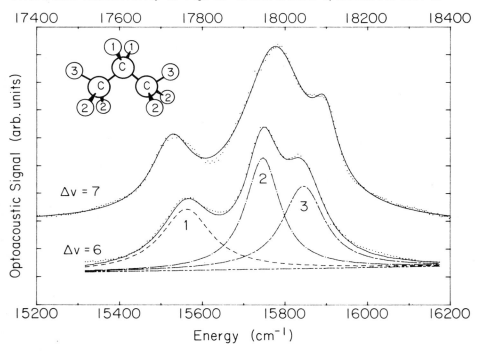

Fig.2 Overtone spectra of gaseous propane. The dots are the digitized experimental data and the solid curves are computer deconvolutions. The upper ($\Delta v = 7$ at 113 Torr) and lower ($\Delta v = 6$ at 117 Torr) spectra are plotted with the results of Analyses III and I, respectively. The $\Delta v = 7$ spectrum has been offset two units vertically for clarity. Note the different wavenumber scales. Peaks 1, 2, and 3 arise from absorptions by the methylene, out-of-plane methyl and in-plane methyl local modes, respectively

placed within the cavity formed by M_1, M_2, and M_3 of an Ar^+-pumped cw dye laser (Spectra-Physics model 375). The relative intracavity power circulating through the cell is monitored by a photodiode detecting the light scattered off a cell window. Both this signal and the optoacoustic signal are phase-sensitively detected and sent to the computer interface. A Commodore PET microcomputer controls the experiment by outputting pulses to a stepping motor, thus rotating the three plate birefringent filter of the dye laser, while simultaneously storing optoacoustic signal divided by the laser power. Spectra are saved on cassette tape and plotted on a Hewlett-Packard 7225A digital plotter.

Three peaks are clearly evident in Fig. 2, the gas phase spectra of propane. From the liquid phase work [9, 10] it is known that the smaller, low energy peak is from the C-H LM's of the methylene group, while the two higher energy peaks arise from the methyl group LM's. These two peaks, previously unresolved, are assigned to the chemically equivalent but spatially inequivalent in-plane and out-of-plane C-H bonds. In the equilibrium configuration, the three carbon atoms and one hydrogen atom from each methyl group are coplanar, hence C_{2v} symmetry. The slight difference in the C-H stretching potential functions [8] splits the methyl LM peaks by about 100 cm^{-1}. Although the methyl groups undergo a hindered internal rotation with a barrier height [11] of approximately 1000 cm^{-1}, this rotation is not fast enough to motionally average the two peaks. Work in progress [12] shows a splitting of the methyl peak in all the n-alkanes from propane to n-hexane.

Since pressure broadening is negligible here, linewidths give dynamical information on collision-free intramolecular relaxation rates [4, 13]. Nonlinear least squares deconvolution yielded positions, widths and areas of the partially resolved peaks assuming Lorentzian lineshapes. Analysis I (Table 1) of the $\Delta v = 6$ spectrum, with parameters freely varying, gave a better fit than II with the relative areas constrained to the stoichiometric

Table 1 Computer deconvolutions of the $\Delta v_{CH} = 6$ (I, II) and $\Delta v_{CH} = 7$ (III) spectra of propane

	Peak	$v(cm^{-1})$	FWHM(cm^{-1})	Area	Variance
I	1	15562.0 ± 3	147 ± 15	1.00 ± 0.11	0.03
	2	15746.0 ± 2	101 ± 7	1.25 ± 0.10	
	3	15845.0 ± 3	123 ± 11	1.13 ± 0.13	
II	1	15559.0 ± 3	124 ± 6	1	0.1
	2	15750.0 ± 1	121 ± 4	2	
	3	15851.0 ± 2	100 ± 5	1	
III	1	17725.8 ± 0.8	118 ± 3	1.00 ± 0.03	0.12
	2	17975.6 ± 0.6	193 ± 2	4.14 ± 0.01	
	3	18097.0 ± 0.6	71 ± 3	0.54 ± 0.05	

ratio of 1:2:1. Peak positions are defined within a few percent of peak
shifts. Intensities and linewidths are correlated and much less well determined by the spectral deconvolution. Analysis of the $\Delta v = 7$ spectrum
gives a broadening of the out-of-plane methyl LM peak and a large change in
the area ratio of the two methyl peaks. Attempts to fit this spectrum with
three Lorentzians with area ratios of 1:2:1 failed miserably. Consecutive
spectra may be superimposed upon each other; thus the observed lineshapes
are accurate. Although the deconvolutions converge well, the fits are not
perfect, particularly in the valley between peaks 1 and 2, and in the wings.
The lineshapes of the individual peaks would not be truly Lorentzian if the
density of states or coupling strengths to the LM's were not uniform. For
propane, the pure overtone, LM transition is nearly resonant with the combinations of the next lower C-H overtone plus two quanta of H-C-H bending vibrations. Indeed the C-H stretch fundamentals are strongly mixed with the
first overtones of the bending motions by Fermi resonance [14]. Such
strong perturbations by specific levels can be expected to distort spectra
from Lorentzian profiles. The 40 to 50 cm^{-1} rotational band structure width
also distorts the lineshape. Thus area ratios and bandwidths from deconvolutions can only be fixed in rather broad correlated ranges.

Non-Lorentzian lineshapes are observed in high overtone spectra of a
number of molecules. For $CHCl_2F$ there is only one pure C-H stretching $v = 6$
level; however, in Fig. 3 there are four vibrational bands of comparable intensity and a number of smaller ones. Thus instead of one Lorentzian peak

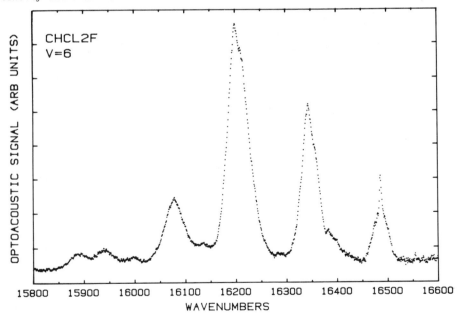

Fig.3 Overtone spectrum of the $\Delta v_{CH} = 6$ region of $CHCl_2F$. The one pure
$\overline{v_{CH}} = 6$ state is mixed with nearly resonant combination states involving
five C-H stretching quanta plus quanta of other vibrational modes. The
shared oscillator strength results in the complex observed spectrum

there are four distinct peaks. This spectrum is an extreme example, but it
clearly shows the importance of strong coupling between specific modes and
the dangers of deconvolutions relying on simple peak assignments and Lorentzian lineshapes.

High Overtone Photochemistry

The observation of different frequencies for chemically equivalent hydrogens
in different steric situations suggests possibilities for bond selective vibrational photochemistry. Significant rate differences have been reported
for excitation of hydrogens at different distances from the site of a heavy
atom rearrangement in the isomerization of allyl isocyanide [6]. Bond selectivity should be marked indeed for a hydrogen transfer reaction if that
hydrogen can be vibrationally excited in the steric configuration most favorable for reaction. Although the absorption oscillator strength of high overtones is derived almost entirely from LM excitation of a single C-H oscillator, the observed broadening shows that there is strong mixing with other
excitations of approximately the same total energy, Fig. 4. A narrow band

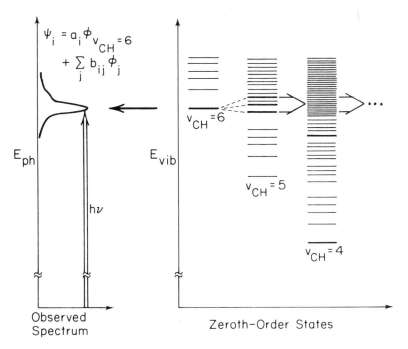

Fig.4 Zeroth order level scheme, coupling scheme and the resulting experimental spectrum. Illustrated is a zeroth order $v_{CH} = 6$ state coupled strongly to nearby states built upon $v_{CH} = 5$ and coupled weakly to the remaining
manifolds of nearly resonant states. Narrow band cw excitation produces a
mixed state of the molecule, ψ_i, corresponding to an eigenfunction of the
molecular Hamiltonian. A sub-picosecond pulse whose bandwidth exceeds the
spectral width is needed to excite the coherent superposition of states which
yields the pure $v_{CH} = 6$ excitation

source excites a mixed state of the molecule; the C-H oscillator and those modes to which it is coupled are excited simultaneously. The entire photon energy may be initially placed into the C-H oscillator only by coherent excitation of the entire band using a mode-locked laser. Thus although the excitation is a linear single photon absorption, a sub-picosecond pulse is required to excite a single bond and the excitation is transferred out of that bond on a timescale given by the inverse linewidth, 0.05 psec for 100 cm^{-1} linewidth. Thus one may expect to see much larger quantum yields for bond selective reaction with mode-locked excitation than with normal monochromatic or random phase polychromatic excitation [2].

Mode selective photochemistry may be especially effective in liquids. Energy may be transferred from the excited molecule to nearly resonant vibrations of neighboring molecules at rates [15] approaching those for intra-molecular energy transfer and much larger than those for reaction after energy randomization (RRKM rates) has occurred. Thus if the reaction involving the optically excited mode does not occur before energy randomization there will probably not be any reaction at all. The vibrational photo-ionization of liquid water which has been observed by exciting levels in the 8600 - 16,700 cm^{-1} range illustrates this possibility [16].

This work was supported by the Division of Chemical Sciences, Office of Basic Energy Sciences, U.S. Department of Energy under Contract No. W-7405-Eng-48 and by the U.S. Army Research Office, Triangle Park, NC, U.S.A.

References

1 P.A. Schulz, Aa. S. Sudbø, D.J. Krajnovitch, H.S. Kwok, Y.R. Shen, and Y.T. Lee, Ann. Rev. Phys. Chem. 30, 379 (1979)

2 C.B. Moore and I.W.M. Smith, Faraday Disc. Chem. Soc. 67, 146 (1979)

3 R.L. Swofford, M.E. Long, and A.C. Albrecht, J. Chem. Phys. 65, 179 (1976)

4 R.G. Bray and M.J. Berry, J. Chem. Phys. 71, 4909 (1979)

5 B.R. Henry, Acc. Chem. Res. 10, 207 (1977)

6 K.V. Reddy and M.J. Berry, Chem. Phys. Lett. 66, 223 (1979)

7 K.V. Reddy, D.G. Lishan, M.J. Berry, and G.S. Hammond, submitted to IX IQEC Conference, Boston MA (1980)

8 D.C. McKean, S. Biedermann, and H. Bürger, Spectrochim. Acta A 30, 845 (1974)

9 M.S. Burberry, J.A. Morrell, A.C. Albrecht, and R.L. Swofford, J. Chem. Phys. 70, 5522 (1979)

10 W.R.A. Greenlay and B.R. Henry, J. Chem. Phys. 69, 82 (1978)

11 S. Weiss and G.E. Leroi, Spectrochim. Acta A 25, 1759 (1969)

12 J.S. Wong and C.B. Moore, in preparation

13 D.F. Heller and S. Mukamel, J. Chem. Phys. 70, 463 (1979)

14 G. Herzberg, Molecular Spectra and Molecular Structure II. Infrared and and Raman Spectra of Polyatomic Molecules (Van Nostrand, Princeton, 1945)

15 W. Kaiser and A. Laubereau, Chemical and Biochemical Applications of La-sers, Vol. 2, C.B. Moore, ed. (Academic Press, New York, 1977), p. 43

16 B. Knight, D.M. Goodall, and R.C. Greenhow, J. Chem. Soc. Faraday Trans. II, 75, 841 (1979)

Generation of UV Radiation (250–260 nm) from Intracavity Doubling of a Single-Mode Ring Dye Laser

C.R. Webster, L. Wöste, and R.N. Zare

Department of Chemistry, Stanford University
Stanford, CA 94305, USA

1. Introduction

Molecules whose electronic transitions lie in the UV portion of the electromagnetic spectrum far outnumber those in the visible. Consequently, it is of considerable chemical interest to extend tunable coherent sources to shorter wavelengths. Potential applications are numerous, including photochemical, kinetic, analytical, and spectroscopic studies. We report here the construction of an intracavity-doubled cw ring dye laser producing single-mode UV output in the range 250-260 nm with a free-running jitter of about ±50 MHz over a time interval of a few seconds. High resolution spectra of the Hg $^3P_1$ - $^1S_0$ transition at 253.7 nm are presented. Fluorescence excitation of natural mercury in a bulb yields a spectrum in which the resolution of the hyperfine structure is Doppler limited, whereas excitation of an atomic beam allows sub-Doppler resolution of these features.

Because dye lasers are limited to wavelengths above 300 nm, the production of tunable deeper UV light requires the use of nonlinear optics such as harmonic generation and sum frequency mixing [1]. To date these methods have provided the lowest wavelength UV radiation but this has been accomplished using pulsed dye laser systems whose theoretical bandwidth is at best the Fourier-transform limit of the laser pulse duration [2].

For narrower bandwidths cw operation is necessary. In 1978 WAGSTAFF and DUNN [3] frequency doubled a rhodamine 6G single-mode ring laser using an ADA intracavity crystal to produce UV output in the range 292-302 nm with good output power and stability. More recently, MARIELLA [4] reached wavelengths near 247 nm by extracavity doubling of a coumarin 480 linear dye laser to produce UV output with a bandwidth of about 2.5 GHz, while CLOUGH and JOHNSTON [5] reported intracavity doubling of a coumarin 535 linear dye laser to produce UV output with a bandwidth of about 4 GHz in the range 257-260 nm. The design which is described below takes advantage of the high circulating power inside a single-mode ring dye laser to generate UV output (250-260 nm) by intracavity frequency doubling. While the present laser system is free running (≤50 MHz bandwidth), actively-stabilized (1 MHz bandwidth) single-mode cw ring lasers are commercially available, suggesting that cw UV radiation with very narrow bandwidths (~2 MHz) will be available in the near future.

2. The Coumarin Ring Laser

The ring laser, shown schematically in Fig.1, is a six-mirror
cavity designed specifically for operation using coumarin 515 dye.
All mirrors are coated for high reflectivity (>99.8% at 507 nm)
in the dye range of 485-550 nm for beams ~15° from normal incidence. Gimble mirror mounts with micrometer adjustment are fixed
on aluminium blocks which define a beam level 12.5 cm above the
supporting NRC optical table having pneumatic isolation mounts.
Excitation·of the dye is provided by a Spectra-Physics model 171
krypton ion laser which produces up to 3.6 watts at 413 nm all
lines. The ion laser output first has its linear polarization
rotated by 90° to match the horizontally polarized dye laser and
is then focused by a 2.5 cm radius of curvature mirror into the
vertical dye jet.

A 1 x 10⁻³ molar solution of coumarin 515 (Exciton) in ethylene
glycol containing a few ml of cyclooctatetraene (COT) is used
in a homebuilt dye nozzle/circulator system. The circulator
is a water-cooled two-chamber design incorporating a 10 μm filter
(Millipore), a 1/8 horse-power motor (GE) and magnetic coupling
assembly (Micropump). A bypass valve allows the dye solution
pressure behind the nozzle to be varied up to 100 PSIG. The
nozzle is made by optically contacting two fused silica blocks
(15 mm x 10 mm x 5 mm) at their largest surfaces and then separ-
ating these surfaces by precision polished fused silica spacers
of ~250 μm thickness. The 15 mm long nozzle has, therefore, a
rectangular cross-sectional area of ~5 mm x 250 μm. The dye jet

COUMARIN RING LASER

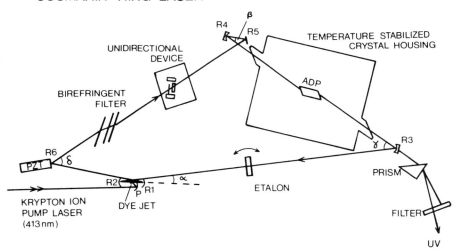

Fig.1 Schematic of the coumarin ring laser: P = pump mirror;
$\alpha \simeq 4°$; $\beta \simeq 25°$; $\gamma \simeq 34°$; $\delta \simeq 42°$; mirrors $R_1 = R_2 = 5$ cm,
$R_3 = R_4 = 50$ cm, $R_5 = R_6 = \infty$; and PZT = piezoelectric translator.

is vertically mounted to cross the plane of the laser ring at
90°, exit through a clearance hole in the table and be gently
collected to avoid acoustic feedback.

The following description refers specifically to Fig.1. The
angle $\alpha(\sim 4°)$ is chosen according to the condition [6,7]

$$R \sin(\alpha/2) \tan(\alpha/2) = Nt \qquad (1)$$

whereby the astigmatic distortions introduced by focusing into
the dye jet are compensated by those introduced by the off-axis
mirrors R_1 and R_2. In (1) R is the radius of curvature of R_1
and R_2, t is the dye jet thickness and

$$N = [(n^2 - 1)/n^4](n^2 + 1)^{\frac{1}{2}} \qquad (2)$$

where n is the refractive index of the dye solution (n = 1.4 for
ethylene glycol).

A second beam waist occurs in the ring laser midway between
mirrors R_3 and R_4 of radii of curvature equal to 50 cm. At this
location is inserted the frequency-doubling unit, the optical
assembly of an Inrad model 5-11 "supercooled" temperature phase-
matching system, containing an ADP crystal close to 5 cm long,
and the heaters and sensors required to set the crystal temperature. The crystal was cut and polished with Brewster-angled
faces to minimize Fresnel losses. It is sealed into a small
Brewster-windowed chamber over pressured with inert gas. In
this way the problems associated with surface adsorption of oil
vapor or other contaminants are avoided. This subunit is partially isolated from a well-insulated massive copper heat sink.
For cooled operation this heat sink is held at 175 K by a closed-
cycle refrigerator, and the crystal temperature maintained at its
set point by bucking the refrigerator with temperature-controlled
heaters. The subunit and blocks sit inside a housing (22 cm x
28 cm x 20 cm) which also has Brewster windows, the space between
these windows and those of the subunit being evacuated to minimize window deposition.

For the purpose of calculating the astigmatism resulting from
focusing into the ADP crystal through the two (each direction)
Brewster windows and balancing this with that resulting from the
off-axis mirrors ($R_3 = R_4 = R = 50$ cm), we use an effective crystal thickness of 5.4 cm and n = 1.52. Then, according to (1) the
angle at which astigmatic compensation is achieved is 12.5°. For
our cavity design the large size of the doubling unit allows $\beta/2$
to be set to 12.5°, but restricts $\gamma/2$ to about 17°. However, the
cavity is essentially astigmatically compensated. As demonstrated
by DUNN and FERGUSON [8], the Z-configuration and the crystal
orientation used mean that coma and astigmatic compensation is
possible both overall and at the focus.

Tuning of the dye laser over the coumarin 515 range is achieved
by rotating an intracavity 3-plate birefringent filter (Coherent
Radiation). The phase matching condition necessary for efficient
second harmonic generation (SHG) is satisfied in ADP by 90° phase

matching [9], i.e. when the fundamental beam makes an angle of
90° with the optical axis of the crystal. For a given wavelength,
this is possible only at a certain temperature. Thus the tuning
of the fundamental and hence the UV output must be accompanied
by an appropriate change in the crystal temperature, which is
maintained at the set temperature to ±0.01 K. During multimode
operation, oscillation occurs simultaneously in both directions
around the ring and UV radiation is coupled out via mirrors R_3
and R_4 (both having greater than 75% transmission in the range
250-260 nm). The longitudinal mode structure of the laser is
monitored using a scanning Fabry-Perot interferometer (Spectra-
Physics 470) with a 2 GHz free spectral range (FSR).

3. Single-Mode Operation

Only a single, uncoated etalon (thickness 6.35 mm, FSR ≃16.2 GHz)
inserted into the ring is necessary to ensure single-mode operation. However, the direction of the travelling wave would suddenly change, the ring oscillating about 50% of the time in either
the clockwise or anti-clockwise direction, spending times in
either given direction varying from a fraction of a second to
tens of seconds. This switching of ring direction is most probably due to microbubbles in the dye jet [3]. The optical bistability is completely removed however by the insertion into
the ring of a homebuilt unidirectional device (UDD).

The unidirectional device comprised a Faraday rotator and a
polarization rotation plate [10,11] both at Brewster's angle.
Each element rotates the plane of polarization of incident radiation by a certain amount, and the sense of rotation of each is
such that a single pass in one direction through both elements
will leave the polarization direction unaltered, while in the
other direction both rotations will add. The single-mode ring
cavity described here has 22 Brewster surfaces. Therefore a
polarization rotation of only a few degrees is sufficient to
introduce enough Fresnel loss in one direction to favor oscillation in the other. The Faraday rotator is a 12.7 mm diameter,
5 mm thick piece of Schott SF 57 glass (refractive index n =
1.8466), flat to $\lambda/20$, held at the center of a 3.6 KG permanent
ring magnet. Compensation is achieved using a 0.1 mm thick crystalline quartz polarization rotator, also flat to $\lambda/20$. The
angular rotation of the linearly polarized beam in the unfavored
direction is 3.5° for each element.

The single-mode dye laser could be scanned either in a mode-
hopping or a continuous-scan configuration. For the former scans,
the etalon is rotated in the horizontal plane by a synchronous
motor drive and the birefringent filter optimized during such a
scan. Altering the inclination angle θ of the etalon with the
ring axis ($\theta = 0°$ at the normal position) changes the transmission
frequencies (where minimum loss occurs) according to the interference condition

$$\nu_\theta = mc/2nL\cos\varepsilon \qquad\qquad\qquad (3)$$

where ε is the angle of refraction, L the thickness of etalon, m an integer, and n the refractive index of the etalon material (1.46 for fused silica). In this way scans over ~1 cm^{-1} in the visible or ~2 cm^{-1} in the UV are readily achieved.

In the continuous scan configuration, the signal from a photo-diode monitoring the visible light (the mirrors transmit 0.15% of the intracavity power) is fed into a Lansing model 80.215 lockin stabilizer consisting of a tuned amplifier, a reference channel, a phase-sensitive detector, an integrator, and a high-voltage dc amplifier. A small modulation voltage at 520 Hz from the stabilizer drives the piezoelectric translator (PZT) (Lansing model 21.938) supporting the cavity mirror R$_6$. The generated correction signal is also applied to this PZT and the cavity is thereby "locked" to the intracavity etalon. Thus, when the etalon is rotated by means of a synchronous motor, continuous single-mode scans are possible. The effective ring length in air is 0.84 m as determined by the measured cavity-free spectral range (c/nL) of $\approx$120 MHz in the visible.

4. Laser Characteristics

In the simple ring arrangement shown in Fig.1 with the crystal, etalon and UDD absent, the laser could be tuned over the range 485-550 nm with a threshold at the peak of this emission requiring only a few hundred milliwatts pump power at 413 nm. The whole doubling unit was measured to have an absorption at 514.5 nm of 4%, and insertion into the cavity raised the required pump power to 0.85 watts for a fresh dye solution (1.4 watts typically after ~10 hours operation). For a fixed crystal temperature of about 234 K, at 508 nm, tuning of the birefringent filter as measured by a spectrometer showed that the UV output had its half-intensity points at ±3 cm^{-1} in the visible corresponding to ±6 cm^{-1} in the UV.

At 508 nm with the laser running multimode the UV output power at mirror R$_3$ is 60 ± 10 μwatts for an intracavity power of about 2.5 watts. A similar output power is observed at mirror R$_4$, and therefore the multimode laser gives 120 μwatts UV power at 254 nm. It should be noted that following its emergence from the crystal the UV output must pass through, in each direction, 2 plates or 4 surfaces oriented at Brewster's angle for the opposite polar-ization, and a mirror only 86% transmitting. Therefore a total of 200 μwatts UV power must emerge from the doubling crystal. When the unidirectional device and the etalon are inserted, the pump threshold is raised to 1.3 watts, and 1.4 watts, respectively, for a fresh dye solution. In this configuration the laser opera-ted single-mode, producing 60 ± 10 μwatts UV output power at 254 nm for a 3.6 watt pump power. With a 2.5 watt pump power, single-mode operation is readily achieved over the range 500-520 nm corresponding to 250-260 nm UV output. The performance of this laser depends critically on the reliability of the crystal doubling unit. The present system has generated UV output during its life (~50 hours used over a period of two months) without any observable degradation. During this time it has not been necessary to change the location of the beam waist inside the crystal, and the crystal temperature has been recycled between room temperature and 200 K several times.

5. Fluorescence Excitation Spectrum of Atomic Mercury

To illustrate the capabilities of the present laser system, the fluorescence excitation spectrum was recorded of the 253.7 nm $^3P_1$ - $^1S_0$ transition of natural mercury (Figs.2-4). As shown in Fig.2 and 3 there are ten hyperfine lines, one from each of the five even-mass isotopes (196, 198, 200, 202, and 204), two from Hg-199 (I = 1/2) and three from Hg-201 (I = 3/2). The differing nuclei cause the lines to be displaced in frequency so that the spectrum under Doppler-limited resolution shows five distinct features (the "hand" of mercury). Because the Hg-196 isotope has a natural abundance of 0.146%, its presence does not contribute significantly to the observed pattern.

The fluorescence excitation spectrum of natural mercury in a bulb (Fig.2) and in a beam (Fig.3) is recorded by scanning the single-mode laser (with mode hopping) and detecting the resultant resonance fluorescence using a RCA 1P-28 phototube in conjunction with phase-sensitive discrimination (PAR Model 124A lockin). In between hyperfine components, scattered light from the cell was about three orders of magnitude lower than the fluorescence signal. The mode-hop positions, occuring every 240 MHz, can be clearly seen in these figures, causing a familiar histogram appearance. Note that the measured Doppler width of ≈1.2 GHz is in good agreement with that calculated.

In Figs.2 and 3 the frequency scales are nonlinear as a result of the angle tuning of the etalon [see (3)]. The UV power is also nonlinear. The visible intracavity power falls as the etalon is scanned away from normal, causing the UV output to drop quadratically with this change. In addition, the crystal temperature is not set for optimum phase matching at the center of the line profile, causing the visible conversion efficiency to depend further on frequency. In both Figs.2 and 3 the UV power drops dramatically with increasing frequency; no correction was made for this variation.

The "bulb" and "beam" experiments are quite different. In the former, the mercury is contained in a quartz cell at a pressure of 2.3×10^{-4} torr, corresponding to 0°C. In the latter, UV output (divergence <1 mrad) was passed vertically through a one-meter baffle arm where it intersected perpendicularly an atomic beam of mercury. The beam source comprised a double chamber stainless-steel oven mounted in a differentially pumped adjoining chamber. Both chambers were evacuated by oil diffusion pumps and liquid-nitrogen-cooled surfaces. A typical mercury backing pressure of 50 torr was used behind the 0.2 mm diameter nozzle. The atomic beam was collimated by a cooled iris.

A comparison of Figs.2 and 3 shows that the latter has a marked increase in resolution resulting from the decreased effective Doppler width of the beam. Indeed, under beam conditions the

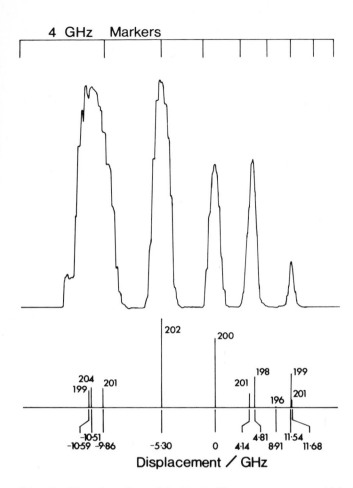

Fig.2 The Doppler-limited fluorescence excitation spectrum of natural Hg vapor in a cell at 0°C. The measured displacements below are from Bitter [12], replotted on our frequency scale. The amplifier sensitivity has increased by x2 after the Hg-200 peak.

line widths are so narrow that it is possible to hop over a line in the mode-hopping scan of Fig.3. This explains the weak appearance of the Hg-201 component at +4.14 GHz. The underlying broader part of the lineshape (obvious for the Hg-202 line which is well off-scale in the figure) is the Doppler-limited contribution from background Hg (5 x 10^{-6} torr) in the beam apparatus.

Figure 4 shows single-mode continuous scans over selected lines, where the resolution achieved, the linewidth, and the background emission contribution are all apparent. We found that the measured linewidth depended on the time taken to scan

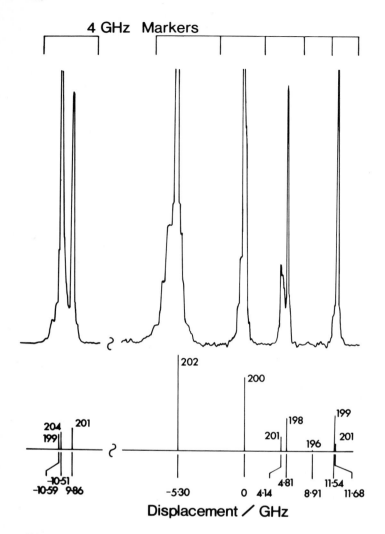

Fig.3 The sub-Doppler fluorescence excitation spectrum of natural Hg in an atomic beam. Note the change in frequency scale between the first and last four components.

over a line, and was a convolution of the contributions from the laser bandwidth and the atomic beam/fluorescence collection geometry. For example, scanning over the Hg-202 line shown in Fig.4b took about 5 seconds and resulted in a measured linewidth of ~100 MHz (FWHM). Scans (a) and (c) were made over slightly longer and shorter times, respectively, (due to the nonlinearity of the etalon scan) and showed slightly larger and narrower linewidths, respectively. We conclude that the free

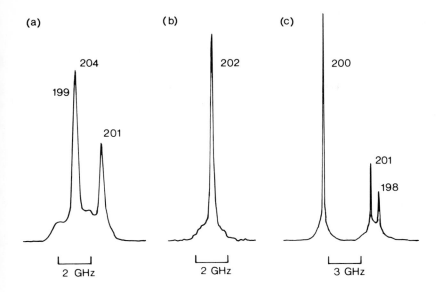

Fig.4 Single-mode continuous scans over selected hyperfine components of the Hg $^3P_1$ - $^1S_0$ transition.

running single-mode laser bandwidth was ≲50 MHz (in the visible) over a period of several seconds. This figure is reasonable for a free-running dye laser using COT additive, no elements to reduce pressure fluctuations in the dye jet, and no 2 μm filter to remove microbubbles. The relative increase in bandwidth due to the modulation applied to the PZT was negligible.

6. Concluding Remarks

An intracavity-doubled cw ring dye laser has been operated single mode to wavelengths as short as 500 nm; the UV output at 254 nm was 60 ± 10 μwatts in a bandwidth of ≲100 MHz (over a few seconds). The coumarin 515 dye laser has been tuned over the 253.7 nm mercury profile to illustrate sub-Doppler resolution and single-mode continuous scanning.

It is well known [13] that crystals with very low absorption and cavities with very low losses are necessary for high efficiency second harmonic generation. In the present laser system having a pump power of 3.5 watts, the power in the single-mode ring cannot exceed ~1 watt. In the future, it should be possible to reduce the present intracavity losses and to actively stabilize the cavity to ≈2 MHz UV bandwidth. Although the present laser system is designed to operate over 250-260 nm, this range can easily be extended to shorter or longer wavelengths by appropriate choice of dye, mirror coatings, and crystal temperature.

It is anticipated that the unique characteristics of this laser will be most effectively exploited in ultra high resolution spectroscopy using Doppler-free techniques.

Acknowledgments

Support by the National Science Foundation under NSF CHE 80-06524 is gratefully acknowledged.

References

1. G. D. Boyd and D. A. Kleinman, J. Appl. Phys. _39_, 3639 (1968); S. Blit, E. G. Weaver, T. A. Rabson, and F. K. Tittel, Appl. Optics _17_, 721 (1978); J. Paisner, M. L. Spaeth, D. C. Gerstenberger and I. W. Ruderman, Appl. Phys. Letters _32_, 476 (1978).

2. R. Wallenstein and T. W. Hänsch, Opt. Communications _14_, 353 (1975).

3. C. E. Wagstaff and M. H. Dunn, J. Phys. D: Appl. Phys. _12_, 355 (1979).

4. R. P. Mariella, J. Chem. Phys. _71_, 94 (1979).

5. P. N. Clough and J. Johnston, Chem. Phys. Letters _71_, 253 (1980).

6. H. W. Kogelnik, E. P. Ippen, A. Dienes and C. V. Shank, IEEE J. Quant. Electron, QE-8, 373 (1972).

7. W. D. Johnston and P. K. Runge, IEEE. J. Quant. Electron. QE-8, 724 (1972).

8. M. H. Dunn and A. I. Ferguson, Opt. Communications _20_, 214 (1977).

9. R. C. Miller, G. D. Boyd and A. Savage, Appl. Phys. Letters _6_, 77 (1965).

10. P. W. Smith, IEEE J. Quant. Electron. QE-4, 485 (1968).

11. S. M. Jarrett, and J. F. Young, Laser Focus _14_, 16 (1978).

12. F. Bitter, Appl. Optics _1_, 1 (1962).

13. A. I. Ferguson and M. H. Dunn, IEEE J. Quant. Electron. QE-13, 751 (1977).

Chemist's Dream About IR Laser Photochemistry

C.T. Lin[*], J.B. Valim, and C.A. Bertran

Instituto de Química, and
Instituto de Física "GLEB WATAGHIN", Universidade Estadual de Campinas
13.100 Campinas SP Brazil

1. Introduction

Since the reports of AMBARTSUMIAN and LETOKHOV in 1974 which demonstrated
that the laser-induced dissociation reaction is isotope selective [1], the
laser photochemists have started to dream about the possibilities of using
this technique in at least two research disciplines. First is the selective
laser-heating of a specific chemical bond (or functional group) in a molecule. This would permit one to carry out the selective bond breakage in an
unimolecular reaction as well as a positionselective chemical reaction in
the bimolecular processes. The second discipline is super-excitation chemical reaction. One hopes that the chemical reaction through a high-energy
reaction channel would have different reaction mechanism, and perhaps would
be more effective than that which occurs via normal low-energy reaction path.

In recent years, the laser photochemistry group at UNICAMP has been actively involved in the research disciplines mentioned above. The late Professor
Sergio P.S. Porto was closely associated with our research group, so it is
proper in this memorial symposium to present some results obtained by my
group with the valuable suggestions and support of the late Professor Porto.

2. Classifications of Infrared-Laser-Induced Unimolecular Reactions

2.1 Center-Atom-Symmetric Molecules: Strong Intermode (or Interbond) Coupling Systems

Infrared multiphoton excitation (IMPE) and infrared multiphoton dissociation
(IMPD) schemes have been proposed and studied [2] by various scientific groups
in the world, specially by Professor BLOEMBERGEN and co-workers [3]. In gen-

---

[*] Author to whom correspondence should be addressed.

eral, one divides the energy levels of a molecule into three regimes of vibrational excitations: discrete levels, quasi-continuum, and true continuum. To compensate the anharmonicity in the IMPE, various effects have been considered such as degeneracy splitting, rotational energy change, coriolis coupling, power broading, pressure broading, etc. [2,3]. Once the excited molecule reaches its quasi-continuum energy regime, the molecular dissociation occurs.

In the intramolecular-heat-bath picture [Ref. 3, Fig. 5], the normal mode of vibrations of a molecule can be divided into the energy levels of active mode, and those of the remaining modes are considered to form a quasi-continuous heat bath. If the rate of laser pumping into the active mode (V_p) is faster than the relaxation rate from the active mode to the heat bath (V_R), then we would have a selective-mode chemical reaction. It is important to point out that most of the work done in this area are for molecules like SF_6, SF_5Cl, BCl_3, NH_3, CH_4, CHFClBr.. etc. [4]. Those molecules are the center-atom-symmetric molecules, and have strong intermode (or interbond) couplings. In fact, Professor BLOEMBERGEN mentioned clearly [3] that this class of molecules can be considered as "hydrogen atom" for multiphoton chemistry and vibrational spectroscopy. Since the active mode for laser excitation is strongly coupled to the heat-bath modes, the selective laser-heating of a specific chemical bond becomes impossible unless one has an ultra-short laser pulse, e.g., picosecond laser pulse. This is due to the fact that when V_p is slower than V_R, the molecule as a whole will have a canonical thermal distribution before the selective bond breakage occurs.

2.2 Non-Center-Atom-Symmetric Molecules: Various Degrees of Intermode (or Interbond) Coupling Systems

From the selective photochemistry view point, the non-center-atom-symmetric molecules, $CF_3{}^{CH} = CHCOCH_3$ (A), should be more interesting than the center-atom-symmetric systems mentioned in the previous section. Compound A is made of 14 atoms and has 36 normal mode of vibrations. The effective couplings vary among the 36 vibrational modes and also among the 13 chemical bonds. In this case, the intramolecular-heat-bath picture should be quite different from that of the center-atom-symmetric systems, i.e., the normal mode of vibrations of a molecule should not be assumed as only "one" laser driven mode and "one" heat bath. Instead, the molecular energy diagram of the molecule A should be considered as an active mode for the laser excitation and "various" quasi-continuous heat baths. Physically, those heat baths are, in fact, corresponding to the "various" portions (functional groups) of the molecules or

the "various" groups of the normal mode of vibrations. The magnitude of couplings varies between the laser active mode and each of the heat baths. A different activation energy is required to overcome the different energy barriers
between any two of the heat baths.

Dynamically, the V_p is still slower than V_{Rn} $(n > 1)$ for the non-center-
atom-symmetric molecules where V_{Rn} are the relaxation rates from the laser
driven mode to the various heat baths of the molecules. Since $V_{R1} \neq V_{R2} \neq V_{R3} \cdots$
$\neq V_{Rn}$, it is probable that the laser excitation energy would accumulate in
one specific heat bath. If V_{Rj} is much faster than any the other V_{Rn} $(n \neq j)$,
then we might have a selective heating of the functional group corresponding
to the heat bath j, i.e., the portion of the molecule corresponding to the
heat bath j is hot while the rest of the molecule is relatively cold.

For larger molecules, e.g., Bis(hexafluoroacetyl-acetonate) uranyl-tetra-
hydro-furan [5], coumarin 6 [6], and sucrose, the laser-excitation mode is
no longer an indivudual normal mode of vibration in the molecule but rather
a group of normal modes that forms an active system for the laser excitation.
This means that the active system is also a heat bath in the intramolecular-
heat-bath picture. Many other heat baths in the intramolecular-heat-bath picture are made of various other groups of normal modes in the molecules. The
fact is that an activation energy is required for passing the laser excitation energy from the active system (itself is a heat bath) to the heat baths
or from one of the nonactive heat baths to the others. Even though V_p is still
slower than V_{Rn} yet the laser excitation energy tends to accumulate within
the heat bath of the laser active system rather than relax its energy to the
other heat baths by passing through a high-energy barrier, i.e., one would
obtain a selective-heating of that specific laser-excited portion of the
molecule.

2.3 Static-Hindered Molecular Structures and Restricted Molecular Relaxations

Systems described in Sects. 2.1, 2.2 are molecules that have free molecular
motions. The degrees of energy transfer from a normal mode to a heat bath or
from one of the heat baths to the others depend on the magnitude of the coupling (the extent of the energy levels overlap) between the active mode and
the heat baths, or one of the heat baths and the others. Presumably, the energy barriers existed for transferring energy from the laser driven mode to
the heat baths are relatively low due to their free molecular motions.

Another class of molecules different from the systems in Sects. 2.1, 2.2
is biphenyl and its group substitutions, e.g., 2-chloro, 2'-iodo, 6-trifluoro-

methyl, 6'-methyl-biphenyl (compound B). The biphenyl molecule composes of two benzene rings which are perpendicular with one another, while compound B offers a static-hindered molecular structure. The energy relaxation from a functional group of one ring to the other functional group of the other ring becomes restricted. This implied that an extremely high energy barrier is posted for the energy transfer to occur. It becomes clear that when CF_3 bond (modes) in compound B is selectively excited by a pulse CO_2 laser, the laser energy should be perserved in the excited portion of the molecule, i.e., the bond breakage can only be either CF_3 or Cl but not the weakest bond, I. Thus one obtains a selective-heating of a specific functional group of the molecule. In short, if the molecular relaxation rate is relatively slow, e.g., the systems in Sects. 2.2, 2.3, one does not need to have a picosecond pulse laser to achieve the selective laser heating chemistry.

3. Experimental Results for Systems of the Second Kind

Figure 1 shows the infrared spectrum of sucrose solid in KBr. The spectrum displays five groups of absorption bands: 3500 cm^{-1} (-OH modes), 3000 cm^{-1} (-CH modes), 1400 cm^{-1} (C = C modes), 1100 cm^{-1} (CO modes), and 950 cm^{-1} (ring modes). One notes that a large spectral overlap is observed for bands 3500 cm^{-1} and 3000 cm^{-1} and also for 1400 cm^{-1}, 1100 cm^{-1}, and 950 cm^{-1}. There is a hole (no absorption bands) observed in the spectral range from 3000 cm^{-1} to

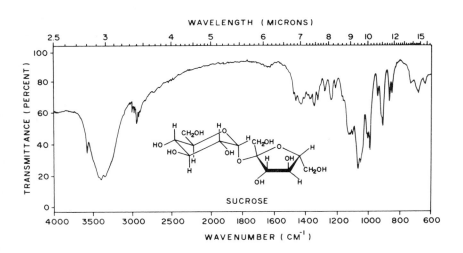

Fig. 1. IR Spectrum of Sucrose Solids in KBr

1400 cm^{-1}. This might suggest that the couplings are strong (energy barrier is low) between -OH modes and -CH modes, and also among C = C modes, CO modes, and ring modes. On the other hand, the group vibrations with vibrational frequencies >3000 cm^{-1} and those of <1400 cm^{-1} should have no (or small) coupling (a high energy barrier).

In the intramolecular-heat-bath picture, each broad IR spectral band for sucrose molecules corresponds to a heat bath in the CO_2 laser induced unimolecular reaction. Since the laser-driven modes ($\nu = 940,56$ cm^{-1}) might also be the most favorable selective-heating bath (a combination of ring modes in sucrose solids), one should see a selective heating of ring vibrational modes for sucrose molecules in our experiment.

The experimental set up for CO_2 laser photochemistry of sucrose has been described elsewhere [7]. The pure sucrose solid sample under vacuum was suspended in the center of the reaction cell. P(24) line ($\nu = 940,56$ cm^{-1}) from a pulse CO_2 laser was grating tuned and focused onto the sample by a GaAs len. The pulse laser was operated at a repetition rate of 3pps for 10 minutes reaction time. The thermal chemistry of sucrose was carried out using a flexible electric heating tape. The temperature attained was 450°C and the heating time for the thermal reaction was also 10 min.

The reaction products of pure sucrose are analyzed by IR spectrophotometer. Figures 2 and 3 display the IR spectra of the reaction products for laser

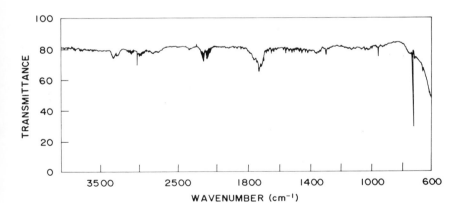

Fig. 2. IR spectrum for the IMPD products of pure Sucrose. $\nu_{laser} = 940-56$ cm^{-1}, laser power density = 1 GW/cm^2 and irradiation time = 10 min

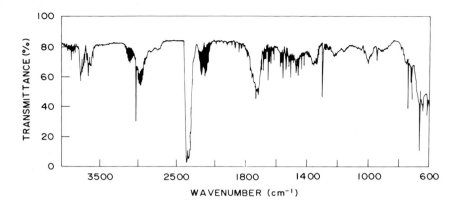

Fig. 3. IR Spectrum for the thermal reaction products of pure sucrose. Temperature = 450°C and heating time = 10 min

photochemistry and for thermal chemistry, respectively. The spectral assignments are: C_2H_2 (3320, 3260, 729 cm^{-1}); CH_4 (3080, 3020, 2960, 1300 cm^{-1}); C_2H_4 (945 cm^{-1}); H_2O vapor (1700-1400 cm^{-1}); H_2CO (1740 cm^{-1}); CO_2 (2360, 2340, 665 cm^{-1}), and CO (2160, 2120 cm^{-1}). The experimental results can be summarized as 1) the relative yields of CH_4 are the same in both cases; 2) C_2H_2 is only produced in the case of laser excitation, and 3) the ratio of the relative yields of $CO/CO_2 = 8$ for the laser photochemistry of pure sucrose and = 0.4 in the case of thermal excitation. Observation 2) suggests that laser photochemistry of sucrose occurs with a different reaction mechanism from that of the thermal chemistry. Moreover, observation 3) might indicate that the laser excitation energy, $\nu = 940,56$ cm^{-1}, is probably locked in the ring modes of the sucrose and thus leads to the specific functional group photochemistry before the thermal equilibrium of the whole molecule is achieved.

4. Experimental Evidence for the "Super-Excitation" Laser Photochemistry

It is known [8-10] that IMPE and IMPD of NH_3 molecules produce NH_2 radicals as their primary reaction products. When IR laser photochemistry of NH_3 is carried out at low sample pressure and low laser power density, the NH_2 radicals are generated in the electronic ground state ($\tilde{X}^2B_1$). On the other hand, the electronic excited NH_2 radicals ($\tilde{A}^2A_1$) were obtained when a high sample pressure and high laser power density were employed.

In our experiment, we used a mixture of 1:1 ratio for NH_3 to O_2 at various total sample pressures. The laser employed was the P(32) line ($00^01 - 10^00$ transition) of a TEA CO_2 laser with a pulse duration of 250 ns and a constant peak power at the focal point of 1 GW/cm^2. The laser was operated at a repetition rate of 3pps and was used to excite the ν_2 band of NH_3. It was found that the IMPD reaction characteristic is very sensitive to the total sample pressure of NH_3 and O_2 used. When the total sample pressure of NH_3 and O_2 is equal or higher than 100 mmHg. The reaction is of an explosive type and accompanied by a yellow flash luminescence. The IR spectrum of the explosive reaction products is shown in Figure 4. The main products observed in the IR spectrum are analyzed as H_2O liquid (broad band ~ 3500 cm^{-1}), H_2O vapor (rotational fine structures 1700-1400 cm^{-1}) and NH_4NO_3 (broad band ~ 1400 cm^{-1}, a distorted band contour ~ 3200 cm^{-1}). When the IR laser photooxidation of NH_3 is carried out at a total sample pressure less than 100 Torr, no apparent luminescence is observed during the reaction. Figure 5b shows the IR spectrum of the reaction products for the nonexplosive type photooxidation of NH_3. The observed products in the IR spectrum are H_2O liquid, H_2O vapor, and N_2O (2211, 224 cm^{-1}).

In order to explain the two distinct reaction pathways mentioned above for laser photooxidation of NH_3, we need to correlate our experiments with the experiments of thermal catalytic reaction of NH_3 and O_2 done by VON NAGEL in

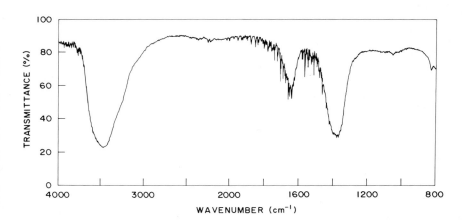

Fig. 4. IR Spectrum for the photooxidation products of NH_3 after a mixture of 60 Torr of NH_3 and 60 Torr of O_2 are subjected to a single pulse CO_2 laser at 10.719 μm and the laser power density = 1 GW/cm^2

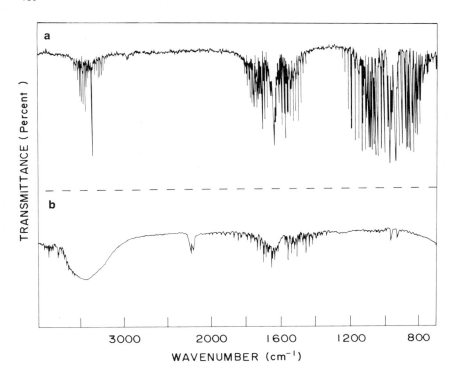

Fig. 5. IR spectra: (a) a mixture of 30 Torr of NH_3 and 30 Torr of O_2 record-ed using a 10 cm optical path; (b) the photooxidation products of ammonia after the mixture (a) is subjected to 1800 pulses CO_2 laser at 10.719 μm and laser power density at the focal point = 1 GW/cm^2

1930 [11]. VON NAGEL observed that the N_2O molecules were the principal prod-ucts when he used a low % weight of NH_3 with O_2 and carried out the reaction at low temperature limits. This result is similar to that of our low sample pressure and low laser power density experiment. Moreover, when the thermal catalytic reaction of NH_3 and O_2 was carried out at a high % weight of NH_3 with O_2 and high temperature, VON NAGEL [11] detected NO molecules as the reaction products. It is known that NO molecule would be transformed into NH_4NO_3 molecule following the reaction.

$$NO \xrightarrow{O_2} NO_2 \xrightarrow{H_2O} HNO_3 \xrightarrow{NH_3} NH_4NO_3 \tag{1}$$

Therefore, it is not surprising that our high gas pressure and high laser power density experiments for the laser photooxidation of NH_3 gives also the reaction products of NH_4NO_3.

From the discussions and correlations above, the probable expalanations for our IR laser photochemistry of NH_3 and O_2 are as follows: I) The reaction mechanism at low sample pressure and low laser power density (low laser energy fluence) is

$$NH_3 + O_2 \xrightarrow{nh\nu} NH_2 \cdot (\tilde{X}^2 B_1) + O_2 \longrightarrow N_2O \tag{2}$$

and II) the reaction mechanism for high gas pressure and high laser power density experiment is

$$NH_3 + O_2 \xrightarrow{nh\nu} NH_2 \cdot (\tilde{A}^2 A_1) + O_2 \rightarrow NO \xrightarrow{reaction} (1) NH_4NO_3 \quad . \tag{3}$$

In summary, our experiments of laser photooxidation of NH_3 have demonstrated that the super-excitation reaction [high-energy reaction channel, $NH_2 \cdot (\tilde{A}^2 A_1) + O_2$] does really have a different reaction mechanism and probably is more effective than the normal low-energy reaction channel [$NH_2 \cdot (\tilde{X}^2 B_1) + O_2$].

Acknowledgments. Financial supports from the FAPESP, Grant No. 74/1334; Convênio CNEN-CTA, Grant No. 103092/74 and the CNPq, Grant No. 40.3304/79 and No. 300.917/79 are acknowledged.

References

1. R.V. Ambartsumian, V.S. Letokhov, E.A. Ryabov, N.V. Chekalin: JETP Lett. *20*, 273 (1974)
2. H. Walther, K.W. Rothe (eds.): *Laser Spectroscopy IV*, Proceedings, Fourth International Conference, Rottach-Egern, Fed. Rep. of Germany, June 11-15, 1979, Springer Series in Optical Sciences, Vol. 21 (Springer, Berlin, Heidelberg, New York 1979)
3. N. Bloembergen, E. Yablonovitch: Phys. Today (May 1978) p. 23
4. Actas del Primer Seminario Latinoamericano Sobre el Laser Y Sus Aplicationes a la Fisica Y a la Quimica, La Plata, República Argentina, 21 al 26 de Agosto de 1978
5. A. Kaldor, R.B. Hall, D.M. Cox, J.A. Horsley, P. Rabinowitz, G.M. Kramer: J. Am. Chem. Soc. *101*, 4465 (1979)
6. J.P. Maier, A. Selmeier, A. Laubereau, W. Kaiser: Chem. Phys. Lett. *46*, 527 (1977)
7. C.T. Lin, S.P.S. Porto, R. dos Santos: "Nonlinear Optics", in Proceedings of the VIth Vavilov Conference, Novosibirsk, USSR, June 20-22, 1979, ed. by V.P. Chebotayev, Vol. 2, p. 31
8. V.S. Letokhov, E.A. Ryabov, O.A. Tumanov: JETP *36*, 1069 (1973)
9. J.D. Campbell, G. Hancock, J.B. Halpern, K.H. Welge: Chem. Phys. Lett. *44*, 404 (1976)
10. C.T. Lin, C.A. Bertran: J. Phys. Chem. *82*, 2299 (1978)
11. A. von Nagel: Z. Electrochem. *36*, 754 (1930)

Multiphoton Ionization Mass Spectrometry and Other Developments in UV Laser Chemistry

K.L. Kompa

Max-Planck-Gesellschaft, Projektgruppe für Laserforschung
8046 Garching/FRG

Summary

This report deals with three topics, namely UV laser-induced multiphoton ionization of polyatomic molecules, collision pair excitation by high power pulsed UV lasers and
stimulated emission processes produced in the products of
iron carbonyl UV laser photolysis. The results under the
first heading are based on work by J.P. Reilly et al. and
P. Hering et al., those for collision pair excitation on
work by H.P. Grieneisen et al. while the last topic relates
to experiments by S.H. Bauer, J. Krasinski et al. All these
authors are associated with our laboratory or held various
guest appointments here during the years 1979 and 1980. The
results show that high power ultraviolet lasers do not just
provide a continuation in the study of molecular processes
along the lines of traditional photochemistry but yield
novel and unexpected results.

1.1 Background

Laser-induced multiphoton spectroscopy (including ionization spectroscopy) is by now a well established area of
study /1/.

A new impetus was provided to this field by the advent of
high power ultraviolet laser sources (rare gas halide, rare
gas excimer and diatomic halogen lasers /2/). By these excitation sources the ionization energy of many polyatomic
molecules can be reached by just two photons going through
a resonant or near resonant intermediate state, Fig. 1.
It has been claimed by several authors /2-6/ that laser-induced ionization shows features which differ markedly from
other ionization schemes (single photon photoionization,
photoelectron spectroscopy, chemical or charge transfer
acivation, electron impact). It is reported that the nature
and relative abundance of secondary ion fragments varies
with laser parameters. Total ion yields can be very high if
resonant or near-resonant intermediate states are available
and if ionization requires not more than two photons. In
such cases, the total ion yield may approach unity for high
laser fluxes. The yield also strongly depends on the nature
of the intermediate states and in particular on their decay

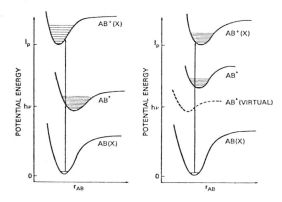

Fig.1. Principal scheme of two step photoionization with resonant intermediate state

times. In favourable cases then this mass spectrometric concept allows for very <u>sensitive</u> trace (potentially single molecule) detection. In addition, the spectral dependence of the ionization probability makes very <u>selective</u> detection of molecules possible. These features combine in the idea of a two-dimensional mass spectrometer whose principal design elements are indicated in Fig. 2.

The benzene molecule has become the "pet" species for this kind of studies. This molecule exhibits resonant intermediate states for both KrF and ArF laser excitation. The magnitude of the transition probabilities coupling the intermediate states to the ground state and to the ionization continuum are known as well as the absolute magnitude of the molecule's ionization potential (9.247 eV). The occurance of relaxation or destructive chemical (isomeri-

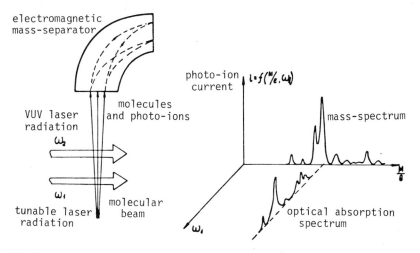

Fig.2. Technical design elements and illustration of two-dimensional mass spectrometer (adapted from /3/)

zation, dissociation) processes affecting these states has been a matter of much speculation /7/. For orientation, Fig. 3 shows the energy relations in some of the ionization sequences of interest.

In addition to the benzene data some discussion is provided for corresponding measurements in a pair of aldehydes in a later paragraph of this paper.

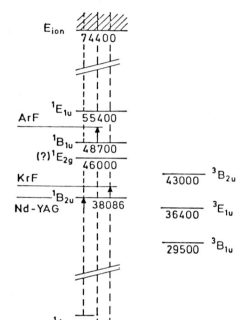

Fig.3. Relevant molecular energy levels in benzene indicating the intermediate and final states which may be reached by the 3 lasers used in this study

1.2 Experimental

Two different experimental set-ups were used both employing a conventional CVC Model MA-3 time of flight mass spectrometer modified to include a laser ionization source (Fig. 4). The source chamber and the flight tube are separated by a metal plate with a thin slit at its center. Across this slit a pressure difference of 200 is maintained leading to typical operating pressures 5×10^{-5} torr in the source and 4×10^{-7} torr in the flight region.

The excimer laser used in this study was homemade providing an output of up to 500 mJ/20 ns as a KrF and up to 200 mJ/12 ns as an ArF laser. The band width of each laser is of the order of 10 A. Only a small fraction of the laser energy passes through a 2 mm aperture and is focused into the source region by a 10 cm focal length CaF_2 lense. It was made sure that the laser light does not produce surface ionization at any point in the spectrometry. The ions generated by the

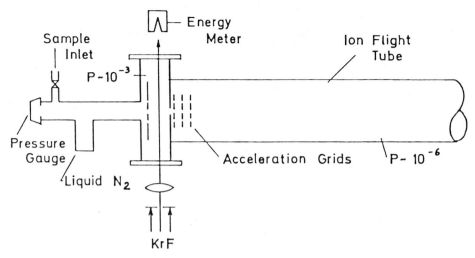

Fig.4. Aparatus used for recording laser induced multiphoton ionization/
fragmentation mass spectra (adapted from /6/)

laser are accelerated by a series of grids into the field-
free flight tube where they separate in time according to
their different m/e ratios. In this way an entire mass spectrum at a given laser wavelength and intensity can be recorded in a single shot. Spectral reproducibilty and mass resolution were at least as good as with the electron impact ion
source previously employed with this instrument (Fig. 5).

Alternatively to the excimer laser a frequency doubled or
quadrupled Nd-YAG laser was used (Fig. 6). This multistage
laser consists of a mode locked oscillator in connection
with a variable optical gate which is used to switch out
single or multiple pulses out of the pulse train leaving
the oscillator. Several amplifier stages bring the energy
up to the desired level. As seen in the figure, two subsequent doubling crystals generate 0.53 µm and 0.26 µm
radiation. Either one of these frequencies or a combination
of the two colours may be incident on the molecular sample
(for further details see /8/).

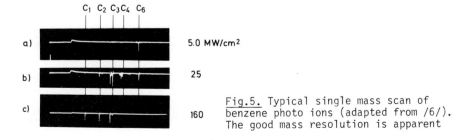

Fig.5. Typical single mass scan of
benzene photo ions (adapted from /6/).
The good mass resolution is apparent

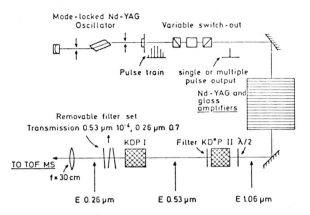

Fig.6. Schematic diagram of picosecond Nd-YAG laser used in multiphoton ionization studies. With the removable filter set in place the 0.53 μm and 1.06 μm laser light is well suppressed while without the filter two-colour experiments can be done. The single pulse duration is 30 ps (FWHM) at 0.53 μm and 20 ps at 0.26 μm. Energies up to 10 mJ in the UV and up to 100 mJ in the visible were used

1.3 Some multiphoton ionization results

Without going into much detail the experimental findings may be summarized as follows. Laser induced ionic fragmentation can be either much more or much less drastic than that produced by electron impact and depends on the light intensity used. The fragmentation observed at the two different excimer laser wavelengths are quite similar (except for minor differences). Up to some MW/cm^2 the benzene molecular ion $C_6H_6^+$ is the dominant species formed. Above that value smaller fragments grow in and while under electron impact conditions fragmentation is essentially complete on the C_4/C_3 level under laser conditions fragmentation appears to continue all the way down to C^+ (compare Fig. 7).

A comparison of electron impact (in parantheses) and MPI is shown in the following scheme.

C^+

C_2^+ C_2H^+ $C_2H_2^+(1)$ $C_2H_3^+(1)$

C_3^+ $C_3H^+(4)$ $C_3H_2^+(5)$ $C_3H_3^+(13)$

C_4^+ $C_4H^+(3)$ $C_4H_2^+(16)$ $C_4H_3^+(18)$ $C_4H_4^+(19)$

$C_5H_3^+(3)$

$C_6H_4^+(6)$ $C_6H_5^+(14)$ $C_6H_6^+(100)$

C6H6 +

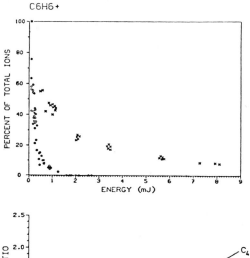

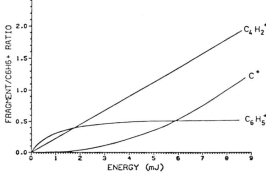

Fig.7. Upper picture: dependence of benzene molecular abundance on laser energy for KrF and ArF irradiation. Lower picture: some representative fragment ion abundencies versus $C_6H_6^+$ formation as function of KrF laser pulse energy (adapted from /6/)

It was first noted with surprise that the total ion yield in the case of ArF radiation is by a factor of 5 lower than that of KrF. A plausible explanation could be provided in /6/ by considering the following rate equation model based on a four-level excitation scheme.

Three or four level model for
C_6H_6 two-step ionization

1. $\dfrac{d(C\ H\ ^+)}{dt} = \sigma_2\ I\ B_{1u,2u}$

2. $\dfrac{d(B_{1u,2u})}{dt} = \sigma_1\ I\ A\ g - (\sigma_2 I + \sigma_3 I + 1/\tau)\ B_{1u,2u}$

3. $A_{1g} = A_{1g}^0 - B_{1u,2u} - C_6H_6^+ - \text{"ISO"}$

4. $\dfrac{d(\text{"ISO})}{dt} = \dfrac{B_{1u,2u}}{\tau}$

An analytic solution (with some simplifying assumptions for
the initial conditions) yielded satisfactory agreement between theory and experiment if an intermediate relaxation
rate (or isomerization rate) is assumed for the ArF case to
amount to 20 ps (Fig. 8).

This result demonstrates that intermediate state dynamics
may be investigated here in a novel way. It also shows that
high total ion yields may only be expected if stable intermediate state exist. While this effect might limit the mass
spectrometric sensitivity in some cases it might also be
beneficial in other cases in comparing molecules with otherwise similar absorption characteristics.

Our results show that the detection sensitivity of benzene
MPI with KrF laser light is very high and may well exceed
10 % of all the molecules in the irradiated sample.

The question of primary versus secondary ionic fragmentation or in other words, the question of "soft" (nonfragment-
ing) versus "hard" (fragmenting) photo ionization has has
received considerable interest both for practical applications in mass spectrometry and with regard to ion spectroscopy.

Secondary fragmentation events may be distinguishable in a
time resolved experiment. For that reason a 20 ps quadrupled Nd-YAG laser was used whose set-up was shown in Fig. 4.
This time resolution should suppress secondary photo dissociation in the photo ions under any reasonable assumptions
for primary decomposition rates based on RRKM estimates
/9/. The experiment also aimed at an investigation of the
effects associated with very high excitation rates (high
field limit). This objective requires a rate of up-pumping

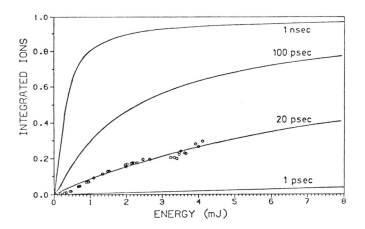

Fig.8. Mass integrated ion yield resulting from ArF laser irradiation of
benzene /6/. Data points are experimental, solid lines model predictions
from four-level rate equations for different intermediate life times

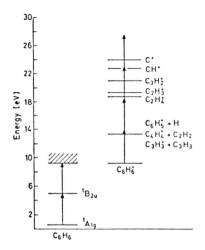

Fig.9. Illustration of energy pass
ways in a polyatomic molecule, e.g.
C_6H_6. The picture shows the compe-
tion between radiative coupling
within the "driven mode" and product states

within the parent molecular ion, which exceeds any coupling
rates to other background states as well as product states.
The principal is illustrated in Fig. 9. It is seen in this
figure that the primary photo ionization occurs through an
autoionizing state at a rate which is essentially instanta-
neous thus suppressing any excitation beyond the ionization
energy in the neutral molecule. The question to be answered
then is that for the further excitation and dissociation
channels. As is indicated in Fig. 9 there are alltogether
four dissociation processes in $C_6H_6^+$ with a threshold of 4
eV which can occur if the primary ion absorbs 1 KrF photon.

The following preliminary conclusions can be reported from
this experiment /7/.
a) Excitation in $C_6H_6^+$ does not go much beyond the energy
required for C_4 and C_3 fragment formation. The generation
of smaller fragments does not occur with single ps- pulse
excitation thus pointing to secondary dissociation pathways. The picture changes at very high power densities
(10^{10} W/cm^2) where one approaches the high field limit and
the excitation rate is less disturbed by lower energy fragmentation channels. At very high energies all molecular
fragments disappear and doubly charged ions are created.
b) The previous KrF/ArF laser data can be reproduced if
multiple 20 ps pulses with t = 6 ns are used. It should
be noted that even at high laser powers molecular resonance
absorption is important and that measurable photo ionization with pure 0.53 µm radiation is not observed here.
These results are in agreement with the picture of limited
secondary fragmentation in the low field case (Fig. 9).

Coroborative evidence is provided by an additional photo
ionization experiment on a pair of aldehydes. The idea of
this measurement is immediately apparent from Fig. 10.
Acetaldehyde has an ionization energy, which is just above
the sum of 2 KrF photon energies while the higher homologue
butyraldehyde has an ionization potential just below this
energy. Consequently, two photon ionization is possible in

190

the latter case while it is not in the former. Indeed no parent acetaldehyde ion is observed and all the ions shown in Fig. 10 are formed after one photon dissociation. The situation obviously changes when an ArF laser is substituted for KrF. Without going into a detailed discussion of these mass spectra (Fig. 10) the conclusion is justified that under the experimental conditions chosen no photo ionization processes occur, which require more than two photons and that fragmentation occurs always once the necessary energy has been accumulated in the molecule.

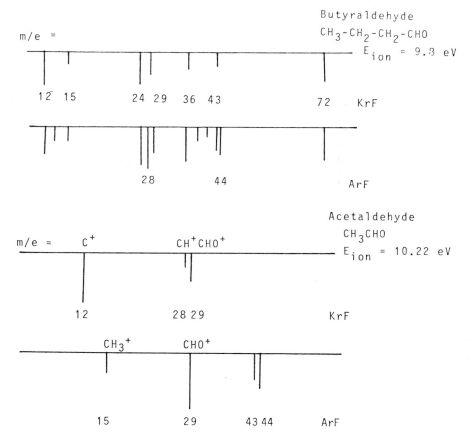

Fig.10. Mass scans of butyraldehyde (upper picture) versus acetaldehyde (lower picture) as seen in KrF and ArF multistep photoionization. While in the upper picture the parent molecular ion is seen in the KrF mass spectrum, the corresponding ion is absent in the lower picture. The mass spectra change considerably if ArF laser light is used. For further discussion see text and /10/

2. F$_2$ laser induced collision pair excitation

The very high photon fluxes available now from high power
UV lasers can be used not study processes which have a low
probability or a low cross section. In a first attempt to
utilize the potential of the molecular fluorine laser at
1580 A the excitation of short-lived collision complexes to
a bound excited state was tried /11/. Xe-Xe collisions were
chosen for this purpose. The experimental set-up is shown
in Fig. 11. There is a shallow van der Waals' potential in

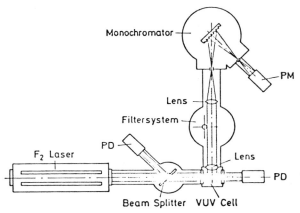

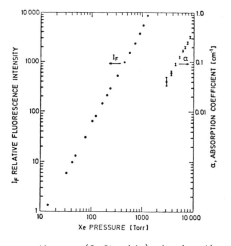

Fig.11. Experimental
arrangement to study F$_2$
laser induced collision
pair excitation /11/

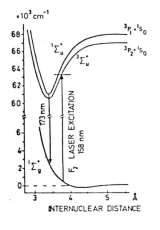

Fig.12. Xenon dimer potential energy diagram (left side) showing the
repulsive ground state and two bound excimer states. In a vertical
transition the F$_2$ laser can excite the quasi molecule only at an in-
ternuclear distance outside the van der Waals' minimum. The right side
of the figure shows that the pressure dependence of the excimer
fluorescence intensity is quadratic over 3 orders of magnitude

the ground state from which however F_2 laser excitation to the excited $_u^+$ state is not possible for energetic reasons. The excitation therefore must proceed as Fig. 12 shows after the collision partners have left the van der Waals' potential and have come to a closer approach. Part of the absorbed energy is reemitted in the form of the well-known 172 nm excimer fluorescence. The pressure dependence of the fluorescence intensity is strictly quadratic as expected for a two-body collision. This may be considered as an example of a laser controlled collision. Besides pure xenon also xenon halogen mixtures were investigated and provided information, which is relevant to the study of kinetic processes in rare gas halide lasers. Obviously it would be more preferable to use tunable laser sources for such experiments as this would yield interesting information on the shape of molecular potentials. Such studies may be attempted in the future.

3. UV laser induced photolytic processes yielding stimulated emission

This last paragraph illustrates again the possibility to produce new effects when the UV lasers are substituted for conventional light sources in photochemical work while $Fe(CO)_5$ normally dissociates into carbon monoxyde and $Fe(CO)_n$ single pass stimulated emissions from Fe* were generated by transverse excitations of iron pentacarbonyl at submbar pressures by KrF pulses /12/. Comparison of fluorescence spectra from atomic iron with respect to both frequency and time evolution showed a very high sensitivity to the excitation conditions, i.e. to the manner in which the excitation was imposed to the molecule. Fig. 13 shows some representative results and gives for comparison also excitation spectra with ArF laser light and with e-beam excita-

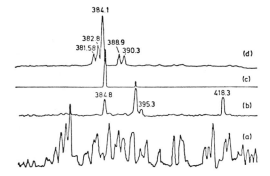

Fig.13. Comparison of fluorescence spectra from $Fe(CO)_5$ excited in four different ways /12/. a) Transversely pumped with ArF laser-unfocused; b) Transversely pumped with KrF laser-unfocused; c) Transversely pumped with KrF laser-focused; d) Electron beam (600 keV; 160 A/cm^2) into 0.1 mbar Fe (CO_5) + 20 bar Ar

tion. The extent of specificity especially when a focused KrF laser is used is simply surprising. Although the detailed mechanism by which the population inversion in atomic iron is produced is yet unknown, this principal may conveniently be used to create new laser lines by excimer laser photo dissociation also in other molecules.

Acknowledgement

This work was supported by the Bundesministerium für Forschung und Technologie and partly by Euratom.

References

1.   see for a general review V.S. Antonov, V.S. Letokhov, Appl. Phys., in print.

2.   D.M. Lubman, R. Naaman, R.N. Zare, J. Chem. Phys. $\underline{72}$, 3034 (1980) and literature quoted therein.

3.   V.S. Antonov, V.S. Letokhov, A.N. Shibanov, Appl. Phys. $\underline{22}$, 293 (1980).

4.   U. Boesl, H. Neusser, E.W. Schlag, J. Chem. Phys. $\underline{72}$, 4327 (1980).

5.   L. Zandee, R.B. Bernstein, J. Chem. Phys. $\underline{71}$, 1359 (1979).

6.   J.P. Reilly, K.L. Kompa, J. Chem. Phys., in print.

7.   P. Hering, A.G.M. Maaswinkel, K.L. Kompa, to be published.

8.   A.G.M. Maaswinkel, K. Eidmann, R. Sigel, Phys. Rev. Lett. $\underline{42}$, 1625 (1979).

9.   F. Rebentrost, A. Ben-Shaul, to be published, J. Silberstein, R.D. Levine, to be published.

10.   J.P. Reilly, K.L. Kompa, to be published in Advances in Mass Spectrometry, Vol. 8, (1980), Heyden & Sons, London.

11.   H.P. Grieneisen, K. Hohla, K.L. Kompa, to be published.

12.   J. Krasinski, S.H. Bauer, Laboratory Report, Max-Planck-Gesellschaft, Projektgruppe für Laserforschung, PLF 34 (1980), J. Krasinski, S.H. Bauer, K.L. Kompa, to be published in Opt. Commun.

Multiphoton Ionization of Atoms

T. Hellmuth, G. Leuchs, and S.J. Smith[1]

Sektion Physik der Universität München
D-8000 München, Fed. Rep. of Germany, and

H. Walther

Sektion Physik der Universität München and Projektgruppe für
Laserforschung der Max-Planck-Gesellschaft zur Förderung der
Wissenschaften e.V.
D-8046 Garching, Fed. Rep. of Germany

1. Introduction

The alkali atoms are especially attractive for the investigation of multiphoton processes. The ionization potentials are small so that the energy of
only a few visible photons is sufficient to ionize the atoms. Therefore, dye
lasers can be used for the experiments and, in particular, make it possible
to investigate, among other things, the wavelength dependence of the photoionization cross sections. On the other hand, the simplicity of the alkali
energy level scheme is an advantage for the theoretical understanding and
interpretation of the results.

Beside the total cross section for photoionization and its wavelength and
polarization dependence, the angular distribution of the photoelectrons is
another experimental result giving important information on the multiphoton
ionization process.

The angular distribution in multiphoton ionization is determined by the
initial and intermediate states involved in the process, the transition amplitudes to the different partial waves of the free electron, the phase difference of the partial waves, the nature of the radiation field and in addition by the final state of the remaining core.

The angular distribution of photoelectrons has, e.g., been investigated
in connection with laser photodetachment [1]. The first experiments with two-
photon ionization have been carried out by Edelstein et al. [7] and later by
Duncanson et al. [2].

In the case of single-photon ionization of atoms or ions with an equal
population of the magnetic sublevels of the initial state, the photoelectron
angular distribution can be described by the general formula

$$\frac{d\sigma}{d\Omega} = \frac{\sigma_{tot}}{4\pi}[1 + \beta_2 P_2(\cos\Theta)]. \tag{1}$$

σ_{tot} represents the total cross section, Θ measures the angle between the
direction of the ejected electron and the polarization of the incident light.
β_2 is the asymmetry parameter and P_2 the second Legendre polynomial. β_2 ranges
from $\beta_2 = 2$ to $\beta_2 = -1$. For a one-electron atom β_2 has been derived by Bethe
[3]. The formula, however, can also be applied to many-electron atoms provided the magnetic sublevels of the intial state are equally populated and
the wavefunctions are represented by antisymmetrized products of spin orbitals.

The angular distribution (1) is also valid in the presence of configuration
interaction and intermediate coupling. In the case of molecules the same form
is obtained when the rotational orientations are averaged [4,5].

---

[1] Permanent address: JILA, National Bureau of Standards, Boulder, CO 80302, USA

In the case of a multiphoton ionization the angular distribution is in addition determined by Legendre polynomials with an order higher than two [6]. For an n-photon ionization P_{2n} is the polynomial with the highest order which may contribute. In general the differential cross section following an n-photon ionization process can be expressed by:

$$\frac{d\sigma}{d\Omega} = \text{const} \sum_{\nu=0}^{n} a_\nu \cos^{2\nu}\theta \qquad (2)$$

where the a_ν depend on the properties of the initial and intermediate states, the radiation field, the partial waves of the emitted electrons etc. The a_ν contain the physics of the multiphoton ionization process.

The number of photoelectrons obtainable in a multiphoton experiment is usually rather small. However, a considerable enhancement can be achieved if the laser frequencies used for the ionization process coincide with atomic transitions. All the experiments on the angular distribution in two-photon ionization have been performed for this quasi-resonant case. The first one on Ti atoms, which were excited by light from a nitrogen laser [7] and the second one on Na atoms which were ionized stepwise by the light of a dye and a nitrogen laser [2]. In the latter experiment the dye laser was either tuned to the $3^2P_{1/2}$ or the $3^2P_{3/2}$ levels of Na.

In the following, experiments on the resonant three-photon ionization of sodium atoms will be described. In addition, the quantum interference effects in the angular distribution in resonant two-photon ionization being introduced by the hyperfine splitting of the intermediate state are studied in detail. The quantum interference effects have also been studied in the total photoelectron current.

2. Three-Photon Ionization

The experimental setup is shown in Fig.1. The atoms of a sodium beam were irradiated by two laser beams originating from two different dye lasers pumped by the same nitrogen laser. The two dye laser beams were linearly polarized in

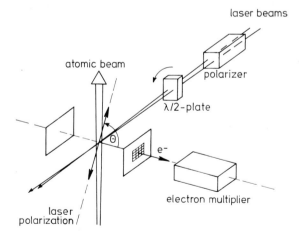

Fig.1 Experimental setup used for the measurement of the angular distribution in three-photon ionization

the same direction. The direction of polarization could be rotated by means of a $\lambda/2$ plate. In this way the angle Θ between the direction of emission of the electrons and polarization direction could be changed. The interaction region between the laser and atomic beams is electrically shielded. The electrons were detected with an angular resolution of 0.35 rad. Behind the aperture, defining the opening angle of the detection system, the electrons were accelerated and detected by means of an electron multiplier.

The excitation scheme used for the three-photon ionization is shown in Fig.2. The first laser beam (ν_1), was tuned either to the $3^2S_{1/2}$ - $3^2P_{1/2}$ or $3^2S_{1/2}$ - $3^2P_{3/2}$ transition of the sodium atom (wavelengths 589,5 nm and 588,9 nm, respectively). The second laser (ν_2) performed a resonant excitation of the n^2D finestructure states.

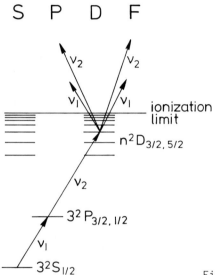

Fig.2 Energy level diagramm showing the transitions used for the threephoton ionization

The duration of the dye laser pulses was 4 ns and the output power between 10 and 50 KW. The spectral width of the laser was about 0.05 Å. In this way the $^2P_{1/2}$ and $^2P_{3/2}$ levels could be excited separately whereas the n^2D states could not be resolved since the finestructure splitting is rather small. The photoionization was performed either by laser ν_1 or ν_2, or by both.

From theory one should expect that the angular distribution in the case of the photoionization via the $^2P_{1/2}$ state is given by the following expansion

$$\frac{d\sigma}{d\Omega} \sim a_0 + a_1\cos^2\Theta + a_2\cos^4\Theta. \qquad (3)$$

This formula corresponds in principle to the result for two-photon ionization. This is due to the fact that the substates of the $^2P_{1/2}$ state are isotropically populated by the excitation with the linearly polarized light. The accurate theory [8], however, gives a contribution for a_3 which is nonzero (Table 1). This contribution is due to the perturbing influence of the neighbouring $^2P_{3/2}$ level. The three-photon ionization via the $^3P_{3/2}$ state contains a contribution up to the sixth power of cos Θ.

The experimental results are shown in Fig.3 in a polar diagram. The solid line was obtained in a least-squares fit of the analytical function (2) to the experimental points. The corresponding coefficients are compiled in Table 1 and compared to the theoretical values which have been calculated by Lambropoulos [8]. The experimental errors of the coefficients are about ± 0.01. There is a rather good agreement between theory and experiment.

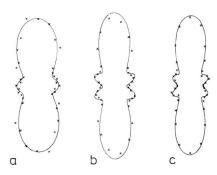

Fig.3 Angular distribution of photo-electrons in resonant three-photon ionization of sodium via the $3^2P_{1/2}$ and 6^2D states (a), via the $3^2P_{3/2}$ and 6^2D states, and via the $3^2P_{3/2}$ and 10^2D states (c). The data were obtained by measuring the signal every 10 degrees over the range of 360 degrees

From Table 1 it can be seen that there is no change in the coefficients when the photoionization is performed either via the n=6 or n=10 state, which is also expected from theory. However, there is significant change when the stepwise ionization is performed either via the $^2P_{1/2}$ or the $^2P_{3/2}$ state. For comparison with theory it would be very interesting to observe the angular distribution when the laser is tuned between both states. Such experiments are presently underway.

Table 1 Coefficients for the angular distribution in three-photon ionization (the coefficient a_2 is normalized to 1)

	$^2S_{1/2}$ - $^2P_{3/2}$ - n^2D - $|1,k>$			$^2S_{1/2}$ - $^2P_{1/2}$ - n^2D - $|1,k>$	
	exp.n=6	exp.n=10	theor.	exp.	theor.
a	0.03	0.03	0.01	0.23	0.19
a_1	0.32	0.30	0.29	-0.48	-0.36
a_2	-1.0	-1.0	-1.0	1.0	1.0
a_3	0.87	0.88	0.90		0.03

3. Quantum Interference Effects in Two-Photon Ionization

The possibility of observing quantum interference effects in photoionization has been proposed in theoretical publications [9,10,13]. The first experimental results have been published recently [11].

In the experiment the two laser pulses used for the photoionization are delayed with respect to each other. The first pulse produces a coherent super-position of closely spaced intermediate levels (e.g., hyperfine levels). This is possible if the duration of the pulse is short compared to $\hbar/\Delta E$ where ΔE is the energy splitting of the intermediate state [11].

The quantum interference effect can be observed both in the total rate of photoelectron emission and in the photoelectron angular distribution if the delay between the two laser pulses is varied. Experiments of both types which have been performed on the sodium atom will be described in the following. As discussed above the angular distribution of photoelectrons for two-photon ionization is given by

$$\frac{d\sigma}{d\Omega} \sim 1 + \beta_2 P_2(\cos\Theta) + \beta_4 P_4(\cos\Theta). \qquad (4)$$

For the case of sodium the angular distribution in two-photon ionization via the $3^2P_{3/2}$ state has been investigated previously by Strand et al. [2]. In this experiment the ionizing pulse was delayed up to 8 ns in order to measure the sensitivity of the angular distribution on the mixing process resulting from the hyperfine coupling in the $3^2P_{3/2}$ level.

In the experiment described here measurements have been performed with delays between 0 and 38 ns, to demonstrate the periodicity, i.e., the quantum interference signal due to the hyperfine levels of the $3^2P_{3/2}$ state and to obtain from it, by Fourier analysis, a spectrum of the hyperfine structure. Furthermore the polarization dependence of the signal has been investigated and the quantum interference effect in the total electron current has been demonstrated.

The level scheme for the two-photon experiment is shown in Fig.4. In an atomic beam experiment a nitrogen laser pumped dye laser, linearly polarized, was used to saturate the transition to the $3^2P_{3/2}$ state. In a second step, part of the 337 nm nitrogen laser radiation, also linearly polarized, was used to ionize the atoms producing photoelectrons with a kinetic energy of 0.6 eV. The nitrogen laser pulse duration was 6 ns and the power density at the atomic beam about 10^7 W/cm^2.

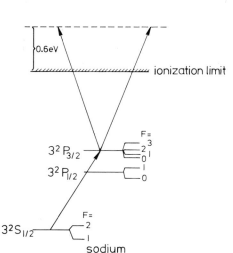

Fig.4 Level scheme of the two-photon experiment

The duration of the dye laser was about 4 ns (Fig.5). The detection of the photoelectrons was performed in a similar way as described in the three-photon experiment (see above). The electrons correlated with the nitrogen laser pulse are measured with a gated integrator capable of resolving single electrons. For an atomic beam density of $10^8/cm^3$ and a collecting solid angle of 0.25 sterad, one to ten photoelectrons were detected per laser pulse for delays not larger than the lifetime of the 3p state. The output of the integrator was averaged for one to three minutes for each angle setting, with the laser pulse repetition rate being 40 pps. The angular distribution was obtained by measuring the signal every 20 degrees over a range of 180 degrees and using the point symmetry of the angular distribution.

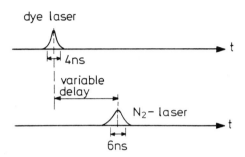

Fig.5 Pulse duration of the laser pulse

By lengthening its optical path, the nitrogen laser pulse was delayed with respect to the dye laser pulse. This was achieved by using up to five plane mirrors to form a simple optical delay line before the Glan-Thompson polarizer and the half-wave plate.

We observed a periodic variation in the total photoelectron current as a function of optical delay. However, these results were erratic since the variation of the time delay always caused a change in the transmission of the optical delay line and required a new adjustment of the laser beams for optimum overlap with the atomic beam. Therefore, the more tedious technique of measuring the angular distribution of the photoelectrons at each optical delay was used instead. This method is sensitive neither to the changes in transmission of the optical delay line nor to changes in overlap of the laser beams at the atomic beam.

The upper part of Fig.6 shows polar diagrams of the angular distributions of photoelectrons measured for various delays and normalized to a constant angle-integrated number of photoelectrons. The linear polarization of the two lasers was parallel. The solid line was obtained by a least-squares fit of (4) to the experimental data, yielding the asymmetry parameters β_2 and β_4. In the lower part of Fig.6, β_4 is plotted as a function of the delay of the nitrogen laser pulse with respect to the dye laser pulse. The delay time was varied between 0 and 38 ns. The statistical errors of the experimental points in the upper part of Fig.6 are at most a few percent.

The observed periodic variation is the quantum beat signal of the four hyperfine levels of the $3^2P_{3/2}$ state. Fig.7 shows the hyperfine structure of the ground state and of the $3^2P_{3/2}$ state of sodium. The arrows indicate transitions allowed due to the angular momentum selection rule $F=0, \pm 1$. Consequently the frequencies that are expected to appear in the quantum beat signal are $\Delta\nu_{32}$, $\Delta\nu_{21}$, $\Delta\nu_{10}$, $\Delta\nu_{32} + \Delta\nu_{21}$ and $\Delta\nu_{21} + \Delta\nu_{10}$. The frequencies of the periodically varying signal in Fig.6 were determined by a Fourier

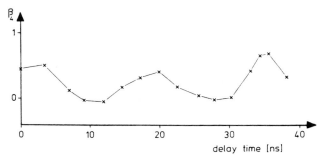

delay time [ns]

Fig.6 Quantum interference effect in the shape of the angular distribution of photoelectrons resulting from photoionization of the $3^2P_{3/2}$ state of sodium

analysis. Two frequencies $\Delta\nu_{32}$ = 59.8 (1.0) and $\Delta\nu_{32} + \Delta\nu_{21}$ = 95.3 (1.0) show up separately in the Fourier spectrum (Fig.8). The frequencies $\Delta\nu_{21}$ and $\Delta\nu_{10}$ are not resolved, a peak at 24 MHz is observed instead. The separations of the hyperfine levels obtained from this preliminary evaluation are in reasonable agreement with previous results on the hyperfine splitting determined, e.g., by Figger and Walther [12].

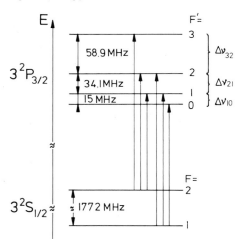

Fig.7 Part of the level scheme of sodium showing the hyperfine splitting of the $3^2P_{3/2}$ and of the ground state. The values for the hyperfine splitting are taken from [12]. The arrows indicate transitions between the two states

With the linear polarization of the two lasers oriented perpendicular to each other the angular distribution of the photoelectrons in two-photon ionization is changed since the influence of the magnetic substates differs in both cases. This can be seen in Fig.9. Therefore, the phase of the quantum interference effect induced by the hyperfine splitting of the $3^2P_{3/2}$ state must also be different if the polarization of the two lasers is perpendicular to each other. A corresponding result is shown in Fig.10.

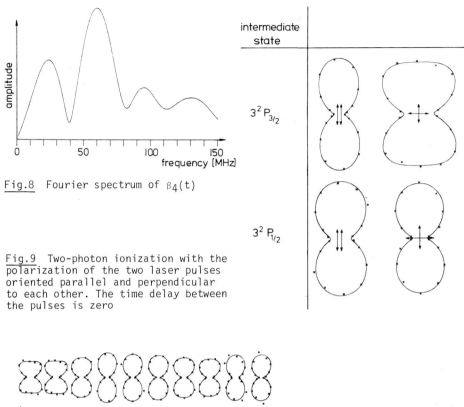

Fig.8 Fourier spectrum of $\beta_4(t)$

Fig.9 Two-photon ionization with the polarization of the two laser pulses oriented parallel and perpendicular to each other. The time delay between the pulses is zero

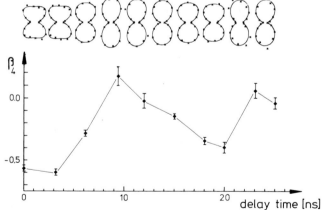

Fig.10 Quantum interference effect with linear polarization of the lasers perpendicular to each other

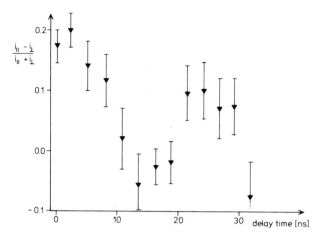

$\frac{i_{\shortparallel} - i_{\perp}}{i_{\shortparallel} + i_{\perp}}$

Fig.11 Quantum interference effect in the total photoelectron current. $i_{\shortparallel}$ and $i_{\perp}$ corresponds to the photoelectron current obtained with the linear polarization of the two laser beams either parallel or perpendicular to each other

The total photoelectron current also depends on the polarization of the two laser beams, therefore the quantum interference effect can also be observed when the following function $(i_{\shortparallel} - i_{\perp})/(i_{\shortparallel} + i_{\perp})$ is evaluated for the photoelectron current i measured with a parallel $i_{\shortparallel}$ and a perpendicular $i_{\perp}$ orientation of the polarization directions of the two laser beams. The result of a measurement is plotted in Fig.11. The signal has the same period as the others shown in Figs.6 and 10. Plotting the function $(i_{\shortparallel} - i_{\perp})/(i_{\shortparallel} + i_{\perp})$ the intensity change is eliminated, being introduced by the optical delay line when the time delay is varied.

Acknowledgements The support by the Deutsche Forschungsgemeinschaft is gratefully acknowledged. One of us (S.J.S.) acknowledges an award from the Alexander von Humboldt Foundation through its program for Senior U.S. Scientist.

References

1. J.L. Hall, M.W. Siegel: J. Chem. Phys. 48, 943 (1968);
    J. Cooper, R. Zare: J. Chem. Phys. 48, 942 (1968)
2. J.A. Duncanson, Jr., M. Strand, A. Lindgard, R.S. Berry: Phys. Rev. Lett. 37, 978 (1976);
    M. Strand, J. Hansen, R.L. Chien, R.S. Berry: Chem. Phys. Lett. 59, 205 (1978)
3. H.A. Bethe: *Handbuch der Physik*, Vol. 24 (Springer, Berlin 1933) pp.483-484
4. J.C. Tully, R.S. Berry. B.J. Dalton: Phys. Rev. 176, 95 (1968)
5. J. Berkowitz, H. Erhardt, T. Tekaat: Z. Physik 200, 69 (1957)
6. M. Lambropoulos, R.S. Berry: Phys. Rev. A8, 855 (1973)
7. S. Edelstein, M. Lambropoulos, J. Duncanson, R.S. Berry: Phys. Rev. A9, 2459 (1974)
8. P. Lambropoulos, private communication

9. R. Zygan-Maus, H.H. Wolter: Phys. Lett. A64, 351 (1978)
10. A.T. Georges, P. Lambropoulos: Phys. Rev. A18, 1072 (1978)
11. G. Leuchs, S.J. Smith, E. Khawaja, H. Walther: Opt. Commun. 31, 313 (1979)
12. H. Figger, H. Walther: Z. Physik, 267, 1 (1974)
13. P.L. Knight: Opt. Commun. 31, 148 (1979)

Part IV

New Laser Devices and Applications

Applications of Tunable Laser Spectroscopy to Molecular Photophysics: From Diatomics to Model Membranes

G.A. Kenney-Wallace and S.C. Wallace

Departments of Chemistry and Physics, University of Toronto
Toronto, Canada M5S 1A1

Recent years have witnessed the development of a myriad of applications of non-linear spectroscopy, through which novel and subtle insights into electronic struc-ture, molecular dynamics and coherent transient phenomena can be gleaned. Many of these nonlinear optical processes such as four-wave mixing, the optical Kerr effect and stimulated Raman scattering, are resonantly enhanced, parametric processes and operate through $\chi^{(3)}$, the third-order nonlinear susceptibility of the system under study.

By studying the amplitude, phase, polarization and k-matching dependencies of the incident and scattered wave in the medium, valuable spectroscopic data are obtained from both time and frequency domain experiments [1,2]. Now these laser techniques can also be applied to monitor the fundamental changes that occur in the electronic properties, and hence in $\chi^{(3)}$, of an atom or molecule as it is taken from an isolated state, to a phase of high number density and collision frequency, namely the condensed phase, where intermolecular forces play a dominant role in modifying both the energetics and dynamics of spectroscopic transitions.

We describe in this paper a new tunable laser spectroscopy multiphoton tech-nique, employing a Raman gas cell. We have applied this technique with two pho-ton absorption to determine for the first time the ionization potential of an impurity molecule embedded in a model membrane. The results of these double resonance experiments are discussed in the context of the influence of the intermolecular forces and fields of the microenvironment on the energetics and dynamics of the guest molecule in comparison to its gas phase behaviour. This Raman cell will be a powerful general tool for probing many investigations into laser-induced molecular photophysics and photochemistry on the nanosecond and picosecond time scale. In order to demonstrate its potential for high resolution spectroscopy we have also included a brief discussion of its application to state-selective studies of the molecules SO_2 and NO when used in conjunction with a tunable dye laser.

Following picosecond laser excitation of an isolated polyatomic molecule, a question of universal interest is how to describe the temporal competition between energy relaxation and photoselective reaction: that is, what is the probability that energy randomization and intramolecular relaxation will depopulate the prepared excited state prior to photoionization, photodissociation, isomerization and so forth? This competition between relaxation and reaction is an intrinsic part of most laser-induced processes stimulated by the selective absorption of one or more photons, and it stems from the availability and coupling of many vibrational modes to provide channels for rapid energy randomization and transfer on a picosecond time-scale or faster [3]. In the condensed phase, the intermolecular interactions frequently lead to changes in the relative spacings between electronic energy levels and thus the overall energetics of a given laser-induced process will vary, as will the coupling between the initial and final states. Symmetry-breaking collisions can facilitate molecular predissociation, while the fluctuating potential wells and barriers in the vicinity of such a reactive event can enhance or impede recombination of neutral or charged fragments. The competition also expands to include collisional and long-range resonance energy transfer effects as well as the possibility that rotational relaxation of the liquid host can determine the equilibrium configurations of the excited guest molecular states and hence control the access to different radiationless relaxation channels.

By exploring the laser-induced photophysics of molecules embedded in membranes, we are introducing a further degree of complexity, namely the chemical and structural inhomogeneity of the microenvironment. However, it is precisely because of the considerations listed above that the optical properties of molecules in biological systems have to be studied from the perspective of the role of the microenvironment, and cannot be extrapolated on the basis of phase density from gas phase data. Such were the motivations of the following experiments, prompted by earlier reports [4] of an unexpectedly high yield of photoelectrons in the fixed frequency, laser-induced ionization of pyrene in the anionic micellar system, aqueous sodium dodecyl sulphate (SDS).

Micelles are thermodynamically stable aggregates [5] of long, hydrocarbon chain molecules with polar end groups, that cluster into spherical and rod-like structures of well-separated hydrophilic and hydrophobic character, and interfacial regions that carry surface charge of varying sign and charge density, the Stern or Gouy-Chapman layer. Micelles are models for lipid membranes, since dynamical effects at the interface mimic many of the biophysical transport processes of importance to living systems. Aqueous SDS micelles have a hydrocarbon interior and a negative surface charge at the interface to the aqueous layer, and are about 30 Å in diameter. Positive sodium ions maintain an overall electrical neutrality in the bulk.

1. Experimental Details of Raman Shifting Techniques

Block diagrams of the apparatus including the Raman cell used in these studies are given in Fig. 1. The basic laser source was a Quanta-Ray Nd YAG oscillator amplifier, which produced up to 160 mJ output energy in the third harmonic at 3550 Å in a 5 ns pulse (fwhm). An uncoated suprasil quartz flat split off a small fraction (7 mJ or 5%) of the UV radiation for the first exciting pulse $\hbar\omega_1$ by stimulated Raman scattering in high pressure H_2 or D_2 [6]. The 80 cm long Raman cell was made using standard, stainless-steel, high pressure fittings, with the only critical part of the design being in the mounting technique for the input and output quartz windows. It is important to avoid strain-induced birefringence, in order to suppress the rotational Raman lines. For H_2, the input radiation was focused into

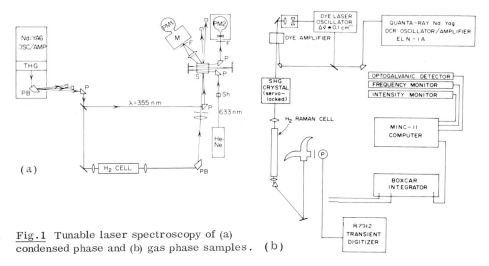

Fig.1 Tunable laser spectroscopy of (a)
condensed phase and (b) gas phase samples. (b)

the cell using a 77.5 cm lens and with either 3550 or 5320 Å beams produced multiple orders of stokes or antistokes radiation, with 20% and 10% conversion efficiencies to the first and second stokes, respectively. Typical H_2 pressures were 10 - 15 atm. The comb of frequencies produced in the Raman cell is spaced by the fundamental molecular vibrational frequencies of 4155 and 2991 cm^{-1} of H_2 and D_2, respectively, so that a very wide range of wavelengths (2,000 - 30,000 Å) can be produced using the combination of Nd:YAG harmonics and H_2, D_2 or CH_4 for Raman shifting. In fact, for spectroscopy and photophysics in the condensed phase, where the intrinsic spectral broadening of electronic transitions is large (~500-1000 cm^{-1}), this latter combination readily provides the necessary laser wavelengths as, for example, in the present spectroscopic study of the photoionization threshold in pyrene.

We have also used the high-order antistokes Raman components [6] for laser spectroscopy in the far UV. However, in this case the input radiation was the narrow-bandwidth output of a tunable dye-laser, which had been frequency doubled into the UV (~ 2900 Å). Since the doubling crystal is servo-locked [7] to the optimum phase matching angle this provides a very flexible, tunable laser source in what was hitherto a relatively inaccessible spectral region (1800 - 2200 Å). With the lower powers in the dye laser, it was necessary to use somewhat harder focusing (40 cm best form lens) and a shorter H_2 cell (50 cm) in order to obtain optimum stability and intensity.

As Fig.1a shows, the absorption of two-photon-induced transient species generated in the sample cell (S) was monitored with a modulated (200 Hz) He-Ne beam, which made up to eight passes across the cell prior to entering the fast photomultiplier, PM2. Emission from the sample was focused onto the slits of a monochromator and detected by PM1. The output signals from the PM were coupled via 50 Ω to a R 7912 Tektronix transient digitizer. Full details of this apparatus have appeared elsewhere [8]. The scheme for the gas phase state-selective studies is shown in Fig.1b, where the output from the Raman cell excited molecules at typically mTorr in a Woods Horn cell. The laser-induced fluorescence was monitored by a photomultiplier P, whose output was directed either into the R7912 or into a 162 PAR BOXCAR and the data stored on a MINC-11 computer.

2. State Selective Studies of Small Molecules

As an initial demonstration of the performance of the dye laser, second-harmonic servo-locking, and the intensity of the antistokes emission from the Raman cell, we show in Fig.2 a portion of the excitation spectrum of the (0,0) band of NO (P_{12} branch) at a pressure of 10^{-3} Torr. Since the exciting wavelength (2270 Å) was only the second antistokes of the second harmonic of the dye laser, it was extremely intense (≥ 1 mJ per pulse) and the resulting laser-induced fluorescence had to be significantly attenuated to avoid saturating the photomultiplier. However, the striking feature of this excitation spectrum in NO is that the spin-orbit splitting (0.2 cm$^{-1} \leq \delta \leq 1$ cm^{-1}) is not resolved even though the measured dye laser line-width $\Delta\nu$ was 0.1 cm^{-1}.

Thus it is clear that there is some spectral broadening of the "antistokes Raman" output which degrades the spectral resolution of the input dye laser radiation. Since the production of the antistokes Raman lines is due to a four-wave mixing process analogous to CARS, the origins of the rather broader linewidth antistokes radiation can be understood. The linewidth of any harmonic radiation is given by the nth order autocorrelation function of the input radiation. Analytical expressions for both lineshapes and linewidths in third-order nonlinear processes have recently been derived [9] and show how both the bandwidth of the input laser radiation and the T_1 and T_2 relaxation times in the nonlinear material contribute to the final lineshape. Thus the full-width half-maximum of the final harmonic wave will increase and the actual lineshape develops Lorentzian wings. Both factors contribute to the broadening of the starting dye laser radiation to ~ 0.8 cm^{-1} in the 2250 Å region. Nevertheless, because of the wide spectral coverage which is possible with the multiple orders of antistokes lines, this Raman shifting technique in H_2 yields a very flexible source for survey studies and exploratory work.

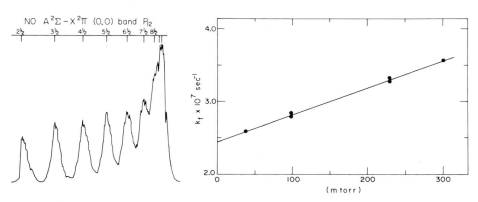

Fig.2 Excitation spectrum of NO   **Fig.3** Self-quenching in SO_2

As an example we show data from some recent studies on the photophysics of the SO_2 molecule in the $\tilde{C}$ ($^1B_2$) state. Figure 3 shows lifetime data as a function of number density for the $^qR_{14}$ sub-band (K states are not resolved) at 42,754 cm^{-1}. The slope of this Stern-Volmer plot (1.1 x 10^{-9} cm^3 s^{-1}) indicates a cross-section $\geq$ 4 times gas kinetic for SO_2 self-quenching, a value which is extremely large for what is expected to be a non-resonant process. Using this Raman source we are currently studying the details of the photophysics and photodissociation of this state in SO_2 in both cells and supersonic nozzle beams [10].

3. Laser-Induced Photoionization of Pyrene in Micelles

The gas phase ionization potential (I_g) of pyrene, a molecule comprising four fused benzene rings, is 7.7 eV. In solution, pyrene molecules absorb in the near UV undergoing a weak $S_0 \rightarrow S_1$ (^1L_b) transition, with vibronic hot bands from 3450 Å < λ < 3720 Å. The double resonance technique was employed to excite pyrene from a hot band into the S_1 state with a laser pulse ω_1 and then to further promote the electronically excited states into the ionization continuum with an optically delayed Raman or dye laser pulse at ω_2 ($\omega_2 < \omega_1$) for which ground state pyrene molecules (S_0) had zero cross-section. When pyrene (S_0) absorbs a 3550 Å photon (ω_1), the S_1 states are formed with ~1270 cm^{-1} excess vibrational energy and display a larger cross-section than S_0 for absorption of a second photon (ω_1) to form ions P^+, e$^-$. Vibrational relaxation occurs on a picosecond time scale [1], but relaxed S_1 states that survive the time interval before the arrival of the second pulse (ω_2) can then absorb a photon from the laser field to produce ions at $\hbar$ ($\omega_1 + \omega_2$). However, for $\lambda_2 \gtrsim 3750$ Å ($\omega_2 \leq 26,667$ cm^{-1}), the absence of real intermediate states and the low power density of the Raman pulse at ω_2 combine to eliminate the probability of observing ions from two-photon photoionization at $2\omega_2$. The mechanism of two-photon laser photoionization of pyrene has been well documented as a consecutive event involving real intermediate excited electronic states in this spectral region [11]. This excited ^1S state has a radiative lifetime of ~300 ns in ethanol and its fluorescence (λ_{max}^f ~ 390 nm) can be readily detected (PM1 in Fig.1a).

We monitor the fluorescence depletion from ^1S upon the arrival of ω_2 and the simultaneous appearance of the optical absorptions of the ions P^+, e$_s^-$ which are well characterized spectroscopically. By knowing the absorption cross-sections and hence population density of the free positive ions and the electrons, which, escaping the coulomb field of the parent ion becomes rapidly ($\leq 10^{-12}$ s) solvated in the aqueous phase [12], we can deduce the relative ion yields as a function of ($\omega_1 + \omega_2$) and the energy per pulse at the individual exciting frequencies.

Figure 4 illustrates the superposition of typical absorption signals for excitation at ω_1 and $\omega_1 + \omega_2$, for (a) pyrene in methanol $\omega_1 = 28,169$ cm^{-1}, $\omega_2 = 24,380$ cm^{-1} (4160 Å), (b) pyrene in SDS micelle, $\omega_1 = 28,169$ cm^{-1}, $\omega_2 = 18,797$ cm^{-1} (5320 Å). The initial absorption constitutes the ion signal from two photon ionization at $2\omega_1$ in the first pulse. Note that the signal is larger in 3b hence more ions are formed for a given incident energy and intensity in the micellar system. The second pulse (ω_2) arrives 17 ns later in each case but only in the micellar system is there an observable increase in the ion yield due to $\omega_1 + \omega_2$. Traces are also shown for just single (ω_1) pulse excitation as a "base-line" for the double resonance experiment.

The absorption and emission data collectively show that (1) the ionization threshold for pyrene in SDS micelles is substantially lower than for pyrene in methanol, and
(2) for a given incident energy ($\omega_1 + \omega_2$) the normalized ion yields are always higher
in the micellar system. Figure 5 shows the frequency dependence of the photoionization efficiency of pyrene, in comparison to the absorption of the electronically excited state ($S_1 \rightarrow S_n$) of pyrene in the hydrocarbon environment of cyclohexane,
measured in independent studies [8].

We may summarize the nanosecond data in the following conclusions. (1)
Pyrene in methanol cannot be ionized at ($\omega_1 + \omega_2$) < 52,549 cm^{-1} placing an upper
limit of I_e = 6.6 ± 0.1 eV. (2) Pyrene in SDS micelles continues to exhibit some
ionization beyond ($\omega_1 + \omega_2$) = 46,966 cm^{-1} but the threshold is extrapolated to be
~ 6000 Å or I_m = 5.4 eV. (3) The frequency dependence of the laser photoionization supports a two-photon photoionization mechanism via the vibrationally relaxed
excited singlet state, S_1. (4) The observed lowering of the pyrene gas phase I_g
by medium-induced effects in methanol is ~ 1.1 eV but, significantly is two-fold
greater in SDS, namely 2.3 eV.

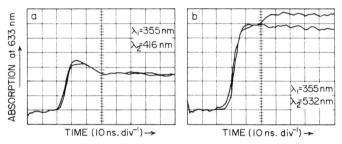

Fig.4 Transient absorption of e_s^- in (a) methanol and (b) micellar system.

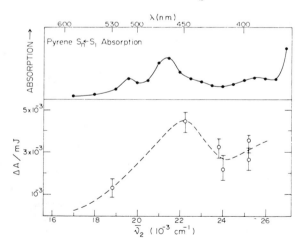

Fig.5 Frequency dependence of photoionization efficiency of pyrene in SDS.

4. Field-Promoted Ionization at Interfaces

The pyrene is located close to the interface of a chemically and structurally inhomogeneous system comprising a hydrocarbon interior, an aqueous and polar exterior
phase and an interfacial region of high charge density. The gradient across the
interfacial region presents a strong local field to the pyrene molecule and it is the
influence of this field that we believe underlies the dramatic lowering of the ionization potential. However, consider first the intermolecular interactions that are
expected to reduce the I value on going from gas phase to liquid phase. Of primary
importance is the work function (V_0) of electron ejection into the medium and the
adiabatic medium polarization (P^+) of the ions. Molecular orientational motion
typically occurs within tens of picoseconds [12] but it is too slow to be included as
polarization energy in the energetics although clearly these molecular dynamics
play a crucial role in establishing the coulomb screening, ionic atmospheres and
solvation shells, all of which govern the probability of ion recombination at picosecond times. Thus rotational relaxation affects the total population density of
ions which escape geminate recombination although not the true threshold for the
appearance of ions. In measuring ion yields and extrapolating to ion thresholds in
liquids, we are also certainly aware of the ambiguity of ionization threshold versus
threshold sensitivity of detection, and we will return to this point at the end of this
paper.

In simple atomic fluids the influence of phase density on I is predicted by [13]:

$$I = I_g + P^+ + V_0 = I_g - \frac{e^2}{2r_+}\left(1 - \epsilon_{op}^{-1}\right) + \left(\frac{\hbar^2 k^2}{2m} + U_P\right)$$

where P^+ is calculated from the Born equation using an effective ionic radius (r^+)
and high frequency dielectric constant (ϵ_{op}) and V_0 is estimated from the electron-
atom scattering potential, which determines the kinetic energy from multiple scattering and the long range electronic polarization acting on the electron, U_p [13].
In molecular fluids the concepts are retained but clearly the necessary approximations decrease the precision of these estimates, particularly for V_0. However, when
we take the V_0 values for water and methanol, which from several experimental
and theoretical reports appear to be -1.3 ± 0.1 eV and -1 eV, respectively [14],
and calculate P^+ for the pyrene positive ion in water, methanol or hydrocarbon to
be -2.25 eV, -2.23 eV and -2.62 eV, respectively, it is not possible to account for
more than 20% of the striking difference between I_ℓ and I_m. In fact, one would
suggest that $I_\ell \approx I_m$, given the uncertainities in P^+ and V_0. For those few molecules studied, the observed I_ℓ for a given impurity molecule upon changing the
liquid environment from polar to nonpolar displays relatively small (10%) changes
in the ionization thresholds. However, the additional variable which must be included in calculating I_m for our data is the influence of the electric field gradient
at the micellar-aqueous interface.

If we consider the (0,0) band for $S_0 \rightarrow S_n$ transitions in pyrene in a hydrocarbon
solvent, the spectrum shows the highest energy peak $S_0 \rightarrow S_5$ at 48,200 cm^{-1} or
5.97 eV. The observed edge of the ionization continuum of pyrene in methanol is
well above this energy but it overlaps both the S_5 and high-lying vibrational levels
of the S_4 for pyrene embedded in SDS. Whether these are the principal precursor

states for the ions, or there are also contributions from underlying Rydberg-like states (which will be significantly broadened in the condensed phase) remains a question for further study. However, in the spectral region where we observe photoionization, the gas phase spectra of other polycyclic hydrocarbons [15] show Rydberg series converging to the first ionization limit, and we therefore expect these states to play an important role, as outlined below.

It is entirely plausible that the electric field gradient at the micellar interface has reduced the barrier to ion formation from these excited states. The radial probability distribution function of an electron in a $n=4$ Rydberg state peaks at 15 Å, giving an effective Rydberg diameter comparable to the average micelle diameter of 30 Å. It is well known that atoms can be ionized in a dc electric field and many detailed studies have been published on multiphoton photoionization and Stark field promoted ionization of excited Rydberg states [16]. The higher the Rydberg state, the more readily ionization proceeds since the Coulomb interaction with the core positive ion becomes progressively weaker. Fields of the order of 10^4 - 10^6 V cm^{-1} exist over the 3 - 5 Å depth of the charged interfacial layer at the surface of the micelle, and thus we expect that the highly excited pyrene molecules, regardless of precise location are significantly perturbed by this field gradient. The situation is shown schematically in Fig.6, in which the superposition of the normal potential well (dotted line) with the Stark field (sloping line) leads to an effective reduction of the barrier by E for an excited state high in the well. Clearly ionization will occur for those states now above the barrier but tunnelling is also anticipated (from r_1 and r_2) for excited states below the barrier.

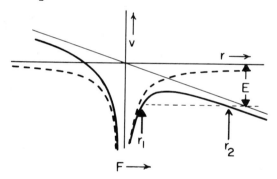

Fig.6 Schematic of Stark ionization.

Thus we propose that the distortion of the Coulomb potential by the dc electric field at the surface of the micelle SDS promotes Stark ionization of the high-lying, Rydberg-like excited states of impurity guest molecules located in the micelle. In micelles of opposite surface charge, the reverse in sign of the field gradient should inhibit photoionization, or at least enhance ion recombination, and preliminary evidence indicates that this is indeed observed. However, since these data were obtained at ns times, and thus precluded <u>direct</u> observation of many of the competing vibronic and vibrational relaxation processes in pyrene, as well as direct observation of geminate-ion recombination, we are now pursuing these ideas with tunable, picosecond laser spectroscopy. Since it is crucial to distinguish between the threshold of ionization (a property of the molecule and its intermolecular interactions) and the threshold of detection of ions (a property of the technique employed) we have spent some effort in ensuring that the smaller number density of ions produced by a picosecond laser pulse, and the temporal and spatial intensity profiles of those laser pulses, will not be limiting factors in the interpretation of data. We conclude by describing a new and versatile autocorrelation technique for measuring the duration of the laser pulses.

5. Future Directions in Tunable Picosecond Laser Spectroscopy

In measuring the picosecond response function of a molecular system to optical excitation with a ps laser pulse, it is essential to know the temporal and spatial intensity profiles of the laser pump and probe pulses, particularly when the relaxation dynamics are those of a molecular state prepared as a consequence of one or more non-linear processes through virtual or real states. In the latter case, the relative cross-sections, competing transition probabilities and saturation in the radiation field can produce apparently simple transient absorption and emission behaviour that masks the true non-linear aspects of the mechanisms involved. Recently real time techniques have emerged [17, 18] to permit continuous display of laser pulses. Figure 7 shows an extension of our earlier work [18] on CW background free picosecond autocorrelation measurements based on a double-speaker configuration, each supporting a corner-cube prism oscillating at audio frequencies (15 - 30 Hz) in the interferometer arms. We have successfully employed these devices as permanent on-line monitors of the pulse diagnostics from a synchronously pumped, tunable, argon ion laser-dye laser system. Only ~10% of the dye laser pulse intensity (nJ pulse^{-1}) is split off into the autocorrelator to obtain excellent signal to noise, while the remainder of the pulse is directed to the chain of dye laser amplifiers pumped by the second harmonic from the Quanta Ray Nd:YAG laser. Gains of ~10^4 are observed from two transversely pumped dye amplifiers, while ~1 mJ pulse^{-1} is anticipated from a third stage (pumped longitudinally) to reach the GWatt powers [19] necessary for nonlinear optical experiments.

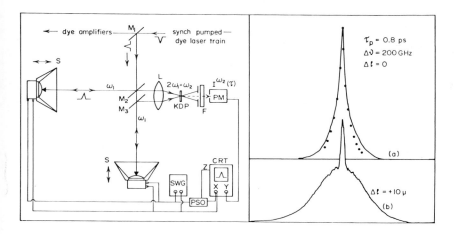

Fig.7 Subpicosecond laser autocorrelation measurements in real time.

The speakers are driven exactly 180° out of phase from a common sine-wave source (SWG) and thus produce a net variation in the relative difference between the interferometer arms which is double the total displacement or "stroke" of one of the speakers. The dynamic range achieved so far is 300 ps, thus permitting auto-correlation measurements of the argon ion pulses to be made as well. The significance of precise cavity matching in synchronous pumping is illustrated by comparison of the subpicosecond laser ($\tau = 0.8$ ps) pulse in Fig.7a with the partially mode-locked pulse in b when the cavity length ℓ is 10 μ too long. The dots in a refer to an $\exp(-|\tau|/T)$ fit to a single-sided experimental pulse with $\tau_p = 0.8$ ps. This autocorrelator in particular offers a large dynamic range (≤ 300 ps) not accessible by other means and a continuously variable range. Furthermore this scheme displays a high degree of linearity. Any asymmetric nonlinearity (i.e., "stretching" in one direction of speaker cone travel and a "compression" in the other direction) in the speaker displacement versus instantaneous voltage behaviour only appears when the speaker is driven at near maximum stroke, since the exact out-of-phase movement of the speakers leads to cancellation of these asymmetric nonlinearities. Thus both the single [18] and double-speaker autocorrelator devices are surprisingly linear, even as the maximum dynamic range is approached, and therefore offer expedient and reliable pulse diagnostics for picosecond and femtosecond tunable laser spectroscopy.

Employing this amplified, subpicosecond tunable laser system and pump-probe double resonance techniques, we plan to not only confirm the ionization threshold of pyrene in the region 5900 Å $< \lambda <$ 6200 Å, as Fig.5 indicates, but also examine the picosecond dynamics of the appearance of ions and fluorescence as we tune ω_2 from well below to just above the ionization continuum. It is also conceivable that these highly excited states just below the continuum have a measurable lifetime before autoionization, and this may influence the time-dependence of the appearance of e_{aq}^- in the aqueous phase. By incorporating electron acceptors onto the surface of the micelles, we should also be able to study electron transfer processes, particularly in those micelles for which the field gradient is expected to accelerate geminate ion recombination leading to a substantially lower quantum yield of free ions.

Finally, it is intriguing to speculate on the role of such field promoted ionization on the energetics of a number of biologically important processes that are initiated by the solar radiation. Clearly these are major questions, but ones which can be addressed in the future with great versatility through the techniques of non-linear laser spectroscopy and Raman spectroscopy.

We gratefully acknowledge the contributions of W. Sharfin, T. Kavassalis, G. E. Hall and K. L. Sala to these diverse projects, the financial support of the Killam Foundation, and the A. P. Sloan Foundation for Fellowships to the authors. Acknowledgement is also made to the Petroleum Research Fund, administered by the American Chemical Society, for partial support of these projects.

1   A. Laubereau and W. Kaiser, Rev. Mod. Phys. 50, 607 (1978).
2   S. C. Wallace and K. Innes, J. Chem. Phys. 72, 4805 (1980).
3   G. A. Kenney-Wallace, Proceedings of Royal Society Discussion on Ultrashort Light Pulses, Phil. Trans. R. Soc. London, in press.
4   S. C. Wallace, M. Gratzel and J. K. Thomas, Chem. Phys. Lett. 23, 359 (1973).
5   N. Mazur, G. Benedek and M. C. Carey, J. Phys. Chem. 80, 1975 (1976).
6   V. Wilke and W. Schmidt, Appl. Phys. 18, 177 (1979).
7   G. C. Bjorklund and R. H. Storz, IEEE J. Quantum Electron. 15, 228 (1979).
8   S. C. Wallace, G. E. Hall and G. A. Kenney-Wallace, Chem. Phys. 49, 279 (1980).
9   N. K. Dutta, J. Phys. B 13, 411 (1980)
10 T. Kavassalis, W. Sharfin and S. C. Wallace, to be published.
11 G. E. Hall and G. A. Kenney-Wallace, Chem. Phys. 28, 205 (1978).
12 G. A. Kenney-Wallace, Adv. Chem. Phys. 47, 628 (1980); E. Ippen and J. Wiesenfeld, Chem. Phys. Lett., in press.
13 J. Jortner and A. Gaathon, Can. J. Chem. 55, 1801 (1977).
14 A. Bernas, D. Grand and E. Amouyal, J. Phys. Chem. 84, 1259 (1980).
15 M. B. Robin, Higher Excited States of Polyatomic Molecules, Academic Press, New York, 1974, 1975, vol.1 and 2.
16 J. E. Bayfield, Multiphoton Processes, ed., J. H. Eberly and P. Lambropoulos, Wiley, New York, 1978, p.191.
17 R. L. Fork and F. A. Beisser, Appl. Opt. 17, 3534 (1978).
18 K. L. Sala, G. A. Kenney-Wallace and G. E. Hall, IEEE J. Quantum Electron. 16, 990 (1980).
19 E. P. Ippen and C. V. Shank, in Picosecond Phenomena, ed., C. V. Shank, E. P. Ippen and S. L. Shapiro, Springer-Verlag, New York, 1978, p.103.

High-Power Picosecond Pulses from UV to IR

F.P. Schäfer

Max-Planck-Institut für biophysikalische Chemie,
Abteilung Laserphysik
Postfach 968, D-3400 Göttingen, Fed. Rep. of Germany

Two-step excitation of atoms and molecules has become of ever-increasing importance during the last few years, e. g. two-step excitation schemes for laser isotope separation or the interesting experiments with Rydberg states reached by two-step excitation. In large molecules as e. g. dyes or molecules of biological interest ps pulses must be used for two-step spectroscopy. But then a great wealth of new information can be obtained as shown by the pioneering experiments of Kaiser and co-workers [1]. For these experiments one needs tunable IR and UV-visible pulses occurring simultaneously.

Unfortunately the present-day methods for the simultaneous generation of tunable UV-VIS and IR pulses are very complex and costly: First, a mode-locked Nd-glass laser oscillator creates a ps-pulse train, from which one pulse is switched out and amplified in several stages to high peak powers. Second, part of the output pumps two lithium niobate crystals, the first acting as parametric fluorescence source, the second as amplifier to create the infrared pulse. Third, the other part of the Nd-laser output is frequency-doubled and -tripled to pump two more crystals for parametric generation of the UV-VIS frequency. There are several severe limitations inherent in this scheme. One is the low pulse repetition frequency of the Nd-glass laser. Another is the limited frequency range of the crystals in the infrared, extending only to about 5 μm. Finally, the complexity of the whole scheme makes it very difficult to achieve and maintain alignment, phase-matching and frequency tuning during the many shots necessary for one experiment.

Considering this, we thought it highly desirable to have an inherently simple system for the generation of high-power tunable ps pulses occurring simultaneously in the UV-VIS and IR. This system should have a repetition frequency of at least 10 Hz, a wavelength coverage from about 350 nm to the near infrared for the UV-VIS pulses and from about 1.5 to 15 μm for the infrared pulses. Such a system is now under construction in our laboratory and even though it is not yet completed, we have done enough measurements on the subsystems to be confident that the design objectives can be met in due course of time.

Figure 1 shows our approach. A short-pulse nitrogen laser pumps one or two distributed feedback dye lasers that produce a single

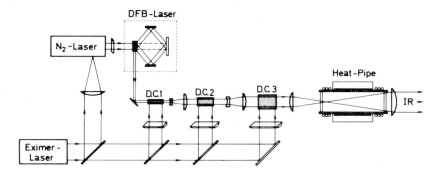

Fig. 1.

ps pulse for every nitrogen laser pulse. These pulses are amplified in three dye cells pumped by a XeCl-excimer laser. The
synchronisation between excimer and dye laser is achieved by
taking a small fraction of the excimer laser beam to fire a
laser-triggered spark gap in the nitrogen laser. The infrared
pulses are produced by stimulated electronic Raman scattering.
This is done be focussing one amplified dye laser beam of suitable wavelength into a heat-pipe oven containing an atomic
vapour.

In the following the distributed feedback dye laser, the amplifier and the infrared-pulse generation will be discussed in detail.

The time behaviour of a distributed feedback dye laser was first
studied, both experimentally and theoretically by Szolt Bor [2].
Normally, an inversion grating is produced in a dye solution by
two interfering beams of equal intensity obtained by passing a
monochromatic laser beam through a 50 % beam splitter. Super-
radiant lasers like nitrogen lasers cannot normally be used for
this purpose because of their large spectral bandwidth and their
incomplete spatial coherence which contribute to a very low visibility of the interference pattern on the dye solution. Bor's
novel pumping arrangement [2], which makes use of a holographic
grating as a beam splitter solved this problem. It is easy to
see that - if certain geometrical relationships are fulfilled -
the different spectral components all recombine on the dye surface to form an inversion grating of visibility one and that
all rays originating in one point of the nitrogen laser near
field pattern are recombined in one point on the dye surface,
thus nullifying the influence of incomplete spatial coherence.
The distributed feedback created in this way is, of course, a
function of time. Instead of the normal relaxation oscillations
of resonators with constant Q, one can obtain much harrower
spikes by the self-cavity-dumping action of the effective cavity
lifetime τ_C, as shwon in the computer solution of the rate equations [2]. The effective cavity lifetime changes drastically
during the pump pulse thus producing short pulses of 41 ps to

112 ps halfwidth when pumping with a nitrogen laser of 2.5 ns
pulse width. Experimentally, this was verified by streak camera
records. If the pump intensity is lowered to less than 20 %
above threshold a single pulse is produced, which according to
the computer solutions should be less than 10 ps wide, if a
half-ns pump pulse is used.

There are many possibilities for tuning the wavelength of such
a DFB laser. Apart from the well-known method of changing the
refractive index of the solution using various solvent mixtures,
a simple method is the rotation of the two mirrors. Since there
are no resonator modes in a DFB laser, because there is no resonator, one can fine-tune the wavelength without any mode-hopping
by applying pressure to the solution. A 100 bar pressure change
results in a wavelength shift of about 10 Å. Since the whole
dye cell contains less than one ml of solution, this is easily
and safely accomplished.

The DFB principle, of course, works in any spectral region and
Bor has shown that one can easily go e: g. from the red to the
blue by simply adding a prism to the dye [2].

The amplifier chain consists of three stages of increasing diameter which are so chosen that the output level of each stage
is well above saturation intensity. The pumping excimer laser
beam is divided into four equal intensity beams which are sent
down parallel to the optical axis of the system. At each stage
beam splitting mirrors direct a suitably chosen fraction towards the dye cell, where four cylindrical lenses focus the
four beams into the center line of the cell. If dye concentrations, diameters and pump intensities are chosen correctly one
can obtain a nearly uniform inversion of circular symmetry
with a slight maximum around the center line. The stages are
transit-time decoupled and ASE will be suppressed by the addition of saturable absorbers. Saturation of the amplifiers will
certainly lead to a reduction of pulsewidth, but how much will
have to be determined experimentally, since this is highly dependent on pulse shape.

The generation of infrared pulses by stimulated electronic Raman
scattering in Cs-vapour is studied extensively by Richard Wyatt
and David Cotter in our laboratory. They first used ns-pulses
from an excimer-laser pumped dye laser of 1 MW peak power in a
nearly diffraction-limited beam. The heat-pipe was of 30 cm
active length and superheated in the middle section in order to
dissociate thermally the strongly absorbing Cs_2-dimers. In this
way threshold for SERS could be reduced by almost an order of
magnitude. Using the 6s-5d Raman transition, they could produce
infrared pulses between 1.6 and 3.2 μm with a quantum efficiency
up to 2 % while scattering to the 7s, 8s, and 9s levels produces wavelengths from 2.2 μm - 8.65 μm, and 10.9 - 16.7 μm,
with quantum efficiency up to 50 % [3]. To study the effect of
high-power ps pulses they used the output of a mode-locked,
frequency-doubled Nd-YAG laser giving 30 ps pulses of 300 MW
peak power. Focussing this radiation into 55 Torr of Cs-vapour
resulted in a 20 % quantum efficiency of conversion to 2'.38 μm
radiation [4]. Threshold lies as low as 200 μJ. Tighter focussing

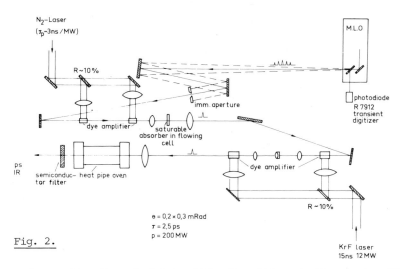

Fig. 2.

reduces the threshold to 100 μJ, although the peak efficiency is reduced.

Recently Wyatt and Cotter applied ps pulses from a passively mode-locked flashlamp-pumped dye-laser system shown in Fig. 2. The ps pulse train from the oscillator is amplified in two stages pumped by a nitrogen laser with 3.5 ns halfwidth so that only one single pulse of the whole train is selected and amplified. After further amplification in two stages pumped by an excimer laser the peak power is 250 MW, the pulsewidth 2 ps, in a nearly diffraction-limited beam, tunable from 460 - 490 nm. Calculations indicate that one should be able to obtain complete coverage of the IR-region between 1.6 and 15 μm using the appropriate Raman transition in caesium vapour. Initial measurements of output energy show quantum efficiencies of ~ 50 % to generate 4 μ. The pulse width of the infrared pulse is expected to be slightly shorter than the pump pulse. An intriguing idea is the possibility of further pulse width reduction using backward Raman amplification in a second heat pipe that is pumped in the opposite direction thus possibly achieving sub-ps pulses, opening up a wide new field for experiments in the physics of large molecules.

References

1. A. Seilmeier, W. Kaiser, A. Laubereau: Opt. Commun. 26, 441 (1978)
2. Z. Bor: IEEE J. Quant. Electron. QE-16, 517 (1980)
   Zs. Bor: Opt. Commun. 29, 103 (1979)
3. R. Wyatt, D. Cotter: Appl. Phys. 21, 199 (1980)
4. R. Wyatt, D. Cotter: Opt. Commun. 32, 481 (1980)

Optically Pumped FIR Lasers

A. Scalabrin, E.C.C. Vasconcellos, C.H. Brito Cruz, and H.L. Fragnito

Instituto de Física "Gleb Wataghin" Unicamp
13100 - Campinas - SP - Brazil

1. Introduction

The generation of far-infrared radiation (FIR) by optically pumping molecular gases was first accomplished by Chang and Bridges in 1970 [1] . They made use of CO_2 laser to pump specific vibrational-rotational states of CH_3F molecule at low pressure, producing population inversion between rotational levels in the excited vibrational state. The importance of this pumping scheme for obtaining FIR laser lines is due to the existence of a great number of polar molecules with vibrational absorption in the 9-11μm region where the CO_2 laser emits, combined with the high efficiency of the CO_2 laser pump. So far more than 50 molecules have been found to lase, yielding over 1000 lines in the 5-250 cm^{-1} spectral range [2] and an intensive search for new and efficient lines is now taking place in several laboratories. A good review of the FIR laser field has been given by Hodges [3] .

The most useful FIR laser molecule is methyl alcohol, CH_3OH, having many efficient lines spread over the submillimeter spectrum. It has received a great attention, most of the potential lines have been discovered, their frequencies measured and many transitions assigned [4] . The deuterated species of methanol have also produced many lines and a considerable amount of new data on them has been taken recently. In this communication we give an up to date summary on FIR laser lines obtained from CH_3OH and its deuterated species pumped by the CO_2 laser.

Optically pumped far-infrared lasers are being used in many interesting applications including the spectroscopy of paramagnetic molecules and free radicals (Laser Magnetic Resonance) [5] , laser frequency synthesis [6] , and plasma diagnostics [7] . The great majority of studies using optically pumped FIR lasers have been done with cw lasers which have narrow linewidths (few kHz) and can be easily stabilized. Pulsed TEA CO_2 lasers are also attractive for optical pumping producing FIR lasers with peak powers of many kW [3] . The main technical limitation is the large linewidth produced by these devices which is typically 500 MHz. Advances have been done in reducing the linewidth of pulsed FIR lasers to 25MHz [8] and still narrower linewidths seem feasible. A practical problem of using TEA CO_2 lasers for pumping is the high cost of operation which is mainly due to consumption of He which is the main component of the laser mixture. In this communication we show that it is possible to substantially reduce the amount of He in the $CO_2/N_2/He$ mixture of the TEA CO_2 laser by the use of low ionization potential dopants without degrading the performance of the laser.

2. Methanol and deuterated isotopes lasers

Methanol is a rich source of FIR lasing lines due to a combination of several factors including the existence of strong absorption bands coincident with CO_2 laser spectrum, large number of potential far-infrared transitions over a wide frequency range and a large permanent dipole moment.

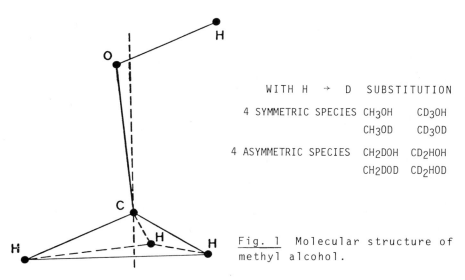

METHYL ALCOHOL

WITH H → D SUBSTITUTION

4 SYMMETRIC SPECIES CH_3OH CD_3OH

                           CH_3OD CD_3OD

4 ASYMMETRIC SPECIES CH_2DOH CD_2HOH

                           CH_2DOD CD_2HOD

Fig. 1 Molecular structure of methyl alcohol.

The molecular structure of CH_3OH is shown in Fig. 1. It is a slightly asymmetric top formed by a methyl group and a hydroxyl group. Its infrared absorption was studied by Woods with a high resolution infrared grating spectrometer [9]. He has done measurements of the infrared absorption with 0.1 cm⁻¹ resolution on CH_3OH, CH_3OD, CD_3OH and CD_3OD, assigning precisely the band centers of most IR vibrational modes. More recently, Serrallach et al. [10] have also measured the IR absorption of all deuterated species of methanol determining the band centers of CD_2HOH, CD_2HOD, CH_2DOH and CH_2DOD. The relevant IR absorption for CO_2 laser pumping comes from the C-O stretch vibrational mode which for CH_3OH is centered at 1033.9 cm⁻¹ [9]. The deuteration of CH_3OH shifts the C-O band by a small amount with the band heads of all isotopes falling within the spectral range of CO_2 laser as shown in figure 2, making them all good candidates for optical pumping with CO_2 laser.

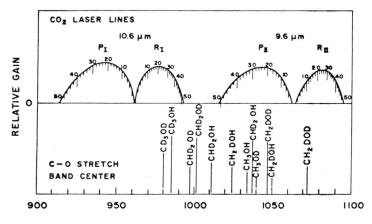

Fig.2 Frequencies of the C-O stretch band center of CH3OH and deuterated isotopes compared to CO2 laser spectrum.

At the working pressure of FIR laser gases (close to 100 mTorr for cw lasers) the absorption lines are narrow and the CO2 laser usually excites only a specific vibrational-rotational transition. To be effective for producing optical pumping the transition frequency must lie within ± 50 MHz of the center of the CO2 laser line, which is the usual tuning range of cw CO2 laser. The requirement of close coincidence imposes severe restriction on how many lines can be pumped. Despite this restriction a surprisingly large number of coincidences exist for methanol and deuterated species.

Table 1 summarizes the observed CO2 laser pump coincidences with normal and deuterated methanol absorptions which produce FIR laser lines. The number of known FIR laser lines for each methanol isotope is shown in the third column of Table 1. CH3OH is the most important of the methanols. This was one of

Table 1   Summary of observed cw laser lines from methanol pumped by regular $C^{12}O^{16}_2$ laser

METHANOL ISOTOPE	NQ of CO2 pump line coincidences	NO of FIR laser lines	Reference
CH3OH	29	107	[4,11,12]
CH3OD	12	28	[2,16]
CD3OH	47	105	[13,14,15]
CD3OD	18	18	[2, 17]
CH2DOH	27	66	[18,19]
CH2DOD	6	10	[21]
CHD2OH	23	46	[18, 20]
CHD2OD	-	-	
Total	162	390	

the first molecules where laser action was achieved by
optically pumping with CO_2 laser [11] . The list of FIR
lines has increased with the development of better laser cavities
and reaches now the number of 107 lines counting only those
pumped by the regular lines of CO_2 laser in the cw mode. A
comprehensive investigation on CH_3OH laser was recently published
by Petersen et al. [4] . Inguscio et al. [12] have reported
three more lines not included in [4]. CH_3OH has lines from $38\mu m$
to 1.2 mm with cw power outputs from less than 0.1 to over
100 mW depending on the line and experimental conditions.

Among the deuterated species CD_3OH is the most important.
Dyubko et al. [13] have reported 70 lines by pumping in the
vicinity of the C-O band. It is not certain that all these lines
come from CD_3OH since the purity of the sample used in [13]
was not known. Many strong short wavelength lines were also
found by Danielewicz and Weiss [14] by pumping in the 10 μm
band. Other lines have been also found by pumping the $9\mu m$ CD_3
deformation vibration band of CD_3OH [15] . CH_3OH [2, 16]
and CD_3OD [2, 17] do not have as many lines but produce
several strong FIR emissions at short wavelengths. The
investigation of the asymmetric methanols is not complete but
the available data shows that they also are good lasers. The
best studied of them is CH_2DOH [18, 19] having many lines as
strong as those of regular methanol well distributed over the
spectrum from 42.5 μm to 616.3 μm. CD_2HOH [18, 20] also
appears to be a good laser and CH_2DOD is under investigation
but does not seem to be as good as the other isotopes [21].

3. Low He consumption CO_2 TEA laser pump.

The achievement of lasing action in CO_2:N_2:He mixtures at
atmospheric pressure requires a uniform transverse discharge
[22] , which is obtained by some kind of preionization. Two
widely used preionization methods the are arc array introduced
by Richardson et al. [23] and the Lamberton and Pearson
[24]wire preionization. Typical ratio of CO_2:N_2:He gas mixture
is 1:1:8 for TEA laser and since open flow is normaly employed
high operation cost result mainly due to He which is the more
expensive gas in the mixture.

Low ionization potential additives can be used to increase
the preionization efficienty [25, 26]. We have observed that
by heavily doping the laser mixture (3%) with dimethylamine or
tri-n-propylamine the optimum mixture for laser operation
requires smaller fraction of He in the mixture. We have observed
this behaviour both in an arc array and in a wire preionized
laser. In Table 2 we give laser parameters and experimental
results for a 130 cm long optical cavity formed by a gold coated
copper total reflector and a germanium spherical mirror with
30 m radius of curvature and 70% reflectivity.

Table 2 — CO$_2$ TEA LASER CHARACTERISTICS

		ARC PREIONIZED	WIRE PREIONIZED
Discharge gap		3.5 cm	3.5 cm
Discharge volume		3.5x3.7x40=518cm^3	3.5x2.0x40=280cm^3
Stored energy		up to 25 Joules	(20nF at 50KV)
Ratio of preionization to stored energy		0.20	0.05
Maximum operating Pressure	a) Heliumless mix (doped)	350 Torr	450 Torr
	b) CO$_2$:N$_2$:He 1:0,6:2 Mix (doped)	> 800 Torr	> 800 Torr
Peak power	a) CO$_2$:N$_2$=1:1 p=300 Torr	2 MW	4 MW
	b) CO$_2$:N$_2$:He=1:0,6:2 p=700 Torr	4 MW	9 MW
Pulse energy		2 Joules typical	
Efficiency		20%	17%
Input energy density(limited by circuit disponibility		48 J/L	89 J/L

Heliumless operation has been obtained for pressures up to 400 Torr. At atmospheric pressure He needs to be added but still at small proportion.

By using a 75 lines/mm diffraction grating instead of the total reflector, oscillation in more than sixty lines was observed over the 9.4μm and 10.4μm bands of CO$_2$. This laser can be used as a powerful and convenient pump source for pulsed FIR lasers. This research was supported by FAPESP, CNEN-CTA and CNPq.

References

1. T. Y. Chang and T. J. Bridges, Opt. Comm. 1, 423 (1970)
2. D. J. Knight, Ordered List of Optically-Pumped Laser Lines

National Physical Laboratory, Teddington-UKNPL № QU45-1979.

3. D. T. Hodges, Infrared Phys. 18, 375 (1978)

4. F. R. Petersen, K. M. Evenson, D. A. Jennings and A. Scalabrin, IEEE J. Quantum Electron. QE16, 319 (1980).

5. J. T. Hougen, J. A. Jennings, and K. M. Evenson, J. Mol

6. D. A. Jennings, F. R. Petersen, and K. M. Evenson, "Direct Frequency Measurement of the 260 THz (1.15 μm) ^{20}Ne Laser: And Beyond", in Laser Spectroscopy IV, ed. by H. Walther, K. W. Rothe, Springer Series in Optical Sciences, Vol. 21 (Springer, Berlin, Heidelberg, New York 1979) pp. 39-48

7. M. Yamanaka et al., Int. J. Infrared & Mill. Waves, 1 , 57 (1980).

8. Z. Drozdowicz, P. Woskoboinikow, K. Isobe, D. R. Cohn, R. J. Temkin, K. J. Button, and J. Waldman, IEEE J. Quantum Electron. QE13, 413 (1977)

9. D. R. Woods, "The High Resolution Infrared Spectra of Normal and Deuterated Methanol", Ph. D. Thesis, U.Michigan (1970).

10. A. Serrallach, R. Meyer, and Hs. H. Gunthard, J. Mol. Spectros. 52, 94 (1974).

11. T. Y. Chang, T. J. Bridges, and E. G. Burkhardt, Appl.Phys. Lett. 17, 249 (1970).

12. M. Inguscio, A. Moretti and F. Strumia, Opt. Comm. 32, 87 (1980).

13. S. F. Dyubko, V. A. Svich and L. D. Fesenko, Radiofisika 18, 1434 (1975).

14. E. J. Danielewicz and C. O. Weiss, IEEE J. Quantum Electron. QE 14, 458 (1978).

15. M. Grinda and C. O. Weiss, Opt. Comm. 26, 91 (1978).

16. Y. C. Ni and J. Heppner, Opt. Comm. 32, 459 (1980)

17. E. C. C. Vasconcellos, A. Scalabrin, F. R. Petersen and K. M. Evenson (To be published).

18. G. Ziegler and U. Durr, IEEE. J. Quantum Electron. QE-14, 708 (1978).

19. A. Scalabrin, F. R. Petersen, K. M. Evenson, and D. A. Jennings, Int. J. Infrared & Mill. Waves 1, 117 (1980).

20. R. J. Saykally, A. Scalabrin, K. M. Evenson (To be published).

21. E. C. C. Vasconcellos, A. Scalabrin, F. R. Petersen and K. M. Evenson (To be published)

22. O. R. Wood, Proc. IEEE 62, 355 (1974).

23. M. C. Richardson, K. Leopold and A. J. Alcock, IEEE J. Quantum Electron. QE-9, 934 (1973)

24. A. M. Lamberton and P. R. Pearson, Electron. Lett. 7, 141 (1971).

25. E. Morikawa, J. Appl. Phys. 48, 1229 (1977)

26. G. Salvetti, Opt. Commun. 30, 397 (1979).

A Direct Observation of Gain in the XUV Spectral Region

D. Jacoby, G.J. Pert, S.A. Ramsden, L. Shorrock, and G.J. Tallents

Department Applied Physics, University of Hull
Hull, HU6 7RX, Great Britain

Lasers operating at short wavelengths in the XUV and X-ray regions have excited considerable interest over the past decade. However, despite the many
imaginative and original approaches suggested, no successful demonstration
of gain in a system which can be scaled to yield laser action has yet been
described. In this paper we report observation of gain on the Balmer α (H_α)
transition of the hydrogen-like carbon ion (CVI) with a gainlength product
in the range 1-5.

These experiments are based on the recombination scheme in which the inversion is formed between the levels n = 3 and n = 2 in the cascade following
recombination of the fully stripped carbon ion. Reports of population in-
versions during the expansion (and therefore cooling) of laser-produced plasmas have been given by several workers [1-3]. In most of these experiments
[1,3] the plasma is formed from a solid target and the inversion generally
forms at too low a density to give practical gain. In our experiments we
have used the expansion of a thin carbon fibre to restrict the number of
carbon ions and increase the rate of expansion [4,5,6]. Preliminary experiments [2] using fibres heated by a Nd:glass laser clearly revealed the presence
of inversion, generating spectra in excellent agreement with a computational
model. These experiments indicated that the bulk of the carbon ions in the
fibre could be heated in this way, and that the laser-fibre energy coupling
was of order 20%.

The present experiments extend these earlier ones by using thinner fibres
and cylindrical focussing. Fibre reduction techniques have been developed to
give carbon fibres with diameters in the range 1.5-4.5μm for these experiments.
The fibres are mounted along the axis of a cylindrical focussing system (Fig.1)
and viewed by two grazing incidence vacuum spectrographs, one viewing the
plasma along its axis, and the second transverse; the complete irradiated
length of fibre being within the field of view of both instruments. The two
spectrographs were carefully calibrated for their response to fibre emission
on the line H_α.

The fibre was irradiated with a Nd:glass laser giving up to 10 j in a pulse
100 ps in duration. As in our previous experiments [2] a nominal 10% pre-pulse
200 ps before the main laser pulse is employed to break the fibre and improve
the laser-plasma coupling. The focussing system produced a spot 2 mm long and
40 μm wide. In order to avoid any absorption at a cold end, the fibre was
mounted with an unsupported end pointing into the axial spectrograph and only
filling 1/2-3/4 of the focal spot.

Gain has been identified in two ways. Firstly the calibration of the two
spectrographs allows a direct measurement of the gainlength product from the

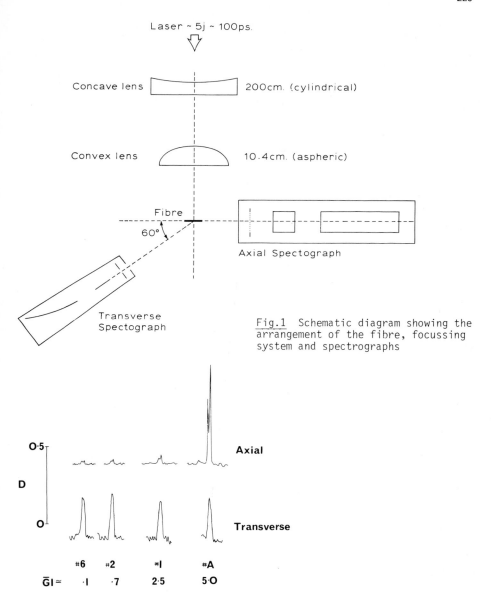

Laser ~ 5j ~ 100ps.

Concave lens   200cm. (cylindrical)

Convex lens   10.4cm. (aspheric)

Fibre

60°

Axial Spectograph

Transverse
Spectograph

Fig.1   Schematic diagram showing the arrangement of the fibre, focussing system and spectrographs

0·5

D

0

Axial

Transverse

#6   #2   #1   #A

$\bar{G}I \approx$ ·1   ·7   2·5   5·0

Fig.2   Microdensitometer traces from a sequence of plates with increasing input energy. The axial and transverse spectra are shown at top and bottom, respectively. Note the progressive increase in the magnitude of the axial signal, whilst the transverse one remains nearly constant. The corresponding increase in resolution of the "doublet" structure along the sequence can be clearly seen

amplification of the spontaneous emission observed in the ratio H_α intensities. This is supported by the observed enhancement of the well-known "doublet" structure of the line along a sequence of shots with increasing input energy, as shown in Fig.2.

Values of the measured gain/length product are given in Table 1 - negative values indicating absorption. There is a progressive increase in the gain/length product with increasing laser energy, but the variation with fibre diameter is relatively weak. The behaviour is in qualitative agreement with our latest computational modelling of the interaction. The values of gain obtained are consistent in order-of-magnitude with those predicted from earlier calculations [6].

Table 1 Listing of the characteristics and measured gains from the spectra obtained. Note the clear sequential behaviour with increasing laser energy

Laser Energy	Fibre Diameter	$\overline{G}\ell$
8.0 ± 1.2	4.1	4.5 ± 0.5
6.7 ± 0.6	2.9	2.5 ± 0.5
6.0 ± 1.3	3.3	3.5 ± 0.5
5.9 ± 2.3	2.2	3.0 ± 1.0
6.0 ± 2.2	4.0	0.1 ± 0.5
5.3 ± 1.0	2.5	0.7 ± 0.5
5.2 ± 1.8	4.2	-0.35 ± 0.5
4.9 ± 2.2	4.3	-1.3 ± 0.5
4.9 ± 2.6	4.6	-2.5 ± 0.5

A direct scaling of these experiments to give a gain/length product of about 10 will allow a direct demonstration of laser action in an amplified spontaneous emission mode. Such experiments will require a focal spot of about 1 cm length of good uniformity, and laser energy density of about 40 j/cm.

References

1. F. Irons, N.J. Peacock: J. Phys. B7, 1109, 1974
2. R.J. Dewhurst, D. Jacoby, G.J. Pert, S.A. Ramsden: Phys. Rev. Lett. 37, 1265, 1976
3. M.H. Key, C.L.S. Lewis, M.J. Lamb: Opt. Commun. 28, 331, 1979
4. G.J. Pert, S.A. Ramsden: Opt. Commun. 11, 270, 1974
5. G.J. Pert: J. Phys. B9, 3301, 1976
6. G.J. Pert: J. Phys. B12, 2067, 1979

Three Layer 1.3 μm InGaAsP DH Laser with Quaternary Confining Layers

F.C. Prince, N.B. Patel, and D.J. Bull

Department of Physics, University of Campinas
Campinas, SP 13100, Brazil

Improvements in the surface morphology of the InGaAsP (DH) laser system were obtained by the addition of small amounts of Ga and As in the confining layers. This method is very promising as a simple and reproducible way to produce high quality wafers giving a very high yield of devices with low threshold current density. The importance of good morphology at the active layer interfaces in reducing optical scattering losses, consequently in reducing threshold current scatter and improving device yield has been demonstrated for GaAs/GaAlAs lasers (1-2-3). The binary InP confining layer typically has poor morphology mainly due to a slight misorientation of the substrate and phosphorous loss. The use of quaternary buffer layer resolves both these problems without recourse to extreme care in crystal orientation and some special means of increasing phosphorous partial pressure near the substrate location.

In the accompanying figure 1 we compare the surface morphology under Nomarsky techniques of several layers that constitute a DH laser system. Pictures (A) and (B) are binary buffer layers while (C) is a quaternary active layer (about 1μm thick) over a binary buffer. Picture (D) is a binary Zn doped top layer over (C). Pictures (A'), (B') are quaternary buffer layers, (C') is a quaternary active layer on quaternary buffer layer and (D') is a complete 3 layer laser system. All the quaternary layers are lattice matched to InP. The wafers (A'), (B'), (C') and (D') have orientations similar to those of (A), (B), (C) and (D) respectively. There is a clear improvement in morphology in changing the confining layers from binary to quaternary. With the quaternary confining layers we can consistently obtain surface roughness of less than (100Å) over the whole wafer at all interfaces.

Lasers devices were fabricated from DH wafers with quaternary confining layers. Typical doping levels and layer thickness were:

First layer: $In_{0.97}Ga_{.03}As_{.08}P_{.92}$ doped with tin at $2 \times 10^{18} cm^{-3}$ 5μm thick;

Second layer: $In_{0.75}Ga_{0.25}As_{0.60}P_{0.40}$ undoped or doped with tin at $3 \times 10^{17} cm^{-3}$, varying in thickness from 0.08 up to 0.75μm;

Third layer: $In_{0.97}Ga_{.03}As_{0.08}P_{0.92}$ doped with Zn at $5 \times 10^{17} cm^{-3}$, 2.5μm thick.

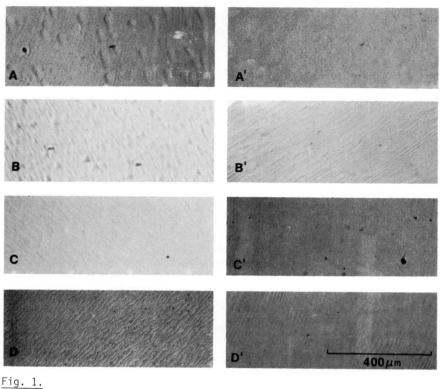

Fig. 1.

A) B/InP   A') Q_1/InP
B) B/InP   B') Q_1/InP
C) Q_2/B/InP   C') Q_2/Q_1/InP
D) B/Q_2/Q_1/InP D') Q_3/Q_2/Q_1/InP

B - Binary InP layer
Q_1- Quaternary Buffer layer $In_{0.97} Ga_{0.03} As_{0.08} P_{0.92}$
Q_2- Quaternary active layer $In_{0.76} Ga_{0.24} As_{0.40} P_{0.60}$
Q_3- Quaternary top layer $In_{0.97} GA_{0.03} As_{0.08} P_{0.92}$

In Fig.2 we show the results of yield and threshold scatter from five different wafers and also the variation of threshold current density as function of active layer thickness (d). In Figs.2c, d, e, and f, which show results for randomly selected broad area lasers, we see that in any given wafer J_{th} maximum is at most only 60% higher than J_{th} minimum, even for wafers with extremely thin active regions. In Fig.2a we show threshold current scatter results of all 70 lasers from a given part of a given wafer. Careful obser-

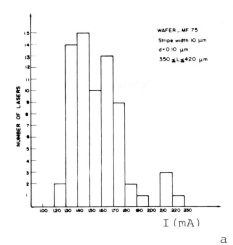

WAFER _ MF 75

Stripe width 10 μm

d = 0.10 μm

350 ≤ L ≤ 420 μm

I (mA)

a

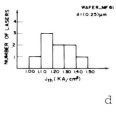

WAFER_MF 61

d = (0.25) μm

J_{th} (KA/cm²)

d

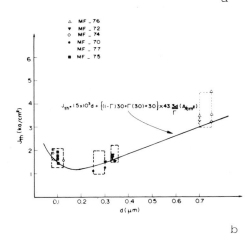

$$J_{th} = 15 \times 10^3 d + \left[(1-\Gamma)30 + \Gamma(30) + 30 \right] \times 43 \frac{3d}{\Gamma} \left(\frac{A}{cm^2} \right)$$

△ MF _ 76
▼ MF _ 72
○ MF _ 74
● MF _ 70
   MF _ 77
■ MF _ 75

b

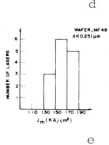

WAFER_MF 48

d ×(0.25) μm

J_{th} (KA/cm²)

e

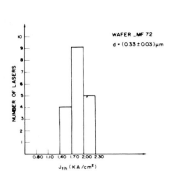

WAFER _MF 72

d = (0.33 ± 0.03) μm

J_{th} (KA/cm²)

c

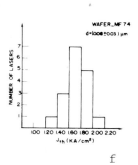

WAFER_MF 74

d = (0.08 ± 0.03) μm

J_{th} (KA/cm²)

f

Fig. 2a-f.

vation showed that the 4 lasers with threshold current between 210 and 230 mA had mirror defects. Discounting these, we again have I_{th} maximum about 60% higher than I_{th} minimum; some of the scatter here may be due to the fact that there is a scatter in laser length (L) between 350 and 420 μm.

In Fig.2b is shown the variation of J_{th} against the active layer thickness with data from different wafers. Also shown is a theoretical fit, in which the values of parameters used are obtained from J_{th} vs 1/L data from lasers from the same wafers.

We would like to point out that from a wafer with d = 0.75 μm, we obtained a laser which gave a normalised threshold current density of 3.4 KA/cm^2/μm, which to our knowledge is the lowest reported to date. The T_0 values of our lasers is in the range 60-70 K and we have obtained cw operation at room temperature.

An additional advantage with this quaternary three layer system is that we can obtain a low resistance contact (differential series resistance of 1 to 2 Ω for 10 μm wide SiO$_2$ stripe devices) directly on the top quaternary layer after a skin diffusion, thus eliminating the need for a cap layer.

Finally, an important advantage of this three layer DH system with quaternary confining layers is that in growing wafers for 1.55 μm lasers we would be able to discuss with the anti-meltback layer.

References

1. J.C. Dyment, F.R. Nash, C.J. Hwang, G.A. Rozgonyi, R.L. Hartman, H.M. Marcos, S.E. Haszko: Appl. Phys. Lett. 24, 481 (1974)
2. G.H. Thompson, P.A. Kirkby, J.E.A. Whiteaway: IEEE J. Quantum Electron. QE-11, 481 (1975)
3. F.R. Nash, W.R. Wagner, R.L. Brown: J. Appl. Phys. 47, 3992 (1976)

Devices for Lightwave Communications*

H. Kogelnik

Bell Telephone Laboratories, Holmdel, NJ 07733, USA

Abstract

A review will be given of recent progress in device research for optical
fiber communications. Recent results include improved growth of epitaxial
layers by liquid-phase, vapor-phase and molecular beam epitaxy in both
the GaAlAs and the InGaAsP materials systems, improved structures, performance and life of laser and detector devices, laser structures for single-
mode operation, and devices operating at the longer IR wavelengths where
fiber losses and fiber dispersion are low. Experimental modulators,
switches and filters in the guided-wave form of integrated optics will also
be discussed.

Introduction

In recent years we have witnessed considerable research effort and rapid
progress in the new technology of lightwave communications (for a recent
review, see [Ref. 1]). This technology is based on generation of signals
in tiny semiconductor lasers or LEDs and on the transmission of optical
signals via hair-thin glass fibers. It has been stimulated by many
promises, including promises of a transmission medium offering larger bandwidth, larger repeater spacings, smaller size, lower weight, lower cross-
talk and potentially lower cost than the traditional media such as copper
wire pairs, coaxial cables or hollow metallic waveguides. Telecommunications engineers are exploring a large variety of applications for this
new technology. These range from short data links interconnecting computers
or other equipment, via systems for television, cable television, and telephone loops, for electric power companies and the military, all the way to
telecommunication trunks between central offices in a city, between cities
and undersea cables linking continents. Characteristic for the progress in
optical fiber technology is the enormous reduction of the transmission loss
of fibers accomplished in the last decade. This is illustrated in Fig.1.

A first generation of lightwave systems has already entered the stage of
field tests and trial systems carrying commercial traffic. There are now
over one hundred installations in Europe, Japan, and North America, and
among them are some major field experiments. Examples for the latter are
the recent Atlanta Fiber System Experiment [2], [3], and the subsequent
Chicago Lightwave Communications Project put in operation by the Bell System
in May 1977, and reporting extremely encouraging results [4].

The first generation of lightwave systems uses multimode fibers of fused
silica (SiO_2) doped with GeO_2, B_2O_3, or P_2O_5. Transmission is mostly in
digital form with data rates between 1 Mbit/s to several 100 Mbit/s, and

236

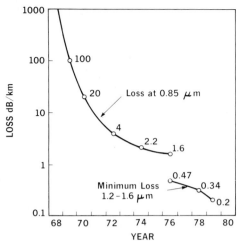

Fig.1 Fiber losses achieved in the research laboratory as reported in the literature in the past 10 years. (Courtesy of T. Li.)

with repeater spacings of several kilometers. The wavelength of the trans-mitted light is near 0.8 μm. The devices used in these systems are AlGaAs junction lasers or LED sources, and Si PIN or avalanche photodetectors.

The transmission speeds and repeater spacings of fiber systems are limited by fiber loss and by pulse broadening due to pulse dispersion in the fiber. Apart from the transmission window near 0.8 μm, fused silica based fibers exhibit two other important windows near 1.3 μm and 1.6 μm. As Rayleigh scattering decreases strongly with reciprocal wavelength, fiber losses can be considerably lower in these "long-wavelength" windows. Recent reports indicate losses as low as 0.6 dB/km near 1.3 μm and 0.2 dB/km near 1.6 μm [5]. The loss spectrum of such a fiber is shown in Fig.2. The dispersive properties of the fiber material contribute importantly to pulse dispersion in optical fibers. This material dispersion is much smaller in the long-wavelength region, where "zero-dispersion" behavior has been predicted for wavelengths near 1.3 μm [1]. Recently the existence of this zero-pulse-dispersion point has been confirmed by pulse-delay measurements using wavelength-tunable radiation generated in a fiber via the stimulated Raman effect [6], and by the transmission of picosecond pulses over kilometer-lengths of single-mode fiber without observable pulse broadening [7]. Apart from material dispersion there is additional pulse dispersion in multimode fibers caused by group velocity differences of the modes. This is absent in single-mode fibers, and Fig.3 shows the dispersion char-acteristics of two single-mode fused-silica fibers with zero dispersion wavelengths near 1.35 μm and 1.5 μm.

There is still a considerable amount of work to be done to improve the properties of first-generation devices. Important issues here include degradation and the operating life of lasers, yield, stability and self pulsations. While this work is proceeding, device research is also giving attention to second generation devices and beyond. One goal is to make available single-mode lasers for single-mode fibers. Another goal is to provide source and detector options for long-wavelength systems that can exploit the favorable fiber properties mentioned above. GaAlAs and Si are no longer suitable for these wavelengths, but considerable success with new materials such as InGaAsP has already been attained.

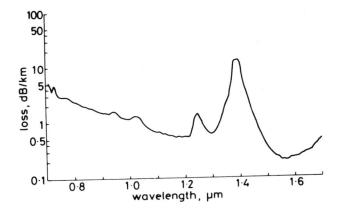

Fig.2 Loss spectrum of a single-mode fiber with SiO_2 cladding and a GeO_2 doped silica core of 9.4 µm diameter. (From Ref. [5].)

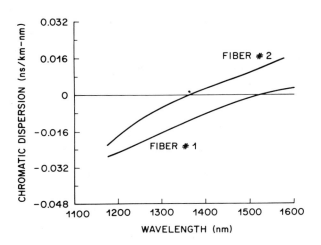

Fig.3 Dispersion spectrum of two GeO_2-doped single-mode silica fibers with core diameters of 5.2 µm (#1) and 7 µm (#2). (From Ref. [67].)

In the following we will attempt to sketch the nature and trend of recent research on devices for lightwave communications. While space and time available for this sketch are limited, the field has grown to such large size, that any selection of illustrations for this work must be subjective and incomplete, and the writer apologizes in advance for all omissions.

The topics selected for discussion include the new semiconductor materials for optical devices, epitaxial growth methods used for device preparation, basic characteristics of new device materials, structures for single-mode lasers, laser degradation, self-pulsations in lasers and new detectors. We will also mention devices that offer possibilities for wavelength multiplexing and switching. These are GRIN-rod devices suitable for multimode fibers systems and integrated optics devices compatible with single-mode fibers.

Semiconductor Materials

The semiconductor lasers of interest for lightwave communications require the preparation of heterostructure crystal layers which are of different band gap but are matched in their lattice constants. Heterostructures are also used for LEDs and some detector devices. Today, by far the best developed materials system is the $Al_xGa_{1-x}As/GaAs$ system, which is providing lasers and LEDs for present lightwave communication systems (for recent reviews see Refs. [8,9]). The bandgap in this material can be tailored to span the wavelength range from about 0.8 μm to 0.9 μm by changing the (AlGa) composition.

In the search for materials suitable for longer wavelengths researchers have investigated a considerable variety of III-V compounds compatible with the substrate materials GaAs, InP, and GaSb. A severe but important materials test was the demonstration of continuous laser operation at room temperature. The break into the long-wavelength range happened in 1976, when three materials passed that test [10, 11, 12]. The first was the $GaAs_{1-x}Sb_x/Al_yGa_{1-y}As_{1-x}Sb_x$ heterostructure system which was graded in lattice spacing to GaAs substrates [10]. CW laser operation was obtained at 1.0 μm and later extended to 1.06 μm. The next was the quaternary system of $In_{1-x}Ga_xAs_yP_{1-y}/InP$ lattice matched to InP substrates [11] and initially operating at 1.1 μm. The third cw room temperature laser operation was demonstrated in $In_xGa_{1-x}As/In_yGa_{1-y}P$ heterostructures graded in lattice spacing to GaAs substrates [12], and operating at wavelengths of 1.06 and 1.12 μm.

At present the focus of research is on the (InGa)(AsP) quaternary system which offers good lattice match (by holding the compositions in the proportion $y \approx 2.2x$). The bandgap in this system can be tailored for operation from about 1.0 μm to 1.7 μm. In the long-wavelength limit we have $y = 1$ and are dealing with the ternary system $In_{.53}Ga_{.47}As/InP$, which is also a desirable detector material.

Another promising detector material is the ternary $Al_yGa_{1-y}As_{1-x}Sb_x$, which can be lattice matched to GaSb substrates [13]. In fact, GaSb appears to be a better quality substrate at present compared to InP. However, much effort is devoted to the improvement of InP, and this should soon bear fruit.

Methods of Epitaxial Growth

There are essentially three different methods which are used for the preparation of epitaxial heterostructures: (1) liquid-phase epitaxy (LPE), (2) vapor-phase epitaxy (VPE), and (3) molecular beam epitaxy (MBE).

LPE is the simplest, least expensive and most commonly used technique. Here, the epitaxial layers are grown from solutions with growth rates of about 100 - 1000 nm/minute. The first cw room-temperature lasers in both the (AlGa)As and the (InGa)(AsP) systems were grown by LPE. LPE-grown lasers in both materials systems can now be made with thresholds better than 1 kA/cm^2.

The VPE and MBE methods promise better uniformity and yield. In the VPE method the epitaxial material is introduced in gaseous form, which offers possibilities for smooth grading of layers and flexibility for the introduction of dopants. Growth rates are of the order of 100 - 500 nm/minute. The most successful VPE technique for the growth of (AlGa)As heterostructures has been the metal-organic chemical vapor deposition (MO-CVD) technique [14]. VPE-grown (AlGa)As lasers have been demonstrated in cw room-temperature operation and with thresholds as good or better than the best LPE material. Recently, VPE-grown (InGa) (AsP) lasers have also been demonstrated to be capable of cw room-temperature operation, and thresholds as low as 2 kA/cm^2 were achieved [15].

At present, MBE is probably the most complex and most expensive of the three techniques, but it offers the broadest range of materials possibilities and thickness control on a 10-Angstrom scale. MBE is essentially an evaporation technique and requires a high vacuum of about 10^{-10} Torr. MBE growth rates are about 5 - 50 nm/minute.

MBE-grown (AlGa)As lasers have been shown to be capable of cw operation at room-temperature [16], and, recently, thresholds as low as 800 A/cm^2 have been achieved [17]. Figure 4 shows a comparison of thresholds achieved in (AlGa)As lasers by the three different methods of epitaxial growth. No cw room-temperature lasers have as yet been made with MBE-grown (InGa)(AsP) material. However, a first step has been made with the achievement of 3 kA/cm^2 room-temperature thresholds in MBE-grown (InGa)As lasers operating pulsed at 1.65 μm [18].

Characteristics of New Device Materials

Experimental efforts are under way to determine the characteristics of the new III-V materials. This is leading to a better understanding of their applicability to optical devices as well as to other electronic devices such as high-speed FETs. Among the results already accomplished is the accurate measurement of the bandgap as a function of composition for (InGa)(AsP) material lattice matched to InP [19]. This is shown in Fig.5.

The temperature dependence of the laser threshold gives indications for the behavior of lasers to be expected at elevated temperatures, and provides clues on mechanism of carrier loss that might be present in the structure. The dependence for the threshold current density $I_{th}(T)$ is expected to be approximately exponential [8,9], i.e.

$$I_{th} \propto \exp (T/T_0),$$

with a characteristic temperature T_0 describing the quality of the material. For (AlGa)As, T_0 is usually in the range between 120° - 165°K [9]. For (InGa)(AsP) heterostructures, T_0 has been measured to be about 70°K near room temperature, both for LPE and VPE material [15], [20], [21]. Measurements of T_0 have given some indications of the presence of an electrically active recombination center in (InGa)(AsP) material [21]. This center affects device operation and should be eliminated if possible.

240

Another materials property that can influence the operational properties of lasers and other devices is due to the presence of an auxiliary indirect bandgap. Electrons are lost from the upper level laser population by intra-band scattering from the direct valley to the auxiliary valley. This scattering is a function of the energy difference ΔE between the direct and the indirect gap. The following is a table of ΔE for a selection of III-V compounds of interest:

GaAs   300 meV

InP   400 meV

GaSb   80-90 meV

InGaAs   800 meV.

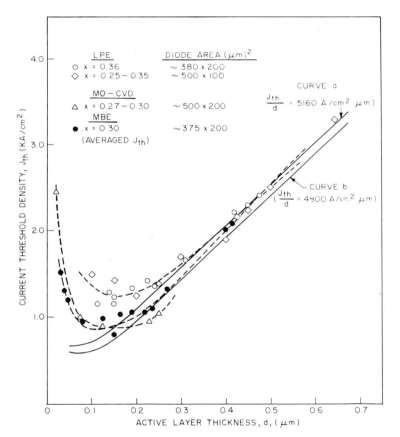

Fig.4 Comparison of threshold currents achieved in (GaAl)As heterostructure lasers as a function of active layer thickness and method of epitaxial growth. LPE data are indicated by open circles and diamonds, MO-CVD data by triangles, and MBE data by dots. (After Ref. [17].)

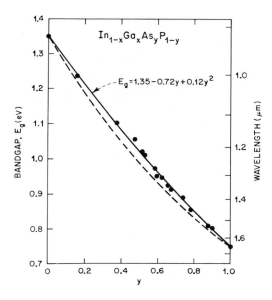

Fig.5 Bandgap energy E_g as a function of the As content y for
$In_{1-x}Ga_xAs_yP_{1-y}$ lattice-matched to InP. The solid line is the best fit to
the measured data, the dashed line is an earlier prediction. (From Ref.
[19].)

The low ΔE-value for GaSb may present a serious hurdle for the accomplishment of cw room-temperature laser operation in that material.

As the new compounds may also be of interest for high-speed electronic
devices, measurements are under way to determine their mobility and similar
characteristics. On the basis of such measurements the mobility of low-
impurity (InGa)(AsP) lattice matched to InP has been projected to increase
monotonically as a function of the As concentration from a value of 4700
cm^2/Vs for InP to a value of 15000 cm^2/Vs for (InGa)As [22].

Single-Mode Laser Structures

For efficient coupling to single-mode fibers one requires junction lasers
that operate in a stable fundamental transverse mode. Frequently single
transverse mode operation brings additional benefits such as kink-free
light-output vs. current characteristics. A considerable variety of single-
mode- structures providing lateral optical confinement have been investigated
for GaAs lasers. Among the more successful ones are the buried heterostructure (BH) laser [23], the transverse junction (TJS) laser [24], the
channeled-substrate planar (CSP) laser [25], and the strip-buried heterostructure (SBH) laser [26]. Cross sections of four laser structures are
sketched in Fig. 6. SBH lasers use a thin and narrow active strip of GaAs.
This strip acts as loading for an adjacent (AlGa)As layer to effect lateral

PROTON-BOMBARDED LASER (BTL)

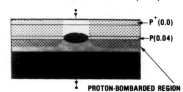

- P⁺(0.0) → $P^+(0.0)$
- P(0.04) → $P(0.04)$
- PROTON-BOMBARDED REGION

BURIED-HETEROSTRUCTURE LASER (HITACHI)

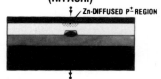

Zn-DIFFUSED P⁺ REGION

TRANSVERSE JUNCTION STRIPE LASER (MITSUBISHI)

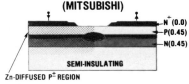

- N⁺(0.0) → $N^+(0.0)$
- P(0.45) → $P(0.45)$
- N(0.45) → $N(0.45)$

SEMI-INSULATING

Zn-DIFFUSED P⁺ REGION

STRIP BURIED-HETEROSTRUCTURE LASER (BTL)

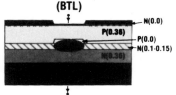

- N(0.0) → $N(0.0)$
- P(0.36) → $P(0.36)$
- P(0.0) → $P(0.0)$
- N(0.1-0.15) → $N(0.1\text{-}0.15)$
- N(0.36) → $N(0.36)$

$$N(x) = N\text{—}Al_x Ga_{1-x} As$$
$$P(x) = P\text{—}Al_x Ga_{1-x} As$$

Fig.6 Cross section of the proton bombarded stripe geometry laser, the BH laser, the TJS laser and the SBH laser. (Courtesy of W. T. Tsang.)

mode guiding. Recent results with SBH lasers using 5 μm to 10 μm strips have exhibited about 85 mA cw thresholds, stable single transverse modes up to 9x threshold, excellent linearity up to 35 mW of cw power output per mirror, and narrow beam divergence [27]. Recent results with TJS lasers on semi-insulating GaAs substrates are also very encouraging. In these structures the junction is made perpendicular to the epitaxial layers. The lasers are showing 12 mA thresholds, up to 15 mW cw output at room temperature, single transverse and single longitudinal modes, and long life [28].

A recent achievement is the single-mode operation in (InGa)(AsP) lasers at a wavelength of 1.3 μm [29]. The structures used in this work were BH lasers of 1-2 μm stripe width having cw room-temperature thresholds of 22 mA, cw output powers of well over 5 mW and demonstrated life of over 1000 hours without deterioration. (Figs. 7 and 8 show the structure and the light output vs. current characteristics of these lasers).

Distributed feedback lasers (DFB) are structures that promise stable output frequencies and single longitudinal mode operation well above threshold. These structures are made by introducing high-resolution (≈ 2000 Angstrom) corrugations into the heterostructure layers. The technology for the preparation of these structures is still rather difficult. However, room-temperature cw operation in GaAs DFB lasers has been demonstrated [30]. Recently a device was demonstrated that insures single transverse modes by use of an SBH structure and provides DFB by means of corrugations in the evanescent field near the strip [31]. Single longitudinal modes and stable output frequencies were obtained up to three times threshold in pulsed operation. Figs.9 and 10 show the structure and the output characteristics of this DFB laser.

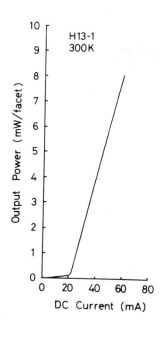

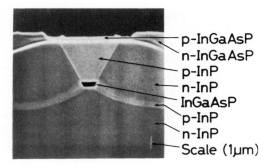

Fig.7 Cross section of the (InGa)(AsP) single-mode laser of Ref. [29].

Fig.8 Output characteristics of the single-mode (InGa)(AsP) laser reported by Ref. [29].

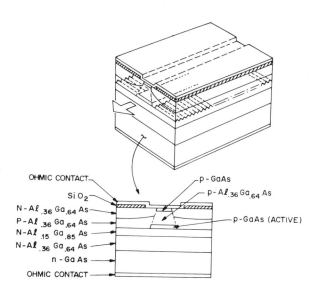

Fig.9 Sketch of the structure of a SBH laser with lateral-evanescent-field distributed feedback. (After Ref. [31].)

Laser Degradation

The operating life of cw junction lasers is, clearly, a vital parameter
for systems applications. At this point it should be recalled that one
deals with current densities of the order of 1 kA/cm^2 in these devices.
In almost 10 years of careful work, the projected mean life of GaAs
junction lasers for cw operation at room temperature has been improved
from a few hours to the order of one million hours as illustrated in
Fig.11. Life projections of this long duration are made on the basis of
temperature-accelerated aging tests in the laboratory. Projections of
10^6 hour mean lifetimes have now been reported for a variety of GaAs laser
structures [32], [33], [34], [35], including a single-mode TJS laser [36].

(InGa)(AsP) junction lasers have been reported to continue operating
after 10^4 hours of cw operation at room temperature. For a new material,
this is extremely encouraging, but much careful work has yet to be done,
and the study of relevant degradation mechanisms is just beginning.

The degradation mechanisms occurring in GaAs lasers are reviewed in
recent text books [8], [9]. They include gradual bulk degradation, facet
damage and contact metallization failure. It has been argued that bulk
laser degradation in (InGa)(AsP) should be considerably reduced as the
photon energy hν is smaller than in GaAs, and degradation times should be

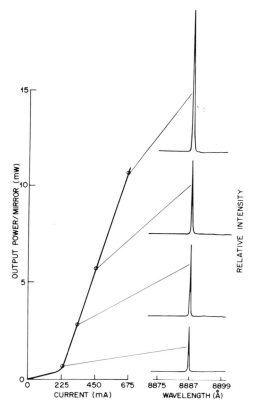

Fig.10 Light-output vs. current
characteristics and associated
output spectra of the DFB laser
sketched in Fig.9. (After Ref.
[31].)

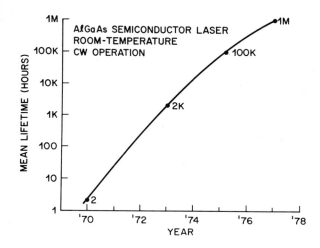

<u>Fig.11</u> Progress in improvement of the reliability of (AlGa)As injection lasers. (From Ref. [1].)

proportional to exp(E_{act}/hν), where E_{act} is the activation energy of the degradation mechanism. However, this argument holds only if there is no new degradation mechansim in the new materials. Indeed, the first studies of optical degradation in (InGa)(AsP) indicate that there is degradation due to a slip-and-glide mechanism that is not normally observed in GaAs [37], [38]. These findings are based on spatially resolved photoluminescence studies and on transmission electron microscopy. They report dark line defects in the <110> direction due to the slip-and-glide mechanism. Dark lines in the <100> direction due to the climb mechanism, which is dominant in GaAs, are virtually absent in (InGa)(AsP).

Laser Pulsations

The occurrence of self-pulsations in junction lasers is one of the remaining problems in (AlGa)As double-heterostructure lasers. These pulsations occur in various laser structures, and under various operating conditions, and they can appear after several 100 hours of aging [39]. Both transient as well as continuous self-sustained pulsations have been observed in the frequency range from 0.2 to 3 GHz. The pulsation frequency is related to the spontaneous recombination lifetime of the carriers of about 2 ns, and to the photon lifetime in the laser cavity of about 10 ps. As they interfere with the high-data-rate modulation of lasers, there are efforts to understand, control and eliminate these pulsations. There are many possible mechanisms for laser pulsations, including filament formation, temperature induced changes and quantum shot-noise effects (for a recent review see Ref. [8]). Phenomenologically, the pulsations in single transverse mode lasers are caused by saturable absorption and/or saturable gain. However, the detailed origin of the responsible saturation effects is still unverified. Recent models trace the pulsations to absorbing dark line defects, to regions of carrier depletion [40], and, possibly, to deep-level traps [41]. Recent experiments with strip-buried heterostructure lasers find a strong correlation between the occurrence of pulsations and the presence of visible defects [42].

A better understanding of the nonlinear absorption and nonlinear gain mechanisms responsible for self-pulsations may not only aid in their elimination, but it can also provide the basis for the controlled generation of pulses in junction lasers (see e.g. Ref. [43]). In fact, some of these mechanisms have been used in recent work on the generation of picosecond pulses in semiconductor lasers. Continuous pulse streams with a 3 GHz repetition rate and pulse duration of 20 ps have been obtained in GaAlAs lasers using an external resonator and active modulation with microwaves at 3 GHz [44]. Pulsewidths as short as 6 ps have been obtained in very recent experiments using SBH GaAlAs lasers and passive modelocking in external cavities with band-limiting elements [45]. Fig.12 shows the cavity arrangement used in these experiments, and Fig.13 shows a correlation function obtained by second-harmonic generation (SHG) and used for the determination of pulse width.

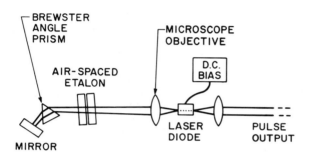

Fig.12 External cavity arrangement used for the generation of picosecond pulses from semiconductor junction lasers. (From Ref. [45].)

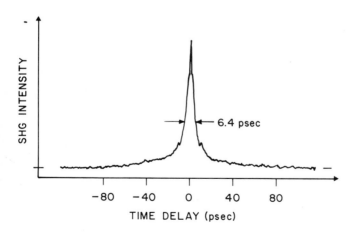

Fig.13 SHG correlation function of semiconductor laser output indicating pulse width of about 6 ps. (From Ref. [45].)

Photodetectors

The detectors required for lightwave communications are room-temperature devices of high sensitivity and fast response, that can detect the attenuated digital optical signal with error rates better than 10^{-9} (for recent reviews see Refs. [46], [47]). First-generation systems use Si PIN photodetectors for lower bit rates and Si avalanche photodetectors (APDs) for higher bit rates. This is illustrated in Fig.14. PIN detectors are followed by low-noise amplifiers such as Si FETs. The internal gain in APDs is optimized to gain values ranging from about 10 - 100 in the presence of excess noise due to the avalanche multiplication process. At bit rates above 10 Mbit/s the APD receivers offer a sensitivity improvement of about 15 dB as compared to Si PIN receivers, for a typical overall sensitivity of 50 dBm (see, e.g., Ref. [1]). However, APDs require relatively high voltages (100-400 V) and temperature compensation circuits to stabilize the gain. There is, therefore, continuing interest in improving PIN-FET combinations, and the recent availability of low noise GaAs FETs has widened the applicability of PIN-FETs to higher speeds (see, e.g., Refs. [48], [49]).

Si detectors are no longer useful at wavelengths longer than 1.1 μm, and device research is busy trying to provide detectors for the long-wavelength

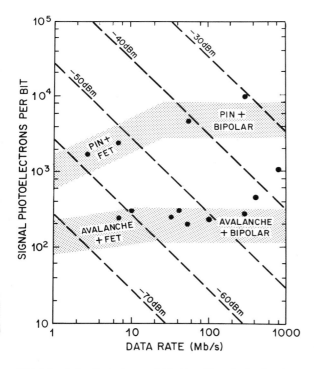

Fig.14 Calculated sensitivity (in photoelectrons per bit) of various optical receivers as a function of bit rate. The large dots represent experimental results achieved to date. (From Ref. [1].)

range (for a recent review see Ref. [50]). Ge is a well developed material that is a good candidate for this purpose. However, Ge is an indirect gap material with a slow fall-off in response towards longer wavelengths and with relatively large dark currents. The search for better materials is focussing on InGaAs, InGaAsP, and GaAsSb, which are direct gap materials and are sensitive up to about 1.7 μm. Fig.15 illustrates the spectral response of these materials and their sharp cut off towards longer wavelengths, which can be moved by changing the materials composition (see, e.g., Ref. [51]). This is a very desirable feature as leakage currents tend to increase exponentially with the cut off wavelength due to the smaller bandgaps. For PIN-FET applications of these materials the requirements include those for low-leakage current, low capacitance, and high-breakdown voltage. Some encouraging results have already been reported where these parameters lie in the range of 10 nA, 1 pF and 100 V (see, e.g., Ref. [52], [53]), but there are also some unexplained phenomena such as an increase of dark current with reverse-bias voltage.

The excess noise due to avalanche multiplication in APDs increases as the ionization rates for electrons and holes become more equal. In Si these ionization rates differ by more than an order of magnitude, but in Ge, InGaAs, InGaAsP, and GaAsSb they appear to be almost the same. Avalanche

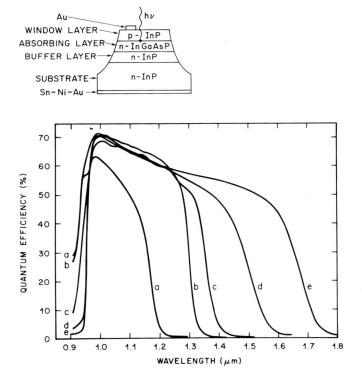

Fig.15 Structure and external quantum efficiency of five $In_{1-x}Ga_xAs_yP_{1-y}$ photodetectors with different composition. The As content y was 0.47 (a), 0.61 (b), 0.66 (c), 0.88 (d), and 1.0 (e), respectively. (After Ref. [51].)

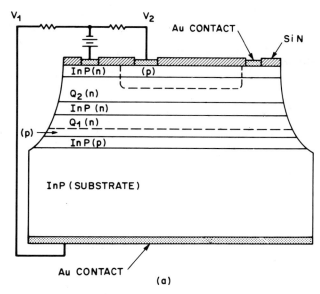

V_1 V_2

Au CONTACT

SiN

InP(n)   (p)

Q_2 (n)

InP (n)

Q_1 (n)

(p)

InP(p)

InP (SUBSTRATE)

Au CONTACT

(a)

Fig.16 Sketch of the structure of a demultiplexing diode. The two (InGa) (AsP) layers are labeled Q1 and Q2. (From Ref. [55].)

gain in the new materials has been observed in several laboratories (for reviews see, e.g., Refs. [46], [50]), but the search for low-noise APD materials still continues. An impulse in the direction may be progress in the understanding of the connection between impact ionization and basic materials properties such as band structure and crystal orientation (see, e.g., Ref. [54]).

The sharp cutoff properties of the new materials are also opening up new device possibilities. A recent example for this is a wavelength demultiplexing (InGa)(AsP) photodiode, that detects and demultiplexes two wavelength bands simultaneously [55]. The structure of such a device is sketched in Fig.16, and its photoresponse is shown in Fig.17.

GRIN-rod Devices

GRIN-rods are glass rods with a parabolically graded refractive index profile similar to that used in graded index fibers. GRIN-rods act like lenses and can be used, in combination with optical elements such as gratings or semitransparent mirrors, to assemble very compact, rugged, and stable devices for the manipulation and processing of optical signals. A variety of devices has been proposed and demonstrated, and most of them are compatible with first-generation multimode fibers, however some applications to single-mode fiber systems appear possible. Examples for GRIN-rod devices are attenuators, directional couplers and switches (see, e.g., Refs. [56], [57]), as well as multiposition switches [58] and wavelength multiplexers [59]. A sketch of a typical wavelength multiplexer is shown in Fig.18, and its demultiplexing filter characteristics are given in Fig.19 for the case of two channels. The latter provides an illustrative example for GRIN-rod

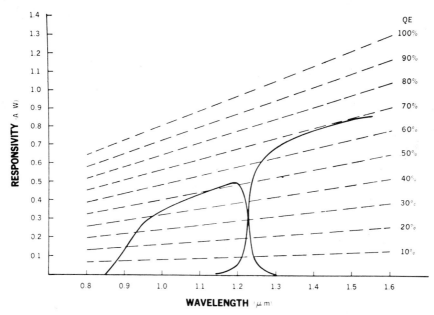

Fig.17 Photoresponse of the two channels of a demultiplexing diode.
(Courtesy of T. P. Lee.)

devices. Illustratively such a multiplexer device might consist of five
input and output fibers, a grin-rod lens and a blazed grating, assembled in
a compact package 1-2 cm long and 2-4 mm in diameter. It would be capable
of multiplexing or demultiplexing four optical channels of different wavelength, with a channel separation of 30 nm and cross talk better than -30 dB.

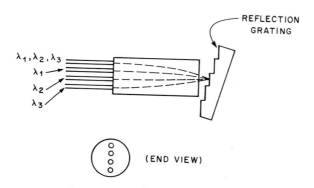

Fig.18 Sketch of a wavelength multiplexer consisting of a GRIN lens and
a grating. Only three input fibers are shown. (After Ref. [68].)

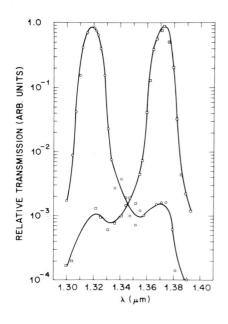

Fig.19 Filter response of a GRIN-rod demultiplexer in the near IR. The response of two channels is shown. (After Ref. [68].)

Integrated Optics Devices

In integrated optics, dielectric waveguides are used on planar substrates to confine the light to small cross sections over relatively long lengths (for recent reviews see, e.g., Refs. [60], [61]). One aims for miniaturized devices of improved reliability and stability, of lower power consumption and lower drive voltages. Devices of interest are couplers, junctions, directional couplers, filters, wavelength multiplexers and demultiplexers, modulators, switches, lasers and detectors. Apart from providing new or improved versions of these devices, a hope of integrated optics is to combine two or more devices on a single chip. Most integrated optics devices and circuits are single-mode devices which are compatible with single-mode fibers. Integrated optics may provide future lightwave communication systems with such sophisticated options as single-mode wavelength multiplexing circuits or electrooptic switching networks. Research in this field is broad and varied, but two examples will have to suffice here to illustrate its nature. Both device examples are based on directional couplers fabricated in the electrooptic material $LiNbO_3$. The coupler consists of two waveguides, about 3 µm wide, which approach each other to a distance of about 3 µm over an interaction length of about 1 mm to 1 cm. The guides are prepared by diffusing Ti into the substrate. Metal electrodes on the surface of the crystal allow the application of an electric field. This induces refractive index changes via the electrooptic effect and allows control of the coupler. Split electrodes and applied voltages of reversed polarity insure that complete crossover of the light from one guide to the other can be achieved by an electrical adjustment [62]. This is called the (alternating) $\Delta\beta$ coupler configuration and has been used in a variety of devices. One of them is a switch and amplitude modulator that includes six sections of alternating $\Delta\beta$, and can operate at data rates in excess of 100 Mbit/s with drive voltages as low as 3 Volts [63]. The structure of this device is sketched in Fig.20. The applied voltage switches the light

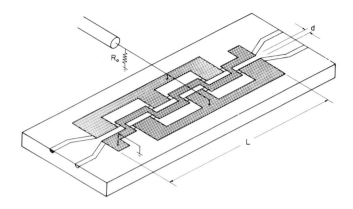

Fig.20 Sketch of alternating Δβ coupler consisting of a Ti diffused directional coupler of interaction length L and guide separation d. The electrodes (shown hatched) provide for six polarity reversals. (After Ref. [63].)

SIX SECTION ALTERNATING Δβ MODULATOR

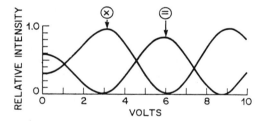

Fig.21 Switching characteristics of a Δβ coupler with six sections of alternating Δβ. Relative output power in the two output guides is shown as a function of applied voltage for light injection at only one input guide. The output guide fed straight through from the input guide is marked =, and the output guide fed by light crossing over from the input guide is marked ⊗. (After Ref. [63].)

from one output guide to the other. The detailed switching characteristics are shown in Fig.21. The second example is a tunable filter device which is very similar in structure to the first. The difference is that the two waveguides are nonidentical and have intersecting dispersion characteristics [64]. To achieve this, the two waveguides are fabricated to different widths (1.5 μm and 3 μm) with different effective refractive index. The device has a measured filter bandwidth of 20 nm and is electrically tunable at a rate of 11 nm/V. The structure and characteristics of this directional coupler filter are illustrated in Fig.22.

Finally, we should mention two examples, where several similar devices have been integrated on a single chip in the research laboratory. The first example is an experimental 4 x 4 optical switching network made by integrating

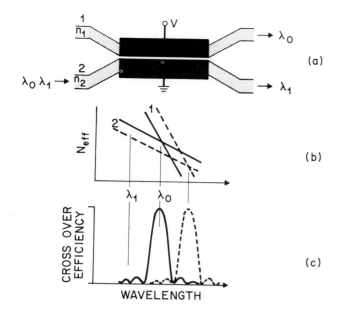

Fig.22 Structure, dispersion and filter characteristics of a tunable directional coupler filter. The device structure (a) consists of a directional coupler consisting of two different guides marked 1 and 2 and electrodes for the application of a voltage (electrode split is not shown). The effective index N_{eff} of the two guides is shown as a function of wavelength λ indicating crossing from one guide (b). The filter response of the light crossing from one guide to the other is shown in (c). (After Ref. [64].)

5 $\Delta\beta$-couplers on a LiNbO₃ substrate [65]. The other example is an experimental wavelength multiplexing chip consisting of 6 DFB lasers operating at different wavelengths with a junction circuit combining the six outputs into one guide on a GaAs substrate [66].

Conclusions

Coupled with the rapid emergence of lightwave communications, there has been a growing research and development effort on optical devices. Device research in this field has seen considerable success, but many exciting challenges remain for solid-state, materials, device, and systems specialists. Research interest is expanding from devices compatible with multimode fibers to those compatible with single-mode fibers, and from devices operating in the 0.8 μm region to devices capable of operating in the long-wavelength region near 1.3 and 1.6 μm.

Because of space and time limitations, the illustrations given in this paper of the diverse trends in device research had to be sketchy, incomplete and subjective. The many necessary omissions include detailed descriptions of several device aspects important in applications such as details on LEDs and the direct modulation of lasers.

References

1. T. Li: IEEE Trans. Commun. 26, 946 (1978).
2. I. Jacobs, S. E. Miller: IEEE Spectrum 14, 33 (1977).
3. Bell System Technical Journal 57, 1717 (1978) (see collection of articles).
4. M. I. Schwartz, W. A. Runstra, J. H. Mullins, J. S. Cook: Bell Syst. Techn. J. 57, 1881 (1978).
5. T. Miya, Y. Terunuma, T. Hosaka, T. Miyashita: Electronics Lett. 15, 106 (1979).
6. L. G. Cohen, C. Lin: Appl. Optics 16, 3136 (1977); C. Lin, L. G. Cohen, W. G. French, V. A. Foertmeyer: Electronics Lett. 14, 170 (1978).
7. D. M. Bloom, L. F. Mollenauer, C. Lin, D. W. Taylor, A. M. DelGaudio: Optics Lett. 4, 297 (1979).
8. H. Kressel, J. K. Butler: In *Semiconductor Lasers and Heterojunction LEDs* (Academic Press, New York 1977).
9. H. C. Casey, Jr., M. B. Panish: *Heterostructure Lasers* (Academic Press, New York 1978).
10. R. E. Nahory, M. A. Pollack, E. D. Beebe, J. C. DeWinter, R. W. Dixon: Appl. Phys. Lett. 28, 19 (1976).
11. J. J. Hsieh, J. A. Rossi, J. P. Donnelly: Appl. Phys. Lett. 28, 709 (1976).
12. C. J. Nuese, G. H. Olsen, M. Ettenberg, J. J. Garmon, T. J. Zamerowski: Appl. Phys. Lett. 29, 807 (1976).
13. H. D. Law, N. Nakano, L. R. Tomasetta, J. S. Harris: Appl. Phys. Lett. 33, 948 (1978).
14. R. D. Dupuis, P. D. Dapkus: Appl. Phys. Lett. 31, 8391 (1977).
15. G. H. Olsen, C. J. Nuese, M. Ettenberg: Appl. Phys. Lett. 34, 262 (1979).
16. A. Y. Cho, R. W. Dixon, H. C. Casey, Jr., R. L. Hartman: Appl. Phys. Lett. 28, 501 (1976).
17. W. T. Tsang: Appl. Phys. Lett. 34, 473 (1979).
18. B. I. Miller, J. H. McFee, R. J. Martin, P. K. Tien: Appl. Phys. Lett. 33, 44 (1978).
19. R. E. Nahory, M. A. Pollack, W. D. Johnston, Jr., R. L. Barns: Appl. Phys. Lett. 33, 659 (1978).
20. Y. Horikoshi, Y. Furukawa: Japan J. Appl. Phys. 18, 809 (1979).
21. R. E. Nahory, M. A. Pollack, J. C. DeWinter: Electronics Lett. 15, 695 (1979).
22. R. F. Leheny, A. A. Ballman, J. C. DeWinter, R. E. Nahory, M. A. Pollack: J. Electronic Mat. (Proc. of Materials Res. Conf., Boulder, CO, June 1979) (to be published 1980).
23. T. Tsukada: J. Appl. Phys. 45, 4899 (1974).
24. H. Namizaki, H. Kan, M. Ishii, A. Ito: J. Appl. Phys. 45, 2785 (1974).
25. K. Aiki, M. Nakamura, T. Kuroda, J. Umeda, R. Ito, N. Chimone, M. Maeda: IEEE J. Quant. El. 14, 89 (1978).
26. W. T. Tsand, R. A. Logan, M. Ilegems: Appl. Phys. Lett. 34, 752 (1979).
27. W. T. Tsand, R. A. Logan: IEEE J. Quant. El. 15, 541 (1979).
28. H. Kumabe, T. Tanaka, H. Namizaki, S. Takamiya, M. Ishi, W. Susaki: Japan J. Appl. Phys. 18, suppl. 18-1, 371 (1979).
29. M. Hirao, A. Doi, S. Tsuji, M. Nakamura, K. Aiki: Appl. Phys. Lett., to be published.
30. M. Nakamura, K. Aiki, J. Umeda, A. Yariv: Appl. Phys. Lett. 27, 403 (1975).
31. W. T. Tsang, R. A. Logan, L. F. Johnson: Appl. Phys. Lett. 34, 752 (1979).
32. R. L. Hartman, N. E. Schumaker, R. W. Dixon: Appl. Phys. Lett. 31, 756 (1977).
33. A. Thompson: IEEE J. Quan. El. 15, 11 (1979).

34. M. Ettenberg: J. Appl. Phys. 50, 1195 (1979).
35. H. Ishikawa, T. Fujiwara, K. Fujiwara, M. Morimoto, M. Takusagawa: J. Appl. Phys. 50, 2518 (1979).
36. S. Takamiya, N. Namizaki, W. Susaki, K. Shirahata: Digest IEEE/OSA Conf. Laser Eng. and Appl. (1979) p. 51.
37. W. D. Johnston, Jr., G. Y. Epps, R. E. Nahory, M. A. Pollack: Appl. Phys. Lett. 33, 992 (1978).
38. S. Mahajan, W. D. Johnston, Jr., M. A. Pollack, R. E. Nahory: Appl. Phys. Lett. 34, 717 (1979).
39. T. L. Paoli: IEEE J. Quant. El. 13, 351 (1977).
40. R. W. Dixon, W. B. Joyce: IEEE J. Quant. El. 16, 470 (1979).
41. J.A. Copeland: Electronics Lett. 14, 809 (1978).
42. R. L. Hartman, R. A. Logan, L. A. Koszi, W. T. Tsang: J. Appl. Phys. 51, 1909 (1980).
43. T. P. Lee, R. H. R. Roldan: IEEE J. Quant. El. 6, 339 (1970).
44. P. T. Ho, L. A. Glasser, E. P. Ippen, H. A. Haus: Appl. Phys. Lett. 33, 241 (1978).
45. E. P. Ippen, D. J. Eilenberger, R. W. Dixon: to be published (1979).
46. H. Melchior: Physics Today 30, 32 (1977).
47. H. Melchior, A. R. Hartman, D. P. Schinke, T. E. Seidel: Bell System. Techn. J. 57, 1791 (1978).
48. S. Hata, K. Kajiyama, Y. Mizushima: Electron. Lett. 13, 668 (1977).
49. D. R. Smith, R. C. Hopper, I. Garrett: Opt. Quantum Electron 10, 292 (1978).
50. H. D. Law, K. Nakano, L. R. Tomasetta: IEEE J. Quant. El. 15, 549 (1979).
51. M. A. Washington, R. E. Nahory, M. A. Pollack, E. D. Beebe: Appl. Phys. Lett. 33, 854 (1978).
52. C. A. Burrus, A. G. Dentai, T. P. Lee: Electronics Lett. 15, 655 (1979).
53. R. F. Leheny, R. E. Nahory, M. A. Pollack: Electronics Lett. 15, 713 (1979).
54. T. P. Pearsall, R. E. Nahory, J. R. Chelikowsky: Phys. Rev. Lett. 39, 295 (1977).
55. J. C. Campbell, T. P. Lee, A. G. Dentai, C. A. Burrus: Appl. Phys. Lett. 34, 401 (1979).
56. K. Doi, S. Nonaka, T. Yunki, M. Takahashi: NEC Res. and Dev. 50, 17 (1978).
57. K. Kobayashi, R. Ishikawa, K. Minemura, S. Sugimoto: Fiber and Integrated Optics 2, 1 (1979).
58. W. J. Tomlinson, R. E. Wagner, A. R. Strnad, F. A. Dunn: Electronics Lett. 15, 192 (1979).
59. W. J. Tomlinson, G. D. Aumiller: Appl. Phys. Lett. 31, 169 (1977).
60. P. K. Tien: Rev. Mod. Phys. 49, 361 (1977).
61. H. Kogelnik: Fibers and Integrated Optics 1, 227 (1978).
62. H. Kogelnik, R. V. Schmidt: IEEE J. Quant. El. 12, 396 (1976).
63. R. V. Schmidt, P. S. Cross: Optics Lett. 2, 45 (1978).
64. R. C. Alferness, R. V. Schmidt: Appl. Phys. Lett. 33, 161 (1978).
65. R. V. Schmidt, L. L. Buhl: Electronics Lett. 12, 575 (1976).
66. K. Aiki, M. Nakamura, J. Umeda: Appl. Phys. Lett. 29, 506 (1976).
67. L. G. Cohen, C. Lin, W. G. French: Electronics Lett. 15, 334 (1979).
68. W. J. Tomlinson: Applied Optics 19, 1127 (1980).

*The material contained in this article was originally published in The Institute of Physics Conference Series Number 53, Solid State Devices, 1979 and is reprinted here with kind permission of the publishers.

Fiber Optics in Brazil

R. Srivastava

Fiber Optics Project, Institute of Physics, UNICAMP
Campinas, SP, Brazil

With the invention of the Laser in 1960, communications engineers became excited at the possibility of using it as a source of carrier waves with the prospect of enormously large bandwidths. Unfortunately, no suitable transmission path was available until a detailed study by Kao & Hockham in 1966 suggested that cladded glass fibers might be used to guide laser light for telecommunications purposes. Commercially available fibers had losses of the order of 1000 dB/km and an improvement of two orders of magnitude was called for to make an economically viable system. A considerable breakthrough occured in 1970 when Corning Glass Works reported fiber losses of 20 dB/km. This result produced a very strong impetus to the field and soon fiber optics became a booming field with research and development activities being started around the globe. Thus although in 1970, the potentialities of optical fiber were seen to be very attractive, no suitable sources were available. The semiconductor laser with its very small size, high efficiency, and capable of being directly modulated at high speed was the obvious choice, but no laboratory as yet had produced a suitable laser for use in optical communications. It was obvious that Brazil could not sit aside and watch the developments. The country could not afford the luxury of importing this new technology when it was available. It was, therefore, deemed necessary to implant a research program at a moderate level to explore the field here in Brazil. So, in 1973, a research program was begun at the Institute of Physics in UNICAMP by a group of scientists to develop a laser source for optical communications. The work had financial support of TELEBRAS - an undertaking of Communications Ministry of Brazil.

In the area of optical fibers, the story was, however, little different. Very optimistic reports were pouring in from many laboratories around the world about the viability of the usage of fiber optical communications within the decade of 70's. So in 1975, we extended the laser program to cover the area of fiber optics. The principal objectives of the project were as follows:
1.   Develop the science and technology necessary for production of optical fibers for communication systems.
2.   Train personnel at technical and scientific level.
3.   Design and construction of basic equipment for production and characterization of optical fibers.
4.   Theoretical and experimental research on fibers to understand physical phenomena involved.

In the area of development, we had short-range as well as long-range goals. The short-range objectives were:
1. Confine efforts to develop step-index multimode fiber using chemical vapor deposition (C.V.D.) process: Ge-doped silica core.
2. Pull the fiber using an industrial CO_2 laser available in the laboratory or with oxy-hydrogen flames.
3. Characterize for:
(a) Attenuation in 500 - 1100 nm range
(b) Numerical Aperture (N.A.)
(c) Geometry - circularity and coeccentricity
(d) Refractive Index Profile
(e) Modal and Material Dispersion
(f) Impurities
(g) Mechanical Strength.
4. Transfer the know-how to TELEBRÁS where it should be perfected and passed to industry.

By the end of 1977, we had achieved many of these short-term objectives and with the creation of the Research and Development Center of TELEBRÁS in Campinas, we started to transfer the development part of the Project to TELEBRÁS. Major part of the laboratories of fiber optics were transferred from the University; along with them went some trained personnel and research staff.

As of today, the fiber optics project has two segments. The one at TELEBRÁS has the objective of carrying out the task set forth earlier in terms of development of fiber for communications whereas we, at the university, are involved in a series of theoretical and experimental research activities. The long-range goals of TELEBRÁS include development of graded index and single-mode fibers. Lately has been necessary to develop large N.A. and large-diameter silica-silicone fibers due to the fact that the hydroelectric plant of Itaipú needs almost two thousand km of this kind of fiber for installation within a year or so. So the group started this program right away and has succeeded in developing this product. This technology is now being transferred from TELEBRÁS to a company in Rio de Janeiro for manufacture and supply of this kind of fiber to Itaipú.

The group of researchers working on optical fibers in the University of Campinas include a total of six physicists and eight graduate students. Most of the laboratories are located in the Institute of Physics in Quantum Electronics Department. This Department was founded by late Prof. Sergio Porto in 1974. It must be mentioned here that Prof. Porto played a key role not only in motivating the young scientists to work in fiber optics but he was helpful in convincing the government and industrial people to finance this Project. Our research activities are in four main areas: theoretical work, propagation work, non-linear optics work and some other related work.

In the following, we summarize the activities of the group.

1. Theoretical work
1.1. Work Completed
(a) Light propagation properties of optical fibers.

(1) Profile design of multimode fibers with and without dispersion.
(2) Dispersion studies of single mode fibers.
(3) Diffusion studies at core-cladding interface.

(b) Fiber fabrication (models)
(1) Automatic control of fiber diameter.
(2) Surface tension effects on coating smoothness.

1.2. Work in Progress
(a) Frequency modulation of signals in optical fibers.
(b) A computer model for degenerate four-wave mixing in a saturable absorbing medium.

2. Propagation in Optical Fibers
2.1. Work Completed
(a) Measurement of mode conversion coefficients
(b) Measurement of modal and material dispersion effects in multimode fibers. (Effect on launching conditions, refractive index
profile, etc.)
(c) Temperature-induced mode cutoffs and mode interference in
liquid-core fibers.

2.2. Work in Progress
(a) Construction of a nanosecond test facility with variable wavelength pulse radiator.
(b) Launching efficiency into optical fibers and methods of increasing coupling efficiency.
(c) Optical Gibb's phenomenon and coherence effects in optical
fibers.

3. Non-Linear Optics
3.1. Work in Progress
(a) Stimulating light scattering in optical fibers.
(b) Construction of a fiber-Raman Laser.
(c) Phase conjugation by degenerate four-wave mixing in fibers.

4. Other Work
(a) Effect of longitudinal tension and temperature on spontaneous
Raman scattering in silica fibers.
(b) Use of holography in determination of fiber diameter.
(c) Detemination of attenuation as a function of tension; localization of defects.

In conclusion, we have tried to communicate the work we have
done, we are doing at present and our plans for the future in the
area of fiber optics. Our success in achieving the goals which we
had set forth reflects the fact that in a developing country, a
university can play a vital role in supporting the technological
needs of industry.

Part V

Laser Biology and Medicine

Laser-Degeneration Study
of Nerve Fibers in the Optic Nerve

N. Carri, H. Campaña and A. Suburo

Instituto Multidisciplinario de Biología Celular (IMBICE),
Casilla de Correo 403, La Plata, Argentina, and

R. Duchowicz, M. Gallardo, and M. Garavaglia

Centro de Investigaciones Opticas (CONICET - UNLP - CIC),
Casilla de Correo 124, La Plata, Argentina

1. Introduction

Knowledge about wiring of neurons is one of the most important goals of
neurobiology. Neuronal processes -axons and dendrites- degenerate when they
are severed from their cell body. Since different staining procedures
distinguish between the degenerating axons and their healthy neighbors,
most neuroanatomical pathways have been mapped through the follow-up of
degenerating axons after spontaneous or experimental lesions at some point
of the pathway. Mapping of neuroanatomical connections has been enormously
enriched during the past few years, thanks to new labelling techniques with
great resolution power [1]. However, the resolution of the older degeneration procedures is only limited by the extent of the lesion and the resolution of the differential staining of degenerating axons. As we will show in
this report, the use of a laser to produce small lesions in the retina of
birds, coupled to the detection of degenerating axons in semi-thin plastic
sections [2] is allowing us to understand the relationship between axons
along the optic pathway with a resolution comparable to that of "in vivo"
labelling techniques.

Wiring between the retina and the brain is established during early
embryonic life, when a population of neurons -the ganglion cells- develops

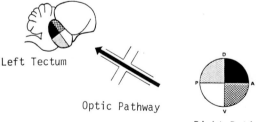

Left Tectum

Optic Pathway

Right Retina

Figure 1. Each point of the retina project to a corresponding point of the
optic tectum. It is not known whether the optic axons maintain a
retinotopic order along the optic pathways or whether they are organized
according to other rules.

in the retina. Each ganglion cell has an axon which grows to the dorsal wall of the mesencephalon -the optic tectum- where it finds an appropriate target neuron. These connections are not formed at random, since each point of the retina is matched to a precise point of the optic tectum (Figure 1). However, we still do not know those factors controlling the organization of these connections. One of the possible explanations is that axons leave the retina in an ordered fashion and that they keep the same neighbors along the optic pathway [3,4]. If this were the case, one should find an ordered distribution of optic axons along the visual pathways. Accordingly, after the lesion of one (or a portion) of the quadrants of the retina shown in Figure 1, one should find a similar distribution of degenerating axons in the optic nerve. Our studies of the optic nerve after laser lesions of the retina indicated that this is nòt so in the visual pathways of quails.

2. Material and Methods

Four week-old quails (Coturnix coturnix) were anesthesized by an intra-peritoneal injection of chloral hydrate (25 mg/100 g. body weight) and procaine (30 mg/100 g. body weight). The right eye was opened with a lid retractor after local anesthesia of the cornea with tetracaine. The beam of a Spectra Physics Ion Laser, model 165, operated at 514,5 nm, was focused to a 100 μm point on the surface of the cornea by a lens system. Various light energies and exposure times were used, and conditions for each experiment are detailed under the corresponding figures. Quails were killed by decapitation one week after irradiation. The position and size of the lesion was determined in flat-mounted retinas which were stained with cresyl violet. Optic nerves were fixed in 2.5% glutaraldehyde in 0.09 M cacodylate buffer with 0.12 M sucrose for 24 hours. After dehydration, nerve slices were embedded in epoxy resins. Consecutive 1.5 μm sections were serially mounted and stained with 1% p-phenylendiamine [2] for light microscopic studies. Ultrathin sections for electron microscopy were stained with lead salts.

3. Results and Conclusions

A week after irradiation, lesions appeared as holes perforating all layers of the retina. Figures 2 and 3 show one of these lesions placed to the nasal side of the pecten, which is the vascular structure lying over the nerve papilla. The retina of birds has no intrinsic vessels and retinal damage was restricted to the beam absorption area. Also affected were those ganglion cells from more peripheral regions whose axons passed through the damaged area in their route to the nerve papilla. Since most axons take a more or less straight pathway to the nerve papilla, a circular sector of the retina was disconnected from brain. The size of this sector depended on exposure times, laser energy output and the distance between the absorption area and the nerve papilla. Obviously, the more peripheric lesions disconnected fewer ganglion cells than those placed on a central position closer to the retina. Thus, laser irradiation made it possible to eliminate as many ganglion cells as it was desired without any alteration of ocular geometry such as would occur after conventional surgical procedures. By the same token, hemorrhagic and infectious complications were minimal.

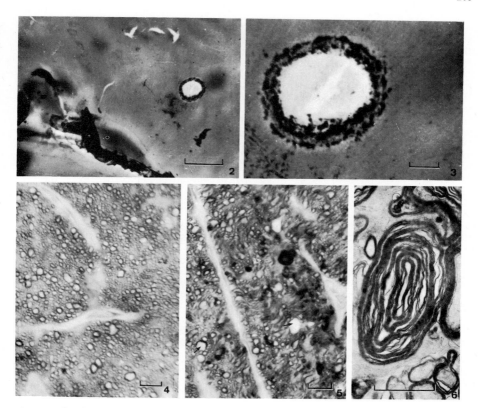

Figures 2.-6.

Figure 2 and 3. A lesion in the ventronasal region of the retina made by an irradiation of 0.5 watts for one second. The lesion appeared as a hole surrounded by reactive pigmented cells. The pecten is the dark structure at the botton of Figure 2. The striated pattern of the retinal surface reflects the pathway of optic axons which leave the eye through the nerve papilla. The latter lies beneath the pecten. Calibration bars 1 mm and 0.2 mm.

Figure 4. Light micrograph of a normal nerve showing myelinated axons of different diameters. Calibration bar 10 μm

Figure 5. A nerve showing degenerating fibers. The arrows indicate axons with degeneration of the clear type. Dark spots correspond to dark degenerating axons. Calibration bar 10 μm.

Figure 6. An electron micrograph showing the disappearance of axoplasm and the disruption of myelin sheath in a degenerating axon. Calibration bar 1 μm.

The optic nerve had an elliptical shape and was surrounded by a thick connective sheath. Connective septa were also found within the nerve but they were placed at random. The nerve contained both myelinated and unmyelinated axons of different sizes (Figure 4). The presence of degenerating fibers could be recognized even in unstained preparations, since many of these dying axons had larger profiles than their normal counterparts. These clear degenerating axons were interspersed with dark degenerating axons -i.e.: those binding more osmium tretoxide- which were clearly recognized by their deep brown colour in p-phenylendiamine stained preparations (Figure 5). Preliminar studies showed that one week survival time gave an optimal picture of degeneration, since axonal fragmentation was almost absent before that period. These findings were confirmed by transmission electron microscopy which showed that both the enlarged profiles and the dark spots of light microscopy belonged to degenerating axons (Figure 6).

Sections of optic nerves were examined with a 1,000 X magnification and the number of dark degenerating fibers per unit area was determined at points separated by fixed intervals of the microscope stage. In this fashion, density maps like those shown in Figure 7 were obtained. The distribution of degenerating fibers varied according to the localization of the lesion on the retinal surface. In some cases (Figure 7 A), the degenerating fibers remained as a group. However, they did not resemble the sectorial shape of the retinal lesion but appeared as stripes across the anteroposterior aspects of the nerve, suggesting that some change in the distribution of the axons had accurred along the pathway. These strip distributions were associated with lesions in the ventronasal region of the retina. When lesions were made in a more dorsal portion of the retina,

```
. . . . . 0 . 0 0 0 0 0 . .        . . . 1 0 0 0 0 0 0 0 . . .
. . . . . 0 0 0 0 0 0 0 0 0        . . 1 0 0 0 0 1 0 1 0 2 . . .
. . 0 0 0 0 0 0 0 2 2 2 3 2        1 2 3 1 1 2 2 2 2 2 0 0 . . .
. . . 0 0 0 0 0 2 4 3 3 4 2        1 2 2 2 2 2 2 2 3 2 3 2 . . .
0 0 0 0 0 0 0 2 4 1 2 2 3 4        0 0 1 0 1 0 1 1 0 1 1 0 0 . .
0 0 0 0 1 2 2 3 2 0 0 2 4 2        1 1 0 0 0 0 0 1 0 1 1 1 0 1 .
1 0 1 1 2 2 2 2 1 0 0 0 1 0        1 1 1 1 0 0 0 0 0 1 0 1 1 1 .
2 3 2 3 2 1 0 0 0 0 0 0 0 0        1 1 1 1 1 1 1 1 0 0 0 1 1 1 0
1 1 2 0 0 0 0 0 0 0 0 0 0 .        1 1 2 1 2 1 2 2 1 0 0 1 1 1 1
0 0 0 0 0 0 0 0 0 0 0 0 0 .        . . 2 1 1 1 1 2 0 1 0 1 1 0 0
1 0 0 0 0 0 0 0 0 0 0 0 0 .        . . . 1 1 1 0 0 1 1 1 1 1 0 0
. 0 0 0 0 0 0 0 0 0 0 0 0 .        . . . . . . 0 . 1 1 2 1 1 0 0
. 0 0 0 0 0 0 0 0 0 0 0 . .        . . . . . . . . . 1 1 1 0 1 .
. . . . 0 0 0 . . 0 0 . . .        . . . . . . . . . . 0 . . . .
              A                                   B
```

Figure 7. Densities of degenerating fibers one week after irradiation with 1.8 watts for 10 seconds. The numbers indicate the amount of dark degenerating terminals on an arbitrarily defined area (0 = background; 1 = 9-20; 2 = 21-32; 3 = 33-44; 4 = 45 or more). (A) Strip patterns found after lesions in the ventronasal region; (B) diffuse patterns found after lesions in the dorsal region of the retina.

degenerating fibers were spread abroad most of the cross section of the optic nerve (Figure 7 B), even though the lesions never affected more than 1/5 of the ganglion cell population. The existence of these diffuse patterns was a clear evidence that at least some of the axons lost contact with their neighbors.

Our observations indicated that there is not a strict retinotopic order in the optic nerve of quails. A lack of such organization has also been observed in cats [5] and it has recently been concluded that the order found in the optic nerve of goldfish is of a chronological nature [6]. Further studies are neccessary to ascertain the biological significance of strip and diffuse degenerating patterns in the optic nerve of quails. However, it can be speculated that they represent two different mechanisms. The guidance of at least part of the fibers would not depend on the maintenance of neighbourhood relationship during the migration of axons.

On the other hand, it should be emphasized that the resolution of this procedure can be further increased, since it is possible to make smaller lesions than the ones reported here. Particularly promising is the study of very small lesions which would not affect axons passing across the absorption area of the laser beam.

4. Acknowledgements

We are grateful to Mr. Robert Kyburz, Cabaña Las Codornices, Escobar, who generously provided the quails used in this study, and to Mrs. Dora de Roth, for her excellent technical assistance.

5. References

1. LaVail, J.H. and LaVail, M.M.: Retrograde axonal transport in the central nervous system. Science, 176, 1416-1417, 1972

2. Holländer, H. and Vaaland, J.: A reliable staining method for semi-thin sections on experimental neuroanatomy. Brain Res., 10, 120-126, 1968.

3. Gaze, R.M.: The problem of specificity in the formation of nerve connections. In: Garrod, D.R. (ed.): Specificity of Embryological Interactions, pp. 51-93, London, Chapman and Hall, 1978.

4. Horder, T.J. and Martin, K.A.C.: Morphogenetics as an alternative to chemospecificity in the formation of nerve connections. In: Curtis, A.S.G. (ed.): Soc. for Exp. Biol. Symp., vol. 32, pp. 275-356, Cambridge University Press, 1978.

5. Horton, J.C.; Greenwood, M. and Hubel, D.H.: Non-retinotopic arrangement of fibers in cat optic nerve. Nature, 282, 720-722, 1979.

6. Russof, A.C. and Easter, S.S., Jr.: Order in the optic nerve of goldfish. Science, 208, 311-312, 1980.

The Argon Laser in the Treatment of Glaucoma

J.A. Holanda de Freitas and J. Quirici

Instituto Penido Burnier, Campinas, and

D.G. Bozinis, A.F.S. Penna, and E. Gallego-Lluesma

Instituto de Física, UNICAMP, 13100 Campinas, Brazil

1. Introduction

The present day ophthalmology is assisting a variety of new surgical techniques intended to aliviate the problem of glaucoma. Most of them, however beneficial, are mutilating to the eye globe and subject to numerous complications. At the same time the clinical treatment of the disease has reached a level of success to the extent that has allowed the control of the disease in most cases and with a large margin of safety. However, as research goes on, new techniques are being introduced and new instruments become available. A Xenon arc lamp was first used by MEYER-SCHWICKERATH [1] in 1956 in the attempt to produce successful iridectomies. ZWENG et al. [2] coupled a Ruby Laser into a direct ophthalmoscope to successfully produce iridectomy in a patient who suffered from pupilar block. KRASNOV [3] used a Q-switched Ruby Laser coupled into a slit lamp and treated 10 patients who suffered from simple chronic glaucoma and which could not be brought under control through clinical treatment. On that occasion the pulse duration was one nanosecond, spot size 250 to 400 microns and pulse energy of 0.2 joules. In all cases he observed reduction of the intra-ocular pressure and improvement of the coefficient of flow of the acqueous humor. More recently, TICHO and ZAUBERMAN [4] , used the Argon Laser in the treatment of 20 eyes suffering from simple chronic wide angle glaucoma and whose pressure was above 40 mm Hg. The anesthesia which they used was 2% xylocaine applied locally for adults, while children were submitted to general anesthesia with Ketalar. In their method, they made a line of 50 point of 50 microns each in the trabecular mesh, while exposing the eye to the laser radiation during 0.2 second pulses, whose intensity varied from 1 to 3 watts of power. After several sessions 16 myotic eyes were brought under clinical control, while in 4 eyes the pressure was reduced to normal. Two of the latter eyes were suffering from congenital glaucoma. POLIACK and PATZ [5] made similar attempts in rabbits and obtained evidence of the presence of iridectomy by histological research. Subsequently they treated 33 eyes suffering from closure angle glaucoma and produced basal iridectomy. In 3 patients, who were previously facetomized and whose iridectomies were incomplete, they succeded in widening them with the Argon Laser. WITSCHEL and RASSOW [6] produced trabeculopuncture in cadaver eyes. The 50 micron

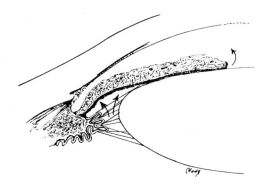

Fig. 1 The circulation
of the aqueous humor is
being hindered by the
close angle.

holes in the Schlemm canal were seen by an electron microscope.
SCHWART et al. [7] used the Argon Laser in the treatment of
closure angle glaucoma of 64 patients. Reportedly, they were
successful in producing iridectomy in 75% of the cases. In
those cases, where pupilar obstruction was present, they
reported 100% successful iridectomy. Subsequently, WAND et
al. [8] studied the effect of panphotocoagulation in 119 patients
suffering from advanced diabetic retinopathy using a Ruby laser.
They were able to conclude that their method was efficient in
controlling the neovascular glaucoma and, at the same time, it
could block the evolution of Rubeosis Iridis. At the same time,
FREITAS et al. [9] studied 18 patients suffering from occlusion
of the central vein of the retina and submitted them to
panphotocoagulation using an Argon Laser. Reportedly, this
method is efficient in treating the retinal bleeding, the
macular edema and the neovascularization of the retina as well
as the anterior segment. They recommend their method in the
prevention of the secondary neovascular glaucoma and the
occlusion of the central vein of the retina. In those cases
where rubeosis was present, they reported reduction of the
neovascularization. More recently, STIEGLER [10] made 176
trabeculectomies using an Argon Laser. Reportedly, in 81%
of the wide angle glaucoma cases the intraoccular pressure
came under control, while in 47% of the closure angle glaucoma
cases the pressure was reduced to normal without medication.
Finally, YASSUR et al. [11] used higher laser powers but shorter
exposure times — 4 Watts and 40 milliseconds. Out of their
53 reported cases 34 were suffering from closure angle glaucoma.
They applied the Argon Laser radiation over the trabeculum and
produced iridectomy. They believe that this method is highly
recommended for the prevention of the closure angle glaucoma.

The purpose of this work is to show the effectiveness of
the use of the Argon laser in the treatment of the glaucoma of
different forms and etiology.

2. Methods

We used a commercial Argon Laser coupled into a slit lamp. The patients who were treated with Argon laser were those who did not respond to clinical treatment. In all cases, the patients were subjected to the routine pre surgical exams such as visual acuity, refraction, ophthalmoscopy, biomicroscopy, tonometry and visual field. Those who were carriers of occlusion of the central vein of the retina, were given special care with emphasis in obtaining more detailed mappings of the retina by the use of color retinography as well as angioretinofluoresceinography. At the same time, in those patients with primary glaucoma, the gonioscopic findings were analyzed in detail. In all cases it was used local anesthesia. The presurgical medication was composed mainly of systemic anti-inflamatory, midriatic and local esteroid drugs. The latter drugs were precribed to patients who were carriers of secondary glaucoma, i.e., uveitis and aphakia. In those cases where the problem was occlusion of the vein we administered plaquetary antiadhesives, fibrinolytic as well as inhibitors of carbonic anidrase. As for the carriers of the simple chronic glaucoma or of closure angle, during the first week were prescribed anti-inflamatory and acetazolamide drugs. The intraocular pressure was controlled every 2 weeks during the 1st month and every 2 months during subsequent six months. After this period the patients were discharged and were considered cured.

3. Results

55 patients were treated and a total of 72 eyes were seen and distributed as follows:

Primary Glaucoma	48 eyes
Secondary Glaucoma or occlusion of central vein	11 eyes
Secondary Glaucoma, uveitis	7 eyes
Secondary Glaucoma, aphakia	6 eyes

Problem eye

Right	20 eyes
Left	16 eyes
Both eyes	36 eyes (18 patients)

Sex

Male	27 patients
Female	28 patients

Visual Acuity	*Before*	*After*
Less than 0.1	33 eyes	35 eyes
0.1 - 0.2	6	4
0.3 - 0.4	5	5
0.5 - 0.6	9	5
0.7 - 0.8	3	4
0.9 - 1.0	10	9

Type of Argon laser surgery

Iridectomy	13 times
Panphotocoagulation	11
Laser puncture	48

Final effect over the intra ocular pressure

Controlled	37 eyes
Controlled with drugs	10
Controlled with surgery	7
Not Controlled	18

4. Comments

In average were applied 1000 shots, which were divided in 3 sessions with a 15 day break. In six eyes the intra occular pressure was reduced to normal, whereas in the remaining six, this was not possible because of the appearance of advanced rubeosis iridis with loss in the transparency of the ocular media. In the uveitis secondary glaucoma we used the technique of iridectomy, with the intent to break up the pupilar block due to the pupilar seclusion. In these cases we used:

Laser Power	1 to 2 Watts
Exposure time	0,2 seconds
Spot Size	500 microns.

Out of the 7 eyes which were treated, the pressure was controlled in 5 of them, while in 2 no result was obtained.

In the aphakic glaucoma of 3 eyes, the purpose was to widen the iridectomy which was incomplete and was responsible of pupilar block and increase of the intraocular pressure. In this case the pressure came under control in all of them. The 3 remaining eyes were carriers of epithelial cyst of the anterior chamber, and as a result of the laser surgery two cysts were destroyed while the third one had to be removed with conventional surgery.

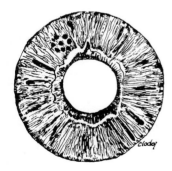

Fig. 2 Initial application of the Argon laser. Subsequently these will become iridectomies.

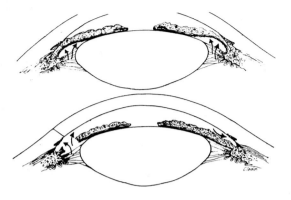

Fig. 3 The iridectomy made with an Argon laser beam allows the free flow of the aquous humor. Lower left arrows indicate the flow of the liquid after iridectomy.

Most of our cases were those of primary glaucoma. We attempted to simplify the surgical technique by the application of the Argon laser over the trabecular mesh. This technique would eventually increase the permeability of the aquous humor and, as a result, it would aliviate the intraocular pressure. It is in this area that specialists in this field have seen the emergence of new techniques and the appearance of sophisticated instruments. It is the purpose of this work to give proper credit to all this effort while offering somo alternatives, aiming towards the reduction of post-operative complications.

In the case of *closure angle glaucoma* one can produce successful iridectomies as long as the Argon laser is used within the following ranges:

Laser Power   1.5 to 2   Watts
Spot Size   200 to 500 microns
Exposure time   0.2 seconds.

During the application of the Argon laser, it is a gratifying feeling to observe the increase of the depth of the anterior chamber, avoiding thus the acute glaucoma while relieving the high intraocular pressure.

In dealing with the *wide angle glaucoma* and with the help of the contact lens, it is possible to direct the laser beam directly over the trabecular mesh and make true holes, not only through the trabeculum but also through the Schlemm's canal. The proper parameters for the Argon laser are as follows:

Power   1.5 to 2 Watts
Spot Size   50 microns
Exposure time   0.2 seconds.

As a general rule, the application of the laser starts

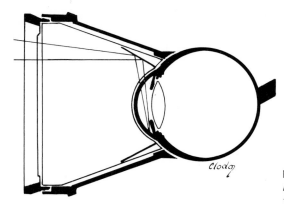

Fig. 4 Ray path of the Argon laser beam directed towards the trabecular mesh

from the lower angle bringing the total of shots to between 60 to 80. Then the patient is kept with anti-inflamatory and acetozalamide drugs during 15 to 20 days. If the pressure continues to be high, a new application of the laser is made over the superior angle with similar parameters and number of shots. The patient is considered cured if, at the end of 6 months, the intraocular pressure is brought under control.

Our statistics show that out of 48 treated eyes, 34 achieved normal pressure. In 24 of these eyes the application of the laser was just enough, while the remaining 10 needed complementary medication. In 8 eyes it was not possible to bring the pressure down to within normal levels, and as a result it was necessary to subject the patients to trabeculectomy following the Cairns technique with complete success. In the remaining six cases it was not possible to reduce the pressure by the sole application of the laser and drugs or because the cases were so bad that some patients rejected the possibility of surgery and the Argon laser treatment was referred to them as a last resource.

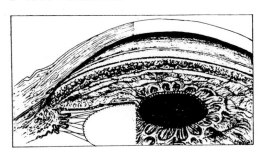

Fig. 5 Laser puncture produced over the trabecular mesh.

During the follow-up period we did not observe any complications. During the first 2 days after the application, a light

Tyndall of the aqueous was observed, in view of the pigmentary dispersion of the iris due to the local heating produced by the laser. Because of the very oblique incidence of the laser beam with regards to the cornea, in 2 cases, right after the application, we noted some opaque points in the corneal stroma which disappeared completely 2 or 3 days after.

Regarding the cases of secondary glaucoma due to vascular occlusions, uveitis and pupilar block, where little progress could be achieved otherwise, the laser proved to be adequate solution.

In the primary glaucoma, when the Argon laser did not produce the desired effect, the patient was operated by conventional trabeculectomy whose success or even failure was not conditioned by the past Argon laser application.

Let us now compare the results of the 48 eyes carriers of primary glaucoma, which were not controlled with the administration of drugs, and which were subjected to filtering *trabeculectomy*

Problem eye		*Sex*	
Right	15 eyes	Male	16 patients
left	5	Female	18
Both eyes	14 patients		

Effect on the intraocular pressure	
Controlled without medication	37 eyes
Controlled with surgery ·and medication	7
Not controlled	4

Complications	
Bleeding of the choroid	1 case
Endophalmitis	1
Iridocyclitis	2
Cataract	4.

In general, the filtering surgical procedures are cataractogenic due mainly to the surgical act. This happened in 4 cases during the period of six months. In the meantime, iridocyclitis, of lower occurence, was controlled completely through anti-inflamatory and atropine drugs. As for the endophalmitis, its occurrence is very rare. Unfortunately, we report one such case with total loss of the ocular globe. The circulatory perturbations of the choroid are justified in view of the surgical trauma itself.

The laser treatment offers the advantage of being a non-traumatic surgery, painless with no risk of infection and non-cataractogenic. It is made with local anaesthesia in outpatient clinic, it does not require hospitalization and it is almost exempt of complications similar to those seen above.

When the two methods are compared hand to hand and the reduction of pressure is considered as the main criterion of success, it becomes evident that the classical trabeculectomy is more efficient. The Argon laser reduced the pressure in 24 of the treated eyes (50%) while the trabeculectomy was successful in 37 eyes (77%). But the laser plus medication brought the figures up to 34 eyes (70%) while trabeculectomy plus medication controlled the pressure of 44 eyes (91%).

The number of operated eyes (44) is not significant. In order to establish proper patterns, we followed up cases during one year subsequent to the operations. It was observed that, when problems do appear, they show up right from the beginning of the application and there was no reduction of the intraocular pressure at any time. On the other hand, in those patients where the intraocular pressure was controlled right after the first or the second application it was observed that the pressure never went out of control thereafter. These facts are leading us to believe that *laser trabeculectomy* is a modern surgical technique, efficient, which can be used equally in the treatment of primary as well secondary glaucoma with very encouraging results.

5. Conclusions

1) The Argon laser is very useful in the treatment of the simple chronic glaucoma.
2) In the case of the secondary glaucoma due to the occlusion of the central vein of the retina, the Argon laser radiation is efficient in inhibiting the neovascularization of the retina and the iris, yielding a reduction of the intraocular pressure.
3) In the closure angle glaucoma, in the pupilar seclusion (uveitis) as well as in the cataract surgery where the iridectomy remained incomplete, it is possible to produce satisfactory iridectomies with the use of the Argon laser radiation.

Acknowledgements

This work was partially supported by Financiadora de Estudos e Projetos (FINEP). We wish to thank Clodomiro Rodrigues and Laercio O. Dias for the art work.

References

1. G. Meyer-Schwickerath, Light Coagulation, 1960.
2. Zweng H.C., Flocks M., Kapany N.S., Silberstrust N. and Peppers N.A., Experimental Laser Photocoagulation, Am. J. Ophthalmol. 58, 353 (1964)
3. Krasnov M.M., Laserpuncture of anterior chamber angle in glaucoma. Am. J. Ophthalmol. 75, 674 (1973).

4. Ticho U. and Zauberman H., Argon laser application to the angle structures in the glaucomas. Am.J.Ophthalmol.94 61(1976)

5. Poliack I.P. and Patz A., Argon iridectomy: An experimental and clinical study. Ophthalmic Surgery 7, 22 (1976).

6 Witschel B.M. and Rassow B., Zur Laser-Trabekulopunktur II: Rasterelektronen, Kroskopische Befunde. Ophthalmologica, 172, 45 (1976)

7 Schwart L.W., Rodrigues M.M., Spaeth G.L. Streeten B.L. and Douglas C., Argon laser iridectomy in the treatment of patients with pupillary block glaucoma. Trans. Amer. Acad. Ophthalmol. Otolaryngol. 85, 294 (1978)

8 Wand M., Dueker D.K., Aiello L.M. and Grant W.M., Effects of panretinal photocoagulation on rubeosis iridis angle neovascularization and neovascular glaucoma. Am.J.Ophthalmol. 86, 332 (1978).

9 Freitas J.A.H., Porto S.P.S., Bozinis D.G. and Piccoli P.M. Tratamento do glaucoma neovascular pelo Laser de Argonio, Arq. Bras. de Oftalmol. 41, 271 (1978)

10. Stiegler G., Laser-Trabekulotomie und Laser-Iridektomie. Drei Jahre Erfahrung mit dem Glaukom. Research-Laser (Brizt.) Klin. Mbl. Augenheilk. 175, 333 (1979).

11. Yassyr Y., Melamed S., Cohen S. and Ben-Sira I. Laser Iridectomy in closed angle glaucoma. Arch. Ophthalmol.97,1920(1979).

Preliminary Evaluation of the Use of the CO_2 Laser in Gynecology

J.A. Pinotti

Fac. de Cs. Médicas, UNICAMP, and

D.G. Bozinis and E. Gallego-Lluesma

Instituto de Física, UNICAMP, 13100 Campinas, Brazil

1. Introduction

From its discovery the CO_2 laser has found many applications in areas of research and has substituted with advantage existing techniques. Because of its
long wavelength it has found immediate application in medicine. Its high efficiency has permitted the reduction of size and adequate power levels are
produced at reasonable cost. A CO_2 laser beam coupled into a microscope is
now used with ever increased frequency in microsurgical procedures such as
removal of malignant tumors of the vocal cords, brain surgery, plastic surgery, dermatology, gynecology as well as otology. As the trend continues almost every other area in medicine will be affected, to the benefit of the
patient who will bleed less, recover faster and eventually will pay less for
his treatment.

KAPLAN et al. [1] have treated a series of 11 patients suffering from
erosions of the uterine cervix and reported that the operation was simpler,
it healed more rapidly, it was less subject to infection and the application
was more accurate. DORSEY et al. [2] have shown the advantages of using the
CO_2 laser in the conization of the Cervix. BELLINA et al. [3] have used the
CO_2 laser in the treatment of Capillary hemangioma and discussed the advantages over conventional surgery. In a previous presentation BELLINA [4] reported the treatment of vaginal adenosis and other vulvar lesions using the CO_2
laser radiation. Our group has treated a large number of patients suffering
from those diseases reported previously and has extended the use of the laser
in almost every other gynecological surgical procedure with the purpose of
appraising the extent of its use.

We intend to present an account of the use of the CO_2 laser in gynecology
and show its advantages as well as its disadvantages when compared to conventional surgical procedures.

2. Materials and Methods

We used a Cavitron AO-300 Surgical laser with rated multimode power of 25 W. The laser was coupled into a colposcope and focused with a 400-mm lens. The minimum spot size was around 1.5 mm. For general surgery we used an articulated arm which delivered the laser power through a sterilizable handpiece to the desired area.

3. Results and Discussion

Table 1 shows the different types of surgery performed. In total we report 90 cases ranging from simple removal of warts to radical mastectomies. The average time of the laser assisted surgery is compared against the conventional surgery.

Table 1. Surgical cases performed with the help of the CO_2 laser and average laser time.

Type of surgery	Number of cases	Average surgery time [min]	Average laser time [min]
Excision of breast nodules	14	15	15
Cauterization of the cervix	32	5	5
Biopsy of the breast papilla	1	5	5
Excision of cervical polyp	4	10	10
Excision of terminal ducts of the breast	4	20	20
Debridment of wound edges	2	10	5
Excision of warts	2	20	20
Excision of skin lesions	2	5	5
Excision of retro areolar breast area	1	25	25
Excision of bilateral ectopic axilar tissue	1	90	70
Marsupialization of the Bartholin gland	1	20	10
Tubal anastomosis	1	180	80
Radical Mastectomy with conservation of the pectoral muscles	2	180	90
Simple Mastectomy	2	145	50
Halsted Mastectomy	2	150	75

Table 1. (continued)

Type of surgery	Number of cases	Average surgery time [min]	Average laser time [min]
Total abdominal Hysterectomy	8	120	75
Hysterectomy + Anexectomy	2	130	100
Oophorectomy	1	60	15
Conization of the cervix	1	45	15
Umbilical hernioplasty	1	60	25
Marchal-Marchetti	2	60	20
Radicalization of Mastectomy	2	75	50
Biopsy	2	75	50

It becomes clear that the use of the CO_2 laser does not extend the average surgery time. However, it should be mentioned that in major surgeries such as mastectomies and hysterectomies the elapsed time is somewhat larger when the CO_2 laser is used as the sole surgical instrument. In these cases, the hemostatic effect of the laser (for vessels up to 0.5 mm) dispenses the use of hemostats and as a result the overall time becomes comparable to the nonlaser surgeries.

The few complications that we observed — 7 in total or about 8% of the cases — amounted almost exclusively to difficulty in healing as well as opening of the scar. HALL [5] in an early paper, has reported this fact and attributed it to the necrosis of the margins of the wound caused by excessive heating effect of the CO_2 radiation. Our response to this problem was the application of some extra subcutaneous sutures, relieving thus the tension on the skin, and prolonging a few days the removal of the stitches.

Since the laser scalpel does not touch the patient, the risk of infection is reduced considerably. Our experience points to this fact as one more advantage of the laser over the scalpel.

What has been a clear advantage in all operations was the reduced bleeding as compared to conventional surgery. Table 2 shows our evaluation in a scale ranging from one to four.

This should not be a surprise since the laser has a hemostatic effect in blood vessels of to 0.5-mm size. In major surgeries, such as mastectomies, this has been a clear benefit for the patient because of the reduced blood

Table 2. Comparison of the bleeding of conventional and CO_2 laser assisted surgery, measured in a scale of one (+) to four (++++).

Surgeries	Conventional surgery	Laser surgery
Excision of breast nodules	+++	+
Excision of polyp	++	+
Biopsy with freezing	+++	+
Excision of ducts	+++	+
Tubal anastomosis	++	+
Mastectomy	+++	+
Hysterectomy	++	+
Conization	+++	+
Oophorectomy	++	+
Marchal-Marchetti	++	+
Radicalization of Mastectomy	++++	+

loss. This simplifies also the performance of small surgeries, which can be done now in an outpatient basis cutting down cost and time of hospitalization.

4. Conclusions

The authors have studied 90 surgeries performed with the help of the CO_2 laser. We concluded that over conventional surgeries the use of the laser:

1) Clearly reduces the amount of bleeding in all surgeries.
2) Some smaller surgeries may be performed in an outpatient basis, due to the simplified surgical procedure.
3) There has been a delay in the healing of the wound, attributed to local necrosis. Additional subcutaneous sutures have overcome the difficulty.

References

1. I. Kaplan, J. Goldman, R. Ger: The treatment of erosions of the uterine cervix by means of the CO_2 laser. Obstet. Gynecol. *41*, 5, 795 (1973)
2. J.H. Dorsey, E.S. Diggs: Microsurgical Conization of the Cervix by Carbon Dioxide Laser. Obstet. Gynecol. *54*, 5, 565 (1979)
3. J.H. Bellina, D.R. Gyer, J.V. Voros, J. Raviotta: Capillary Hemangioma Managed by CO_2 Laser. Obstet. Gynecol. *55*, 1, 128 (1980)
4. J.H. Bellina: Gynecology and the laser. Contemporary Obstet. Gynecol., set. 1974
5. R.R. Hall: The healing of tissues incised by a Carbon Dioxide Laser. Brit. J. Surg. *58*, 3, 222 (1971)

Application of Vertical Brackets
in Orthodontic Treatments: A Laser Speckle Study

M. Abbattista

Facultad de Odontología, Universidad Nacional de La Plata (UNLP)
La Plata, Argentina

L. Abbattista*

Facultad de Odontología, UNLP and Hospital Subzonal Infantil "Adolfo
M. Bollini", Ministerio de Salud, Provincia de Buenos Aires
La Plata, Argentina, and

N. Rodríguez**, R. Torroba**, L. Zerbino**, M. Gallardo, and M. Garavaglia

Centro de Investigaciones Opticas (CONICET - UNLP - CIC)
Casilla de Correo 124, La Plata, Argentina

1. Introduction

This paper reports the application of a laser speckle technique to orthodontic studies. The purpose of these studies is to certify the validity of
a proposal related with a new type of apparatus, and the methodology to be
used in orthodontic treatments.

As it is well known, man's evolution shows a reduction in maxilla size.
Such reduction can be observed by comparing the size of maxillas with the
size and distribution of teeth in some members of the filum that, hypo-
thetically, was initiated in the final period of the Third Era and continued up to the man of present times.

In Figure 1 a) it is compared the maxillas of chimpanzees, gorillas,
orangoutans, and men. As one can see, the number of teeth is the same in
the four cases, but their sizes and distributions are different, depending
upon food regimen, and in the case of men, it is very important to add,
upon its cultural behaviour. Besides, Figure 1 b) emphazises the comparison
of teeth profiles between gorillas and men. Gorillas and other primates
have prominent canines, and their diastemas are fit for providing the best
occlusion for mastication. Such prominent canines play an important role
for their strong, hard, and dry mastication. In the fossil rests of all the
primitive men we know, as those of the "Pithecanthropus Erectus", the size
and position of canines are almost the same as in maxillas of present times,
as shown in Figure 1 c). However, their teeth were stronger than ours, to
provide for a mastication like that of primates. It produced a permanent
horizontal and vertical migration of teeth, so preventing caries and
periodontal illness and, at the same time, the muscular excercises contributed to the growth of maxillas.

* Fellow of the Comisión de Investigaciones Científicas de la Provincia
  de Buenos Aires (CIC), Argentina.

** Fellow of the Consejo Nacional de Investigaciones Científicas y Técnicas
   (CONICET), Argentina.

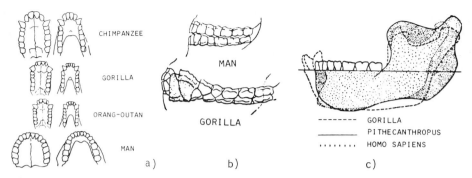

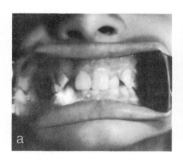

Fig.1. Comparison of maxillas and distribution and profiles of teeth in man and primates

The soft alimentation of men, which appeared with cooking and, afterwards, with the use of fork and knife-that is, with human civilization — produced a reduction of those muscular movements, starting an involutive process of maxillas. This process was also accelerated by the diminution of the suckling period of babies. These are the main arguments to justify that at least a 40% of patients between 10 and 15 years old suffer from markedly discrepancies between the distribution of clustered teeth and the size of maxillas.

2. Orthodontic apparatus and techniques. Proposed modifications: Vertical brackets versus horizontal brackets.

Figure 2 shows a clear example of the situation represented by the patient case N° 88. Its part a) is a living and impressive picture of the mouth of a girl 12 years old, while part b) is a gypsum model made at the initial state of the treatment, showing malocclusion, and part c) is a view of the occlusal plane of both maxillas.

Fig.2. Initial state of the treatment of patient N° 88

This kind of malocclusions can be orthodontically treated by using several clinical techniques which, in general, follow the basic principles stated by Edward Angle in 1928 [2]. The pioneering technique developed by

Angle is based on the bracket edgewise appliance, properly cemented to
teeth, which serve for applying forces or couples to them. They are applied
by using stainless steel arch wires or rubber bands, as it is shown in
Figure 3 b). All techniques developed until now employed horizontal
brackets cemented to teeth, and as a result of the application of forces in
a point of teeth crowns, their roots pivot on gingival and apical fulcrums,
producing rotations of the entire dental piece around an axis located
somewhere between, as shows Figure 3, a). Then, the employment of horizontal brackets produces a combination of rotations and translations of teeth.

Using horizontal brackets, the position of the crowns at the end of
treatment can be good from the aesthetic point of view, but the radiological study shows anomalies in root orientation, as observed in Figure 3 c).
This is the origin of serious damage of the masticatory function and of
periodontal illness.

To solve these questions, a new type of brackets was introduced and various modifications in well-known apparatus made, as in the Mershon arch. The
most important innovation is the development of the technique that uses
vertical brackets instead of the horizontal ones. This technique is called
distal-corrector action technique employing combined gentle forces [2].

By using vertical brackets it is possible to charge the crown with a
system of forces whose resultant is applied in a appropriate point, as it
is shown in Figure 3 d), to produce the displacement of teeth parallel to
themselves.

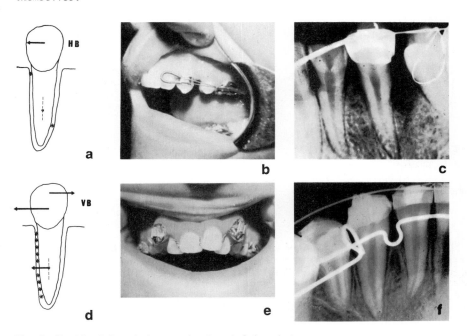

Fig.3. Vertical brackets vs. horizontal brackets

Figure 3 e) shows a clinical case with vertical brackets applied in the upper canines. This kind of vertical brackets are recommended to be applied in those clinical cases where dental pieces must be moved parallel to themselves, to their right places, or to produce a careful treatment finishing, with a good occlusion.

As we said before, vertical brackets allow orthodontists to apply to the crown a system of forces instead of a force in a point. Furthermore, it is possible to give to vertical brackets the proper angle of orientation, and to associate another type of forces as shown in Figure 3 e). In this case, the vertical bracket applied to the upper canines is 10° from its axis, to compensate the influence of the second incisor, and in addition, an extrusion force will be applied on it, to pull it down to the occlusal plane.

3. Laser speckle measurements.

A Type-O-Dont simulator was employed for modelling the physical situation of the mouth to prove that the action of vertical brackets produces movements of teeth that are perpendicular to the axis of their roots. In the same laboratory simulator was also modelled the action of horizontal brackets. Then, the results obtained serve to compare both techniques. Figure 4 a) shows details of the simulator. Observations were devoted to both canines and to the second premolar. The Mershon arch was modified, and its end was turned out around the mesial face of the second premolar, to prevent its mesioversion, as it is shown in Figure 4 b). The first premolar was extracted.

Fig.4. Type-O-Dont laboratory simulator for modelling the action of vertical brackets

Laser speckle interferometry was employed for measuring the movements in artificial teeth. Two simultaneous perpendicular planes of observation were illuminated with an ion argon laser at 514.5 nm wavelength and 1 watt power. One of them was the occlusal plane, and the other was a tangential plane to the dental arch in the canine position. As it is known, specklegraphy can be used for measuring the magnitude and angular orientation of small movements, but not the sign of the displacement vector. In the experiments we performed, signs were defined according to physical considerations. In order to obtain better resolution, holographic film was used for recording double exposure specklegrams. The first pair of perpendicular exposures was photographed with the dental system unloaded. After that, forces were applied to the model, following the orthodontic fashion, and the second pair of perpendicular exposures was then photographed. In order to apply the appropriate set of forces with well defined angular orientation and intensity, an adequate stainless steel arch wire was developed, which is shown in Figure 5.

Magnitude and angular orientation of dental displacements were electronically determined by illuminating the specklegrams with 2 mW power 632.8 nm wavelength He-Ne laser.

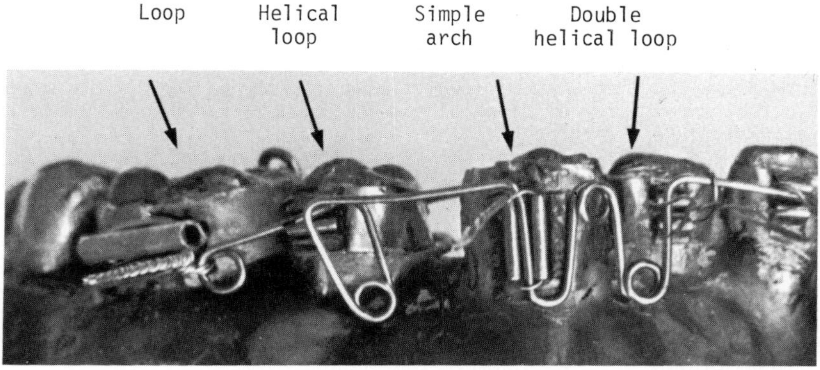

Loop
Helical loop
Simple arch
Double helical loop

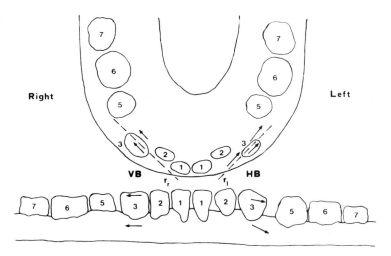

Fig.5. Arch wire employed with vertical brackets

A summary of the results with the movements of canines produced by the action of vertical and horizontal brackets is shown in Figure 6. We can

Right
Left
VB
HB
7
6
5
3
2
1
1
2
3
5
6
7

Fig.6. Displacement vectors determined by laser speckle interferometry in the cases of horizontal (HB) and vertical brackets (VB)

observe that the displacement of canines and the gum surrounding them appeared to be parallel to the respective dental arch tangent, if one observes the actions of the vertical and horizontal brackets in the occlusal plane. But, if observations are made in the planes tangent to dental archs, a pure translation is measured in the case of vertical brackets, and a combined roto-translation in the case of horizontal brackets.

Figure 7 shows a summary of quantitative results of one of our experiments. They confirm what we said before related to the aesthetic point of view in crown positioning. If we observe the lines of forces and the displacement in the occlusal plane, the results are almost the same for both cases, that is, by using horizontal brackets as well as vertical brackets. But, attending to the efficiency of the masticatory function, orthodontic results obtained with vertical brackets are incomparably better than those obtained with horizontal brackets, as we can corroborate by radiological examination of clinical cases. Figure 3 f) is a reproduction of a clinical case radiography at the final stage of treatment.

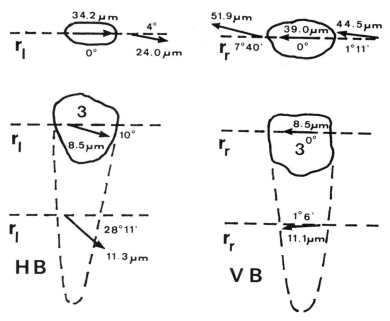

Fig. 7. Quantitative results from laser speckle interferometry in the cases of horizontal (HB) and vertical brackets (VB)

Clearly shown is the degree of parallelism obtained between roots of incisors, canines, and premolar. Besides, Figure 8 shows the gypsum model of the clinical case N° 88 at the end of the successful orthodontic treatment.

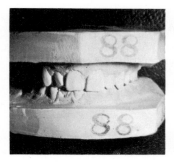

Fig.8. Gypsum model at the final stage of the treatment of clinical case
N° 88

4. Conclusions.

In conclusion, in our opinion the advantages using vertical brackets instead of horizontal brackets, for those clinical cases where such orthodontic technique can be recommended, are supported by the results of optical experiments, clinical treatment, and radiological studies. These
results are good enough from functional, aesthetical, and psychological
points of view. Besides the metallic elements and components to be employed in the vertical brackets appliance are the same as than those ortho-
dontist uses in applying the previous techniques. Then, our proposal is not
expensive or, at least, not much more than previous techniques. The time of
treatment with vertical brackets is comparatively equal or even shorter
than that with horizontal brackets.

5. Acknowledgments

The invaluable support provided by the Secretaría de Estado de Ciencia y
Tecnología, Argentina, and the Organization of American States is
gratefully acknowledged.

6. References

1. Angle, Edward H.: The latest and best in orthodontic mechanism, Dental
   Cosmos 70, 1143, 1928; 71, 164, 1929; 71, 260, 1929; 71, 409, 1929 and
   71, 416, 1929.

2. Abbattista, M.: Técnica distaladora y correctora simultánea con fuerzas
   combinadas. Nuevos conceptos. Doctoral Thesis, Universidad Nacional de
   La Plata, 1977.

3. Françon M.: Granularité Laser (speckle) et ses Applications en Optique,
   Masson Editeurs, 1975.

4. Erf. R.K., Editor: Speckle Metrology, Academic Press, 1978.

Lasers in Biology:
Fluorescence Studies and Selective Action

A. Andreoni, R. Cubeddu, S. De Silvestri, P. Laporta, and O. Svelto

Centro Elettronica Quantistica e Strumentazione Elettronica - Politecnico
Milano - 20133 Milano, Italy

1. Introduction

Likewise the case of chemistry, lasers can be used in biology in either one
of the following two ways: (i) To probe a biomolecule, in which case techniques such as Raman scattering, fluorescence, flash-photolysis etc. can
all be used to get structural information about the given biomolecule; (ii)
To act on a biomolecule so as to induce an irreversible change in its structure i.e. to perform photobiology. In this case, the possibility of performing selective photobiology, using schemes which are either similar or different to those already used to perform selective photochemistry appears to
be a very challenging one.

In this paper, we will first briefly review the work that our group has
been doing in the last few years on point (i) of above, i.e. to probe a
given biomolecule. The technique that has been used here is based on laser-
induced fluorescence   [1] . The paper will then more throughly consider
and discuss the results that we have recently obtained on topic (ii), i.e.
laser selective photobiology. We will show that workable schemes of laser
selective photobiology have indeed been conceived and made to operate, thus
opening new interesting possibilities in fields such as genetic engineering
and laser phototherapy.

2. Laser Microfluorometry Experiments

To perform these experiments a laser microfluorometer, sketched in Fig. 1,
has been developed [1] . The output beam of a nitrogen-pumped dye laser
of special design [2,3] (giving pulses of $\sim$ 100 ps duration, $\sim$ 100 kW
peak power and 100 Hz repetition rate at any wavelength in the visible
range) is sent to a microscope and focused by its objective to a spot approximately equal to the resolving spot of the microscope ($\sim$ 0.5 μm). In
this way, excitation can be provided to a small area in the biological specimen. The fluorescence light emitted is then sent to a fast photomultiplier
(Varian type 154 M) and its time behavior is observed. On account of the
weakness of the signal, which is emitted from a sample less than 1 μm^3 in
volume, a home-made averager has been developed that provides both fast
response ($\sim$ 35 ps) and high accuracy [1] .

This system has been used to perform fluorescence studies of the complexes
that are formed when a dye, belonging to the acridine family, is bound to
DNA. Acridines form a rather large class of dyes [4] which absorb in the
blue and fluoresce in the green region of the spectrum [5] . The chemical

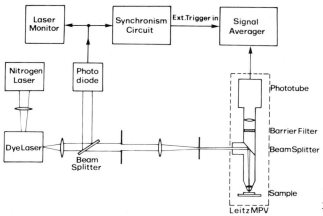

Fig.1 Block diagram of the laser microfluorometer

structure of some of these dyes is shown in Fig. 2. Besides the widely used dye Acridine Orange, we show in the figure also the structure of Proflavine (henceforth denoted by PF) and Quinacrine Mustard (QM) that have mostly been used in our experiments. Note that the neutral, monocationic and dicationic forms are respectively indicated for AO, PF and QM [4-6] . These are in fact the forms that are found for these dyes at the pH value used in our experiments,i.e. pH = 4.6 [7] .

When a dye of the acridine family is made to interact with DNA, an acridine-DNA complex is formed. The complex is readily formed on account of the very high binding constant of the process. The structure of this complex is sche-matically shown in Fig. 3 (according to the so-called modified Lerman's mod-el [8] . The acridine ring intercalates two adjacent base pairs of the DNA, the binding being predominantly due to the van der Waals force between the π-electrons of the acridine ring and the π-electrons of the purine or

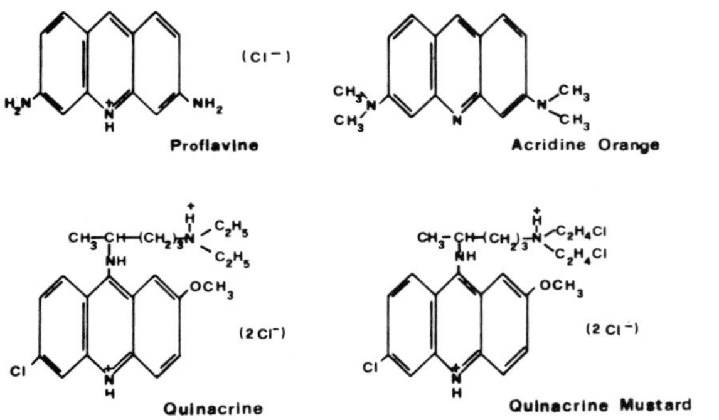

Fig.2 Chemical structure of a few acridine dyes

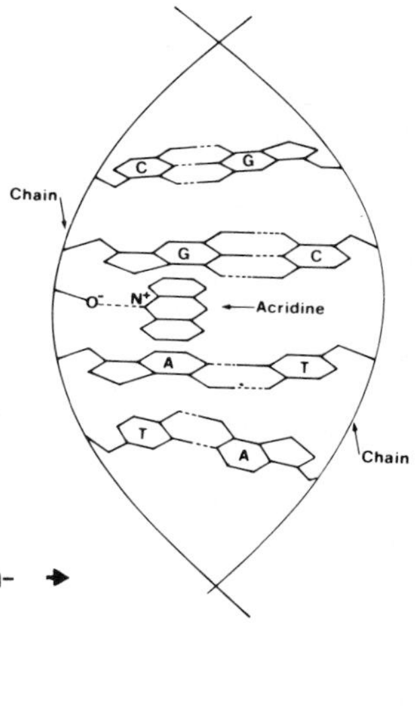

Fig.3 Modified Lerman's model of acridine-DNA com-➔
plex

pirimidine bases of the DNA. The binding is further strengthened by the ionic interaction between the basic Nitrogen atom of the acridine ring and the oxygen ion of the phosphate group of the DNA backbone [9] .

The acridine-DNA complexes have interesting photophysical and biological properties. In particular, we wish to mention that the fluorescence decay time of the complex is strongly dependent on the base-pair sequence that the dye intercalates [1] . The lifetime is quite long (10-20 ns) when the dye intercalates two Adenine-Thymine base pairs (AT-AT sequence) while it becomes much shorter (1-5 ns) when it binds to sequences containing at least a Guanine-Cytosine base pair (i.e. GC-AT and GC-GC sequences). From now on, as short-hand notation, we will denote as complex 1 and 2 the complexes with short and long fluorescence decay time, respectively.

Using the fluorescence microfluorometer in Fig. 1, experiments have been performed in the last few years by our group on acridine-DNA complexes. Note that, when the whole sample, rather than a small spot, was excited, the microscope was not used in the apparatus in Fig. 1. On account of the relatively large amount of the results obtained, we limit ourselves here to summarize the most relevant ones.

The mechanisms that are responsible for the fluorescence quenching in the case of complex 1 have been investigated in some detail. On account of the more recent results which have been obtained [7] , the quenching seems to arise from phenomena of proton transfer in the acridine dye in the excited state. The acridine seems, in fact, to jump into states of different protonation (neutral, monocationic or dicationic) depending upon the environment, i.e. upon the base pair sequence, that have different fluorescence decay times.

Studies on the base-pair sequences of DNA have also been performed [10] . If the average concentrations C_1 and C_2 of complexes 1 and 2 are, at the same time, present in the volume excited by the laser beam, the fluorescence time behavior is then expected to consist of the sum of two components, the fast one arising from complex 1 and the slow one from complex 2. If we now assume that the spontaneous lifetime is the same for the two complexes, we can then say that the extrapolations of the two components at time $t = o$ give two quantities which are proportional to C_1 and C_2 , respectively. If we further assume that the dye has the same probability of intercalating any base pair sequence [11] , the quantity C_2 will be proportional to the concentration of AT-AT base-pair sequences, while the quantity C_1 will be proportional to the concentration of AT-GC plus GC-GC base-pair sequences. In this way, information on the DNA base-pair sequences can be obtained. It must be noted, however, that this will be an information averaged over the excited sample volume, whose dimensions cannot be smaller than the wavelength of the excitation light, i.e. ~ 4000 Å , while the distance between two adjacent base-pairs in the DNA is ~ 3 Å. Nevertheless, valuable information has been shown to be obtainable in cases such as bacteria [10] and chromosomes [1] .

The technique of laser microfluorometry has also been applied to differentiate between different phases of a cell (e.g. human lymphocytes) . In this case, the fluorescence decay curve has been found to depend on the phase (e.g. G_1 , G_2 of human lymphocytes) in which the cell is actually found [12] .

3. Laser Selective Action

Consider two biomolecules A and B on which we want to perform a selective action. This means that we want to produce an irreversible change of the chemical structure of either biomolecule A or B. We will consider the very common case in which the two biomolecules are so similar that the corresponding absorption spectra do not differ enough to allow a selective excitation of one of them. In such a case, a suitable dye is added to the biomolecules so that two complexes with biomolecules A and B are formed. Since the lifetime of the dye is the parameter most sensitive to the environment, it usually occurs that the lifetimes of the two complexes are appreciably different. This difference allows for a selective photodamage of either one of the two complexes, resulting in selective action on biomolecule A or B [13] .

If we call again 1 and 2 the complexes with shorter (τ_1) and longer (τ_2) lifetimes respectively, selective action on complex 2 can be performed as it follows (see Fig. 4). The complexes are made to interact with two laser

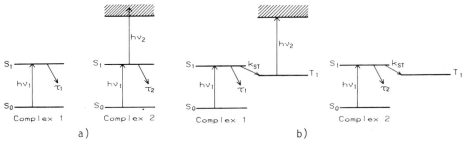

Fig.4 a) Selective action on complex 2. For the sake of simplicity, the less efficient ionization step of complex 1 is not indicated. b) Selective action on complex 1. For the sake of simplicity, the less efficient ionization step of complex 2 is not indicated

pulses of frequencies ν_1 and ν_2 respectively. The pulse at frequency ν_1 promotes the dye to its first excited singlet state S_1 . The pulse at frequency ν_2 ionizes the dye starting from its excited singlet state S_1 [14] . If the second pulse is sent with a delay τ_d after the first one, so that $\tau_1 < \tau_d < \tau_2$, the second pulse will find an S_1-state population of complex 2 appreciably larger than that of complex 1. This population will have, in fact, decayed from the S_1 state to a new state, which may be the dye triplet state T_1 (or some other lower-lying state, such an exciplex etc.). Complex 2 will thus be predominantly ionized. Similarly, selective ionization of complex 1 can also be performed. The frequency ν_2 is now selected in such a way as to perform ionization from the T_1 state of the dye. The time delay between the two pulses τ_d is now chosen so that $\tau_2 < \tau_d < \tau_T$, where τ_T is the lifetime of the T_1 state (usually much longer than τ_1 and τ_2). In this case, the second pulse will find a T_1-state population of complex 1 appreciably larger than that of complex 2, and complex 1 will be selectively ionized.

The proposed schemes present the following advantages: (i) They have rather general applicability, since usually suitable dyes can be found that bind specifically to any given biomolecule. (ii) Visible or near-uv photons are required for the excitation and ionization steps of the dye; moreover, at these wavelengths the biomolecules are usually transparent, thus avoiding possible complications arising from direct absorption. (iii) Since the lifetimes τ_1 and τ_2 fall in the nanosecond range, nanosecond or subnanosecond pulses are required. This is to be compared with picosecond or even subpicosecond pulses which are required for the scheme proposed by Kryukov et al. [15] . (iv) The selective action on complex 2 can also be performed by using either CW or long-pulse irradiation. In this case, the photons at the frequencies ν_1 and ν_2 are acting at the same time. On account of the difference in lifetime, the steady state S_1 population of complex 2 will be larger than that of complex 1 and the selective ionization of complex 2 will be produced.

Experiments have been performed to test some of the ideas of selective action, which have been discussed. Acridine-DNA complexes have been used, since they meet the requirements needed for selective action [16] . In the first experiment, which was done using QM, a Nitrogen laser (giving pulses of $\sim$ 7 ns duration, 200 kW peak power, 100 Hz repetition rate) provided both the excitation and the ionization pulses. Since the pulse duration was comparable with τ_2 the scheme discussed in connection with the case of CW or long-pulse irradiation holds in this case [17] . Moreover, the scheme works with $\nu_1 = \nu_2$. The desctruction of the QM molecules following the photoionization was monitored by the reduction in the QM fluorescence as excited by probe pulses at 430 nm from an auxiliary dye laser. Indeed a substantial fluorescence fading of QM was readily observed after $\sim$ 10^4 shots (corresponding to an irradiation time of a few minutes). To estimate quantitatively the results, we define a damage probability per laser shot p as

$$p = \Delta N/N \quad , \tag{1}$$

where $N = N(n)$ is the concentration that is present at a given laser shot and ΔN is the corresponding change due to the next laser shot. From (1) by an iterative procedure, we obtain that the population $N(n)$ left out after the n-th laser shot is given by

$$N(n) = (1 - p)^n N_0 \quad , \tag{2}$$

where N_0 is the population before irradiation. From (2) we have

$$\ln \{ N(n)/N_0 \} = n \ln \{ 1 - p \} \quad . \tag{3}$$

Since $N(n)$ is proportional to the detected fluorescence signal, the quantity $N(n)/N_0$ is readily obtained as the ratio of the corresponding fluorescence signals. Equation (3) shows that a plot of $\ln\{N(n)/N_0\}$ vs n should yield a straight line whose slope gives p, i.e. the damage probability per laser shot. This quantity can then be used to get quantitative evaluation of the selective action.

A first set of measurements has been performed to show that the observed
destruction of the sample was indeed arising from a two-step photodamage
of the dye. To this purpose, Fig. 5 shows the plots of $\{\ln N(n)/N_0\}$ vs n
for several values of the laser peak power, while the irradiate-volume cross
section was kept constant (1 mm^2). For these measurements, a solution of
QM in 0.2 M acetate buffer (pH 4.6) was used. At each intensity the best
fitting straight line is also shown in the figure. According to Eq. (3), we
obtain the damage probability per laser shot p. This quantity is then plotted as a function of the laser peak intensity in Fig. 6. The dots in Fig. 6
can be reasonably well fitted to a square-low dependence (solid line), while
the slight roll-off at higher intensities comes from the saturation of the
S_0- S_1 transition. The results show that the observed dye dissociation arises

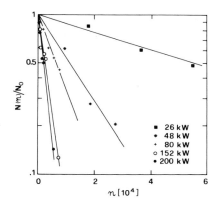

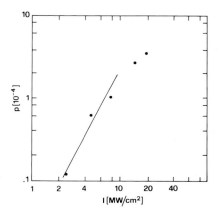

Fig.5 Fraction of QM population
$N(n)/N_0$ as a function of laser shots
n at different N_2-laser peak power

Fig.6 Damage probability p per laser
shot versus irradiation intensity I.
The straight line has a slope of 2

from a two-step process. Once this was established, measurements have been
done on QM bound to Poly dG-Poly dC (a synthetic polynucleotide in which
all base pairs are GC) and to Poly dA-Poly dT (a synthetic polynuclotide
in which all base pairs are AT), in order to show the selectivity of the
two-step photoionization process. In fact, according to the different lifetimes of QM bound to these polynucleotides [1] , they provide good models
for complex 1 and complex 2 of the above discussion. Figure 7 shows the
experimental results. The damage probability for QM bound to Poly dA-Poly
dT (complex 2) turns out to be 4.7 times larger than that for QM bound to
Poly dG-Poly dC (complex 1). Results of thin-layer chromatography on silica
gel of the hydrolyzed polynucleotides before and after irradiation have shown
that extensive damage of the polynucleotide bases occurred. Thus the photoionization of the QM molecules has indeed induced irreversible damage of the
polynucleotide bases [18-20] . However, no quantitative evaluation of this
damage has been possible so far by this technique.

A second set of experiments has been performed to test that the idea of
using suitably delayed pulses was indeed working. To this purpose, the acridine dye Proflavine (PF) and the experimental set-up of Fig. 8 have been

292

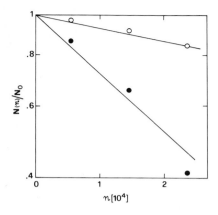

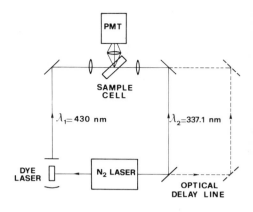

Fig.7 Fraction of QM population $N(n)/N_0$ versus the number of laser shots n: (●) QM in Poly dA-Poly dT; (o) QM in Poly dG-Poly dC

Fig.8 Block diagram of the experimental apparatus

used. An atmospheric pressure N_2 laser [2] giving two oppositely directed laser beams (each with ~ 500 ps duration, 100 kW peak power and 100 Hz repetition rate) has been used. One beam pumps a dye laser oscillating at λ = 430 nm (120 ps duration, 100 kW peak power and 100 Hz repetition rate). The dye laser beam is sent to the sample cell to perform the excitation step of PF. The second beam of the N_2 laser is sent through a suitable optical delay-line to the sample cell and performs the ionization step. The amount of damage was monitored by the reduction in sample fluorescence when excited by the 430 nm beam alone and observed by a photomultiplier (PMT). Damage of the dye was again readily observed when the two beams were present, while no damage could be detected when either one of the two beams was blocked. This showed the biphotonic character of the process. The 337.1 nm beam was then delayed as compared to the 430 nm beam. Figure 9 shows the damage probability p per laser shot vs the delay τ_d between the two pulses. It refers to the case of PF in 0.2 M acetate buffer solution (pH 4.6). The probability p decreases with increasing delay on account of the decay of the S_1-state population. Furthermore, the best fit to the measured points (solid line) gives quantitative values of the excited state photoionization cross-section. Proflavine bound to either Poly dG-Poly dC (complex 1) or Poly dA-Poly dT (complex 2) was then irradiated in the same experimental set-up in Fig. 8. The corresponding dissociation probability per laser shot p as a function of the time delay τ_d is plotted in Fig. 10. The damage probability of PF in Poly dA-Poly dT is always larger than that in Poly dG-Poly dC. While the ratio of this quantities (which may be called the selectivity of the photodamage) is already large enough for moderate time delays (~ 2 ns), the selectivity becomes much larger at a time delay of ~ 8 ns. Indeed, no damage was observed in the case of Poly dG-Poly dC at this delay. On account of the uncertainty of the detected undamaged-PF fluorescence, we then obtain a value of 40 as the lower limit of selectivity at this delay. The result appears to be quite remarkable and deserves some comments. In fact, since the lifetime of complex 2 (PF bound to Poly dA-Poly dT) is ~ 4.5 ns [21], at 8 ns delay most of the excited state population of both complexes is expected to have decayed from the S_1-state. In the case of PF bound to Poly

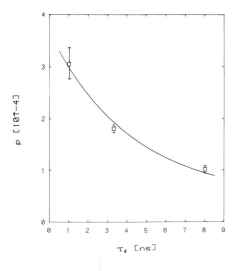

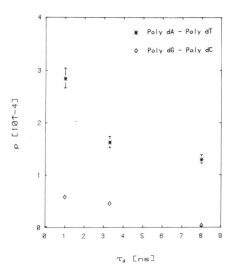

Fig.9 Damage probability per laser shot p of free PF in buffer solution at three different delays τ_d (square dots) between the two laser pulses. The solid line is the theoretical best fitting

Fig.10 Measured values of the damage probability per laser shot p of PF bound to Poly dA-Poly dT (asterisks) and to Poly dG-Poly dC (diamonds) at three different delay τd between the two laser pulses

dA-Poly dT, however, the damage probability p is still appreciable and is likely to arise from two-step ionization starting from the triplet state. Since no comparable damage is observed for PF bound to Poly dG-Poly dC, we are led to the tentative conclusion that this complex decays to a short-lived and/or deeper-lying state. This conclusion is consistent with the assumption made in previous works [1,13] that this state could be a non-fluorescent exciplex.

4. Conclusions

In this paper a review is presented of our work based on the technique of laser-induced fluorescence to obtain structural information on dye-biomolecule complexes. Results on the form of binding, on the DNA base-pair sequences and on the phase of human lymphocytes have been obtained in the case of acridine-DNA complexes. Schemes of laser selective action on dye-biomolecule complexes are also discussed. Experimental results, showing a highly selective action arising from the application of these schemes, are presented.

References

1. A.Andreoni, C.A.Sacchi, O.Svelto: In *Chemical and Biochemical Applications of Lasers*, Vol. 4, ed. by C.Bradley Moore (Academic Press Inc., New York, 1979) p. 1

2. R.Cubeddu, S.De Silvestri: Opt. Quant. Electr. **11**, 276 (1979)
3. R.Cubeddu, S.De Silvestri, O.Svelto: to be published in Optics Comm.
4. A.Albert: *The Acridines* (E. Arnold Ltd, London 1966)
5. A.R.Peacocke: In *Heterocyclic Compounds: Acridines*, Vol. 9, ed. by R.M. Acheson (Interscience, New York 1973) p. 723
6. A.C.Capomacchia, S.G.Schulman: Anal. Chem. Acta **77**, 79 (1975)
7. A.Andreoni, R.Cubeddu, S.De Silvestri, P.Laporta: Optics Comm. **33**, 277 (1980)
8. N.J.Pritchard, A.Blake, A.R.Peacocke: Nature (London) **212**, 1360 (1966)
9. R.K.Selander: Biochem. J. **131**, 749 (1973)
10. G.Bottiroli, G.Prenna, A.Andreoni, C.A.Sacchi, O.Svelto: Photochem. Photobiol. **29**, 23 (1979)
11. O.F.Borisova, A.P.Razjivin, V.I.Zaregorodzev: FEBS Lett. **46**, 239 (1974)
12. G.Bottiroli, P.G.Cionini, F.Docchio, C.A.Sacchi: In *Proceedings 1980 International Conference on Lasers*, Shanghai May 5-8 (1980) and Beijing May 19-22 (1980), in press
13. A.Andreoni, A.Longoni, C.A.Sacchi, O.Svelto, G.Bottiroli: In *Tunable Lasers and Applications*, Proceedings of the Loen Conference, Norway, 1976, ed. by A.Mooradian, T.Jaeger, and P.Stokseth, Springer Series in Optical Sciences, Vol. 3 (Springer-Verlag, Berlin, Heidelberg, New York, 1976) p. 203
14. A.Andreoni, R.Cubeddu, S.De Silvestri, P.Laporta, O.Svelto: to be published in Phys. Rev. Lett.
15. P.G.Kryukov, V.S.Letokhov, D.N.Nikogosyan, A.V.Brodavkin, E.I.Budowsky, N.A.Simmkova: Chem. Phys. Lett. **61**, 375 (1979)
16. R.Rigler: In *Chromosome Identification. Nobel Symposium XXIII*, ed.by T.Caspersson and L.Zech (Academic Press, New York, 1973) p. 335
17. A.Andreoni, R.Cubeddu, S.De Silvestri, P.Laporta: Chem. Phys. Lett. **72**, 448 (1980)
18. J.Piette, C.-M.Calberg-Bacq, A.Van de Vorst: Photochem. Photobiol. **26**, 377 (1977)
19. J.Piette, C.-M.Calberg-Bacq, A.Van de Vorst: Photochem. Photobiol. **27**, 457 (1978)
20. J.Piette, C.-M.Calberg-Bacq, S.Cannistraro, A. Van de Vorst: Intern. J.Radiat. Biol. **34**, 213 (1978)
21. S.Georghiou: Photochem. Photobiol. **22**, 103 (1975)

Time-Resolved Resonance Raman Techniques for Intermediates of Photolabile Systems*

M.A. El-Sayed

Department of Chemistry, University of California, Los Angeles
Los Angeles, CA 90024, USA

This paper summarizes the work of my group over the past four years. Different resonance Raman techniques are described which are useful in studying intermediates of photolabile systems in the millisecond, microsecond, nanosecond, and picosecond time domains. These techniques are used to study two important photobiological systems: bacteriorhodopsin (bR) and carbonmonoxy-hemoglobin (HbCO). The summary of the results of applying these techniques to study the retinal system in bR is given and discussed in terms of what is known about its photochemical proton pump cycle. The main results are 1) the largest retinal configurational changes occur in the first step (the absorption step) and 2) the Schiff base proton in

$$-\overset{|}{C} = \overset{+}{N}H-$$

ionizes in 40 μs (in the $bL_{550} \to bM_{412}$ step).

The picosecond Raman spectra of the porphyrin system of the intermediate produced in the photolysis of HbCO in the picosecond time domain suggest that photodissociation of CO might proceed from a quintet state of HbCO to give a high spin Fe(II) with the Fe(II) being coplanar with the porphyrin plane. The iron moves out of the plane in the microsecond time scale, suggesting that its motion is probably controlled by the protein motion.

---

* Modified versions of this paper are being submitted to three different invited publications: a) The Proceedings of the Sergio Porto Laser Memorial Meeting, Rio de Janeiro, Brazil, June 29-July 3, 1980; b) The Proceedings of the VIIIth International Congress on Photobiology, Strasbourg, France, July 21-25, 1980; c) a volume edited by Lester Packard on *Visual Pigments and Purple Membranes*. It is also being submitted as an abstract to the International Congress on Raman Spectroscopy, Ottawa, Canada, August 4-9, 1980.

1. Introduction

It is proper in a meeting held in tribute to Sergio Porto that I discuss my group's activities in time-resolved Raman spectroscopy. Furthermore, in a laser meeting, I would like to discuss our recent results on resonance Raman spectroscopy of transients formed in the picosecond time scale.

While picosecond lasers have been used previously to obtain the resonance Raman spectra of stable molecules [1] (in particular to reduce interference from fluorescence radiation), the first resonance Raman spectra of transients formed in the picosecond time domain have just been reported this year by our group [2] and others [3]. Of course, optical absorption and emission spectroscopy of picosecond transients has been useful during the past decade in determining the number, rise time, and decay time of picosecond intermediates in photolabile systems. However, due to the broad nature of the optical absorption of the system studied, these spectra did not yield the kind of structural information one would like to have in order to identify the exact structural changes taking place in these processes. Vibration spectra, as obtained from the Raman scattering process, are expected to give more structural information. For this reason, as well as the development of cavity-dumped picosecond lasers and the imaging detection systems, time-resolved Raman spectroscopy is expected to be an active field of research over the coming decade or two.

2. Time-Resolved Resonance Raman (TRRR) Techniques [4]

2.1 The Millisecond, Microsecond, and Nanosecond Transients (with J. Terner, A. Campion, C.L. Hsieh, A. Burns)

Different time-resolved resonance Raman techniques have been developed [4] for determining the resonance Raman spectra of the intermediates formed from photolabile systems, e.g., the proton pump system of bacteriorhodopsin [5-8] and the CO hemoglobin [2]. The method used varied and depends on the time scale in which the photointermediate builds up in concentration. All the methods that we have used have the following features in common:

1) One laser is used which acts both as the photolytic as well as the Raman probe light source. This is especially applicable to systems of broad absorption bands that one expects to have on overlap of the absorption band of the photolabile parent compound and that for the intermediate whose resonance Raman is being examined.

2) The experiment is carried out in a time scale appropriate for the rise and decay times of the intermediate being studied.

3) The laser used is adjusted at a wavelength which gives high photolytic probability and large R. Raman enhancement for the intermediate examined. Furthermore, the scattered Raman radiation should have minimum overlap with any fluorescence present.

4) For obtaining the spectrum of a certain intermediate, chemical or physical perturbations are used, if possible to maximize its concentration.

5) Satisfying the above conditions, two Raman spectra are then recorded using the optical multichannel analyzer for detection; one spectrum is obtained at very low powers (to obtain the spectrum of the unphotolyzed parent compound at minimum photolysis), and the second spectrum is obtained at high powers (to maximize the concentration of the photoproduct studied).

6) Computer subtraction techniques are then used to subtract out the low-power spectrum from the high-power spectrum to obtain the Raman spectrum of the intermediate, having maximum concentration in the time scale of the experiment used and which has the maximum enhancement at the wavelength of the laser used.

In order to satisfy condition 2) above, i.e., adjust the time scale of the experiment to maximize the concentration of a certain intermediate, one of the following techniques is used:

I) *Pulsed Lasers* [5,9,10]

Only intermediates appearing in a time equal to or shorter than the pulse width could be detected by Raman spectroscopy if the wavelength is adjusted for maximum enhancement. Intermediates with rise times shorter, and decay times longer than a few nanoseconds can be studied [5,9,10] by using the N_2-pumped (e.g., Molectron) or Nd YAG-pumped (e.g., Quanta-ray) dye lasers which have few-nanoseconds pulse width. Chromatics could be used for intermediates with decay times in the microsecond time scale. In principle, picosecond intermediates could be detected by using the high-power pulsed picosecond lasers. However, the low duty cycle of these lasers, from multiphoton processes, could hamper the observation of good signal-to-noise values in these experiments.

II) *Modulation of cw Lasers* [6,7]

Electric or mechanical modulation of cw lasers can produce pulses with different pulse width and at a given modulation frequency. Electric modulation could give short pulses with high duty cycles. Mechanical modulation could give longer pulses with lower duty cycles. We have used mechanical choppers (rotating disks) fixed with variable size slits. A cw laser could function as a pulsed laser with variable pulse width. The fact that the laser can be brought up to a small focus in the micron range makes slit width of a few microns usable in these experiments. With the available practical motors usable in this experiment, intermediates in the 50-100 nanosecond time scale could be detected [6]. Of course, with slow motors and large slit width, millisecond intermediates could be easily observed. Two slits, one for the photolysis and one for the probe laser could also be used [7] with variable time delays (i.e., separation between the slits). The duty cycle in these experiments is determined by the number of slits in the rotating disk as well as the motor speed. In any case, they could be much better than in the pulsed laser experiment.

III) *Flow Techniques* [4,8]

Instead of pulsing the laser, the sample could be "pulsed" by flowing it across a focused cw laser beam. Actually the flow technique was first used by MATHIES et al. [11] to determine the Raman spectrum of the unphotolyzed rhodopsin. MARCUS et al. [12] were the first to use it for kinetic studies in bacteriorhodopsin by varying the flow rate of the sample (which could be changed by a factor of ten) to obtain different time scales for different intermediates. We have extended the time scale of this method by realizing [4,8] that, for the same flow rate, the time scale of the experiment could be varied by varying the laser focus itself. By using a microscope objective and a flow rate of 10-40 m/s, the experimental time resolution (determined by the time it takes the flowing sample to cross the focused laser beam) could be in the 50-ns time scale! More importantly, the scattered Raman relation is being collected continuously in this experiment.

2.2 Resonance Raman of Picosecond Intermediates [2] (with J. Terner, T.G. Spiro, M. Nagumo, M.F. Nicol)

By replacing the cw laser in experiment III) above with a mode-locked cavity-dumped Argon or Krypton ion-pumped dye laser, the resonance Raman spectrum of intermediates in the picosecond time scale could be recorded. In this case, the time resolution of the experiment is no longer determined by the sample residence time in the laser beam as in III), but rather by the pulse width of the picosecond laser used. This is true only if the time between the laser pulses ($\sim$1 μs) is longer than the residence time of the sample in the beam ($\sim$0.1 μs).

3. Resonance Raman Results on the Proton Pump System of Bacteriorhodopsin
[6-8,13] (with J. Terner, C.L. Hsieh, A. Burns)

Bacteriorhodopsin absorbs visible light ($\lambda_{max} = 570$ nm) and passes through
a number of intermediates (at least four) before the bR returns to its initial form [14]:

$$bR_{570} \xrightarrow{h\nu(ps)} bK_{590(610)} \xrightarrow{2 \mu s} bL_{550} \xrightarrow{40 \mu s} bM_{412} \xrightarrow{ms} bO_{640} \longrightarrow bR_{570} \quad .$$

As a result of this cycle, (one or two) protons are pumped out of the cell,
thus creating proton gradients across the bR cell membrane [15]. It is this
electric free energy that is believed to be used in the synthesis of the high-
energy molecules (ATP).

We have attempted to use the TRRR technique in order to examine two problems concerning the above photocycle. The first one is concerned with the
retinal conformation changes. Based on chemical reconstitution studies [16],
it is believed that the retinal in bR_{570} is all-trans retinal. In rhodopsin,
resonance Raman studies showed that [11] the retinal is in the 11-cis configuration and it has been the common belief that the absorption process leads
to isomerization to the all-trans form in the first step. This change in the
retinal configuration leads to changes in the retinalprotein interaction
energy as well as entropy. It is the change in the free energy upon the absorption that leads to the storage of the free energy necessary to drive the
system through the latter process. The question then arises: If the retinal
in the bR_{570} already contains an all-trans retinal, what is the mechanism of
converting solar energy into free energy in the absorption act of the bR
system? Could it be that even if retinal in bR_{570} is in the all-trans configuration, further configuration changes take place in the first step of the
cycle? (It should be pointed out that the all-trans retinal *inside the protein*
might not necessarily have the lowest value of the free energy of the system.)
Another possibility is that when the all-trans retinal combines with the bacteriorhodopsin to form the retinalprotein complex it does not retain its all-
trans configuration. In support of this is the difficulty in comparing the
fingerprint region of the retinal in bR_{570} with that of all-trans model compounds [13], unlike the rhodopsin system when good agreement between the
11-cis and the rhodopsin system is obtained [11]. In any case, one would like
to investigate whether or not the first step in the cycle involves a change
in the retinal configuration. By using TRRR one can examine the fingerprint
region (1000-1400 cm^{-1}) (sensitive to retinal configuration changes) for the

$^{bK}_{590(610)}$ intermediate and compare it to that for $^{bR}_{570}$. The $^{bK}_{590(610)}$ intermediate is formed in the picosecond time domain; however, it lasts a few microseconds. Thus, the TRRR experiment can be carried out in the 0.1-1 μs time scale.

The second problem concerns the origin of the protons pumped during the photochemical cycle. The retinal in $^{bR}_{570}$ is bonded to the lysine in the protein via a Schiff base linkage:

$$\begin{array}{c} | \quad + \\ -C = N- \\ | \\ H \end{array} \quad .$$

While not definitely proven, it has been assumed [15] that one of the protons pumped out in the $^{bR}_{570} \rightarrow ^{bM}_{412}$ transformation comes from this group. We have followed, by TRRR techniques, the

$$\begin{array}{c} | \quad + \\ C = N- \\ | \\ H \end{array}$$

vibration at $\sim$1640 cm^{-1} to determine the step at which this band disappears (due to the formation of the unprotonated Schiff base):

$$\begin{array}{c} | \\ -C = N- \end{array}$$

which has a frequency below 1620 cm^{-1}.

The important results in the fingerprint region (which is sensitive to the retinal configuration) and the $C = N$ stretching region (to follow the deprotonation of the Schiff base) can be summarized as follows:

1) Similar to rhodopsin [17], large changes in the fingerprint region are observed during the first

$$^{bR}_{570} \xrightarrow{h\nu} \; ^{bK}_{590}$$

   transformation.

2) Like in rhodopsin [17], large enhancement is observed for the low-frequency vibrations in the $\sim$980 cm^{-1} region for $^{bK}_{590}$. This might suggest twisting of the polyene system.

3) Unlike in rhodopsin [11], a complete identification of the isomeric form of the different intermediates is difficult to achieve by comparing our

spectra with those of model compounds in solution. This could result from one or more of the following reasons: a) Some distortion of the spectra obtained by subtraction techniques. b) Larger perturbation of the spectra of retinal by the retinal-opsin interaction in bacteriorhodopsin than that present in rhodopsin. c) The small difference between the all-trans and 13-cis fingerprint spectra in solution.

4) In the $bK_{590} \rightarrow bL_{550}$ transformation, the 980 cm^{-1} region becomes normal. This might suggest a relaxation of the twisted polyene structure.

5) Small changes in the fingerprint region take place in the $bL_{550} \rightarrow bM_{412}$ process, the process which is accompanied by the largest change in the position of the optical absorption maximum in the cycle. It also involves the deprotonation of the Schiff base. This might suggest [13] that the theories based on ionic interactions [18] rather than retinal configurational changes [19] might be the correct ones in explaining the origin of the red shift in the retinal absorption upon combining with the opsin.

6) The CN stretching vibration is greatly reduced in frequency during the first step ($bR_{570} \rightarrow bK_{590}$). However, it is found that it is reduced further in D_2O solvent. This might suggest that in the first step of the cycle, the interaction between the Schiff base nitrogen and the proton has been reduced. This is unlike the results in the rhodopsin [17] system, in which the CN frequency in the batho form has remained similar to that in the parent compound.

7) The CN frequency in the bL_{550} form is found to be similar to the parent compound (and is thus protonated). This suggests that deprotonation of the nitrogen Schiff base takes place during the $bL_{550} \rightarrow bM_{412}$ process, i.e., in ~40 μs.

4. The Carbonmonoxyhemoglobin Results [2]

Using optical picosecond spectroscopy [20,21], it is found that HbCO dissociates in the picosecond time scale and the intermediate that appears lasts for a longer time than [22] 680 ps. What we [2] wanted to investigate was to determine the vibration spectrum of this intermediate. The Fe(II) in HbCO is in the low-spin state and is coplanar with the porphyrin plane. In deoxyhemoglobin (Hb), however, the Fe is off the porphyrin plane and is in the high-spin state. In the high-spin state, an electron occupies the d_{x2-y2} orbital which forms σ^{*} antibonding molecular orbitals with the pyrrole nonbonding atomic orbitals. This leads to expansion of the porphyrin ring and to a large reduction in the frequency of some of the Raman active $C = C$ stretching vibrations in the Hb molecule as compared to the HbCO molecule. Thus, these bands could be used to label the spin state of the Fe in any Fe-porphyrin system [23]. Of course, the more in-plane the Fe becomes, the stronger the σ^* repulsion becomes and the further is the reduction in the $C = C$ stretching frequency.

Carrying out the picosecond experiment described in Sect. 2.2, it is found that when the laser is focused (high-intensity condition), three new bands appear in the $C = C$ stretching region at 1603, 1552, and 1540 cm^{-1} for the picosecond intermediate; with the 1552-cm^{-1} band being anomalously polarized. These bands are red shifted greatly from the corresponding ones in HbCO molecules and are further reduced by only 4-6 cm^{-1} from those for the Hb molecule itself. This suggests that in this intermediate, Fe(II) is in the high-spin state (to explain the fact that these bands are closer in frequency to Hb than to HbCO) with the Fe being more in the porphyrin plane in this intermediate than in Hb (to account for the observed small frequency reduction from those for Hb).

Low-spin to high-spin conversion *in the ground state* of Fe porphyrin is known to take place in the microsecond [24] time domain. The fact that the low-spin (in HbCO) to high-spin (in this intermediate) conversion takes place in the 25 ps time domain (the pulse width of our laser) suggests that the conversion must have taken place in the excited states. This then suggests that in HbCO, $S_0 \to S_1$ absorption is followed by $S_1 \to Q$ (quintet state) intersystem crossing process. Dissociation might then take place from the quintet state itself. Some quintet states are theoretically predicted to fall below S_1 in Fe porphyrins [22].

Acknowledgment. The author wishes to thank Drs. Stoeckenius and Bogomolni for supplying some of the bacteriorhodopsin samples. The financial support of the U.S. Department of Energy (Office of Basic Energy Sciences) is gratefully acknowledged.

References

1. M. Bridoux: C. R. Acad. Sci. *258*, 620 (1964);
   M. Bridoux, A. Chapput, M. Crunelle, M. Delhaye: Adv. Raman Spectrosc.
   65-69 (1973);
   M. Delhaye: in Proceedings of the Fifth International Conference on Raman

Spectroscopy, ed. by Schmid et al. (1976) pp. 747-752;
M. Bridoux, A. Deffontaine, C. Reiss: C. R. Acad. Sci. *282*, 771 (1976);
M. Bridoux, M. Delhaye: in *Advances in Infrared and Raman Spectroscopy*, Vol. 2, ed. by R.J.H. Clark, R.E. Hester (Heyden, London 1976) p. 140;
P.P. Yaney: J. Opt. Soc. Am. *62*, 1297 (1972);
R.P. Van Duyne, D.L. Jeanmaire, D.F. Shriver: Anal. Chem. *46*, 213 (1974);
F.E. Lyttle, M.S. Kelsey: Anal. Chem. *46*, 855 (1974);
M. Nicol, J. Wiget, C.K. Wu: Proceedings of the Fifth International Conference on Raman Spectroscopy, ed. by Schmid et al. (1976) pp. 504-505

2. J. Terner, T.G. Spiro, M. Nagumo, M.F. Nicol, M.A. El-Sayed: J. Am. Chem. Soc. *102*, 3238 (1980)
3. M. Coppey, H. Tourbez, P. Valat, B. Alpert: Nature *284*, 568 (1980)
4. For a previous review see "Time-Resolved Resonance Raman Spectroscopy in Photochemistry and Photobiology", in *Multichannel Image Detectors in Chemistry*, ACS SYMPOSIUM SERIES Bk. 102, Chap. 10 (1979) pp. 215-227
5. A. Campion, J. Terner, M.A. El-Sayed: Nature *265*, 659 (1977)
6. A. Campion, M.A. El-Sayed, J. Terner: Biophys. J. *20*, 369 (1977)
7. J. Terner, A. Campion, M.A. El-Sayed: Proc. Natl. Acad. Sci. USA *74*, 5212 (1977)
8. J. Terner, C.L. Hsieh, A.R. Burns, M.A. El-Sayed: Proc. Natl. Acad. Sci. USA *76*, 3046 (1979)
9. W.H. Woodruff, S. Farquharson: Science *201*, 831 (1978)
10. K.B. Lyons, J.M. Friedman, P.A. Fleury: Nature *275*, 565 (1978);
R.F. Dallinger, J.R. Nestor, T.G. Spiro: J. Am. Chem. Soc. *100*, 6251 (1978)
11. R. Mathies, T.B. Freedman, L. Stryer: J. Mol. Biol. *109*, 367 (1977)
12. M.A. Marcus, A. Lewis: Science *195*, 1328 (1977)
13. M.A. El-Sayed, J. Terner: J. Photochem. Photobiol. *30*, 125 (1979)
14. R.H. Lozier, R.A. Bogomolni, W. Stoeckenius: Biophys. J. *15*, 955 (1975)
15. D. Oesterhelt: Angew. Chem. Int. Ed. Engl. *15*, 17 (1976)
16. M.J. Pettei, A.P. Yudd, K. Nakanishi, R. Henselman, W. Stoeckenius: Biochemistry *16*, 1955 (1977)
17. G. Eyring, R. Mathies: Proc. Natl. Acad. Sci. USA *75*, 4642 (1979)
18. B. Honig, A.D. Greenburg, V. Dinur, T. Ebrey: Biochemistry *15*, 4593 (1976)
19. R. Kornstein, K. Muszkat, S. Sharafy-Ozeri: J. Am. Chem. Soc. *95*, 6177 (1973)
20. L.J. Noe, W.G. Eisert, P.M. Rentzepis: Proc. Natl. Acad. Sci. USA *75*, 573 (1978)
21. C.V. Shank, E.P. Ippen, R. Bersohn: Science *193*, 50 (1976)
22. B.J. Greene, R.M. Hochstrasser, R.B. Weisman, W.A. Eaton: Proc. Natl. Acad. Sci. USA *75*, 5255 (1978)
23. T.G. Spiro, J.M. Burke: J. Am. Chem. Soc. *98*, 5482 (1976)
24. J.K. Beattie, N. Sutin, D.H. Turner, G.W. Glynn: J. Am. Chem. Soc. *95*, 2052 (1973);
J.K. Beattie, R.A. Binstead, R.J. West: J. Am. Chem. Soc. *100*, 3044 (1978)

Part VI

Picosecond Bistability

Optical Bistability in Semiconductors

S.D. Smith

Department of Physics, Heriot-Watt University
Edinburgh, United Kingdom

1. Introduction

I dedicate this paper to the memory of Sergio Porto, a much loved and
respected colleague, with whom I shared an enjoyable first half in a football
match in Novosibirsk.

The first observations of optical bistability in a nonlinear Fabry-Perot
interferometer fashioned from a parallel sided crystal of InSb were made in
1979 [1] independently and almost coincident in time with observations in
epitaxial films of GaAs [2]. The work on InSb was rapidly extended, using
the differential gain mode to two beams clearly showing the "optical transistor" effect in which one optical beam was amplified by the second with
gains of up to 10 being observed [3]. The large size of the bandgap-resonant
nonlinearity has enabled optical thickness changes as large as $5\lambda/2$ to be
induced allowing the observation of bistability in fourth and fifth orders
in both transmission and reflection. The early work was conducted at 5 K;
we here report extension to 77 K and time-resolved operation.

2. Characteristics of Semiconductor Optically Bistable Devices

The particular characteristic of semiconductor devices lies in the fact that
very large dispersive nonlinearities can be obtained which facilitates the
creation of intrinsic one-element devices of small size and power requirement.
The required optical feedback is provided very simply by using either the
natural reflection of the crystal surface (0.36 for InSb) or two layer
reflection coatings. The devices therefore have relatively modest finesse
which is consistent with a fast intra-cavity field build-up time. A second
advantage of the large nonlinear refraction is that intensity tuning is large
and the interferometer can therefore be operated at relatively low power and
the thickness can be small. In the first InSb devices, 500 μm thickness was
used, the second generation interferometers are $\sim$ 130 μm thick and future
thicknesses are likely to be $\sim$ 50 μm or less. Thus cavity field build up
times can be restricted to be of the order of picoseconds. Power densities
in the first observations at 5 K were of the order of 10 μW/μm² and results
reported here now reduce this by a further order of magnitude. It is
relatively easy to construct a device $\sim$ 1 cm² in area so that the potential
two-dimensional capability for all-optical data storage and beam amplification
is quite considerable with limiting element dimension of a few μm. Ultimate
speeds will depend only upon the speed of the nonlinear response: theoretical
mechanisms for the giant nonlinear effect exploited in this work are discussed
in a later section.

3. Experimental

The experimental realisation of a series of new effects observed in InSb is achieved by exploiting the coincidence of the 60 or 70 CO laser lines from 1930 cm^{-1} to 1660 cm^{-1} with the bandgap of InSb at 5 K ($\sim$ 1900 cm^{-1}) and at 77 K ($\sim$ 1840 cm^{-1}). The development of a continuously variable attenuator which, combined with a spatial filter, enables the beam intensity to be controlled over at least four orders of magnitude whilst retaining a near-perfect Gaussian profile [4] has been a key technique in uncovering a new series of low power nonlinear optical effects [5]. The first of these effects to be discovered [5] was beam broadening at intensities greater than 10 W/cm^2 ($\sim$ 10 mW incident). A macroscopic analysis of the near- and far-field beam patterns [6] enabled us to conclude that this effect was due to self-defocusing with a nonlinear refractive index n_2 defined from $n = n_1 + n_2 I$, with a value of $n_2 \sim -6 \times 10^{-5}$ cm^2/W at 1886 cm^{-1} and 5 K. The sign of this effect also indicated that the origin was unlikely to be thermal as did a series of time-resolved experiments.

The output beam from the grating tunable CO laser (Edinburgh Instruments PL3) is passed through the attenuator, through an electro-optic modulator for time-resolved measurements and then through the sample held in a cryostat with anti-reflected ZnSe windows. A detector then detects the transmission of the interferometer as the intensity is changed either by the attenuator (that is, for a steady state measurement) or the modulator (for a time-

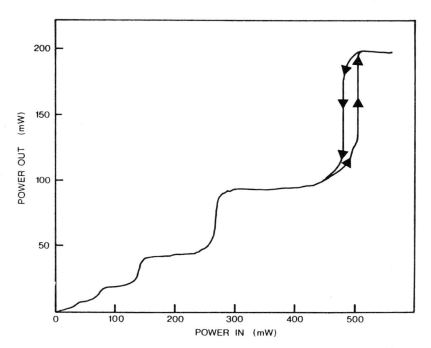

Fig.1. Optical bistability in InSb at 5K. Crystal thickness 560 μm with natural reflectivity (36%) faces. Incident cw CO laser wavenumber 1895 cm^{-1}, spot size 180 μm

resolved measurement). The output from the interferometer is illustrated in
Figure 1 for 1895 cm^{-1} and with the sample held at 5 K. It shows five
distinct steps corresponding to the successive changes of optical thickness
of $\lambda/2$. It is interesting to note that the increment of power between the
steps increases as the intensity increases indicating a saturation in the
refractive nonlinearity. The entire Gaussian beam power was observed at the
detector and despite the inevitable incident Gaussian intensity variation
across the specimen, clear optical bistability is observed in fifth order.
Figure 2 shows further observations in both transmission and reflection
showing bistability in both fourth and fifth orders, i.e. multistability.
The results also show that the background absorption in the interferometer
is low, i.e. $\sim$ 50% transmission or reflection is observed.

The conclusion that the basic effect is a fundamental electronic process
led us to predict that the intensity change can be provided by two beams
rather than one. Thus, the steps in Figure 1 which show differential gain
(that is, the output changes more than the input) can be induced with a
second beam. The results of this experiment are demonstrated in Figure 3.
At each step the second beam modulates the transmission of the first beam
and signal gains of up to 10 are demonstrated. This is a clear demonstration
of 'optical transistor' action and since it is caused by the change of phase
thickness induced by one beam, we term it, by analogy with the transistor,
"transphasor" action. It should be noted that the physics of such a two-
beam device is, however, not identical to differential gain with a single
beam. When a second beam is coincident with the standing wave field inside
the Fabry-Perot cavity the resulting conditions are exactly those required
for degenerate four-wave mixing (DFWM). (In the case of a single beam this
DFWM term vanishes into the nonlinear refraction). Indeed we have observed

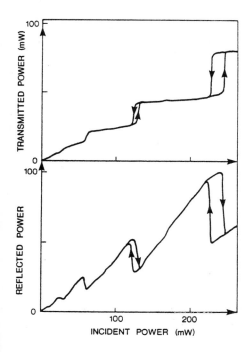

Fig.2. Multistability in transmission and reflection. Sample and
laser line as for Fig.1 with different cavity tuning (by lateral
displacement of sample)

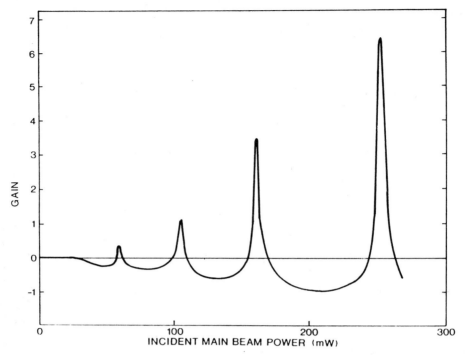

<u>Fig.3.</u> Two-beam differential gain — 'transphasor' action. The gain is the ra-tio of the power change in the weak signal beam to the induced change in the main pumping beam, showing clear peaks corresponding to the one-beam differ-ential gain of Fig.1. Sample and laser beams as in Fig.1; laser beams, from separate lasers, focused coincidentally onto the sample face

DFWM in InSb at power levels well below the levels used in the "transphasor" experiments [7]. In the "transphasor", therefore, new beams are generated by DFWM which must affect the detailed operation of the device, and it would not be justifiable to claim 'optical transistor' action simply on the basis of single beam differential gain measurements.

4. Mechanism of Nonlinearity

Since the large nonlinear effect is very effective in a device sense we are concerned to understand the microscopic origin of the nonlinearity. This requires considerable explanation since the value of $\chi^{(3)}$ $(\omega:\omega,-\omega,\omega)$ which corresponds to the intensity dependent refractive index is $\sim 10^{-2}$ esu at 5 K and as high as 10^{-1} esu at 77 K at frequencies near the bandgap. Previously familiar values for $\chi^{(3)}$ in semiconductor materials away from the bandgap are in the range 10^{-8} to 10^{-11} esu. Resonance near the bandgap can increase these values to about 10^{-6} esu at best but it is seen that a further four or five orders of magnitude are required to explain the present experimental results. From Figure 1 we can plot the variation of n_2 with intensity (or otherwise expressed, there exists a $\chi^{(5)}$ as well as a $\chi^{(3)}$). This plot, Figure 4, indicates that the refractive nonlinearity saturates. Saturation

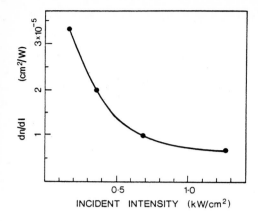

Fig.4. Intensity dependence of dn/dI, the nonlinear refractive index, derived from the separation of nonlinear interference orders in Fig.1

of the absorption occurs in the band-tail region [4]; we have shown that most of this effect disappears before the onset of the nonlinear refraction responsible for the device operation [6]. Photoconductivity measurements [8] indicate that mobile carriers can be created by laser irradiation in the same region. It is therefore reasonable to associate the excitation of carriers and possible saturation of transitions with the existence of these highly sensitive nonlinear effects. The mechanism of excitation is not so far established (we refer to this below) but we can nevertheless deduce from the measured absorption coefficient at the appropriate laser power that betwee 1×10^{14} and 3×10^{15} free carriers would be created optically if we assume a recombination time of about 200 ns (a reasonable scaling from experimental numbers). The excitation of such a carrier density might then lead to the following effects:

1) Screening or saturation of exciton effects.

All absorption near the bandgap frequency in a semiconductor is in principle modified by Coulomb effects. It is known that the screening of discrete exciton absorption is associated with the refractive non-linearity in GaAs [2]. This mechanism, however, seems unlikely in InSb where the Bohr radius of the exciton is ~ 700 Å and excitonic effects are totally screened at carrier densities of around 1×10^{-14} cm^{-3}, the lowest initial value of the electron density in our material. In practice, exciton absorption spectra are not seen in InSb and both single and two photon absorption spectra near the bandgap energy are better fitted to experiment if excitons are neglected. We therefore consider significant exciton effects unlikely.

2) Nonlinearity caused by free carrier plasma creation.

This effect has been invoked by Jain and Klein [9] to explain a value of $\chi^{(3)} \sim 10^{-7}$ esu in silicon. The effect is described by the equation

$$n_2 = -\frac{2\pi\alpha e^2 \tau}{n m_r \hbar \omega^3}$$

(in esu) where α is the linear absorption coefficient, τ is the recombination time, n is the linear refractive index, m_r is the reduced effective mass, and h, e and ω have their usual meanings. Inserting the values quoted above and noting that the absorption coefficient $\alpha \sim 1$ cm^{-1} this yields a value of n_2 around 3×10^{-6} cm^2/W - 10 or 20 times too small to explain our results at 5 K [6].

3) Excitation of carriers has been shown to saturate the absorption above the bandgap in InSb [10,11]. This has been attributed to a dynamic Burstein-Moss shift caused by filling the bottom of the conduction band with electrons. Equally, such an effect removes states contributing to dispersion and could cause nonlinear refraction. The published absorption work relies upon an excitation at energies above the bandgap and subsequent thermalisation of the carriers. We can estimate [12] the relative values of this effect and the free carrier plasma

$$\frac{n_2 \ (\text{Burstein-Moss})}{n_2 \ (\text{Plasma})} = \frac{1}{4} \left(\frac{\omega_G}{\omega_G - \omega} \right)$$

where $\hbar\omega_G$ is the bandgap energy.

In our case, however, the excitation energy is less than the bandgap and it is not clear that within the timescale of the excitation the carriers will distribute themselves in the same way. Thus it may not be correct to interpret this formula too literally.

4) In direct analogy with discrete two-level systems, broadened by a dephasing time T_2, it is possible to consider the band-tails of a set of two-level oscillators as capable of exciting electrons for laser photon energies less than the bandgap. This could be described as 'off-resonant pumping' and would be significant within a frequency range of $1/T_2$ of the bandgap. The suggestion of Javan and Kelley [13] that saturated absorption leads to nonlinear refraction, in principle by causality, applied to nonlinear processes can then give a contribution to nonlinear refraction. For the case of a semiconductor there is already evidence that a two-level model can reasonably explain saturable absorption in p-type Ge [14]. The Bloch states which describe interband transitions constrain the absorption to k-conserving vertical transitions so that a set of two-level oscillators is a quite reasonable model. At least it gives us a complete a prior model to compare with experiments. An expression for n_2 can then be obtained by summing the dispersive contributions using a density matrix formulation put forward by Miller, Smith and Wherrett [12]. The expression for n_2 is as follows:

$$n_2 = \frac{-2\pi}{15} \ \frac{\hbar}{n^2 c} \ \left| \frac{eP}{\hbar^2 \omega} \right|^4 \ \left(\frac{2m_r}{\hbar} \right)^{3/2} \ \frac{T_1}{T_2} \ (\omega_G - \omega)^{-3/2} \ .$$

Here all the parameters e, h, ω, c take on their usual meanings, with n as the linear refractive index, m_r the reduced effective mass and P the momentum matrix element. $\hbar\omega_G$ is the bandgap energy. The parameters in this theory which require careful interpretation are the energy relaxation time T_1 and the phase relaxation time T_2, assumed constant throughout the bands for simplicity in this expression. We can interpret T_2 as being the time between collisions of excited electrons or holes giving randomisation of the quantum-mechanical phase of the wavefunctions; we therefore expect T_2 to be related to the speeds of intraband scattering mechanisms which can be on a picosecond timescale (the momentum relaxation time obtained from mobility data is typically picoseconds or less in cooled InSb). In atomic systems T_1 is normally the time taken for the excitation to relax to the ground state; in the semiconductor we might then interpret T_1 as the band-to-band recombination time. While this recombination time must set an upper limit on T_1, it is not the only process by which the excitation of a particular two-level system can be effectively relaxed. After excitation, the electron(hole) can be scattered to another state inside the conduction (valence) band by the fast intraband processes; the excitation of this state is then effectively relaxed because we now find this state in the condition it was initially - namely, with the electron (i.e. no hole) in the valence band and no electron in the corresponding conduction band state. Intraband energy relaxation has been measured at about $\sim$ 100 ps [15] in InSb albeit in a magnetic field. Because of the difficulty of interpreting T_1 and T_2, their ratio must remain a fitting parameter in our theory. However, we would expect to see a response of the nonlinearity on both a picosecond and a nanosecond timescale. The comparison between n_2 resonance measurements and theory with T_1/T_2 = 10 and 100 is shown in Figure 7 for a temperature of 5 K. Such values for the ratio T_1/T_2 are not inconsistent with the result of Gornik et al. [15]. Recent measurements at 77 K of $n_2 \simeq 3 \times 10^{-4}$ cm^2/W at 1830 cm^{-1} suggest that using this model T_1/T_2 is as high as 10,000 at this temperature. In this case, the absorption edge is substantially more broadened, implying a shorter T_2 than the 5 K case.

All the processes discussed above are in principle capable of giving large contributions to $\chi^{(3)}$. We have undertaken some experiments which give some indications of the relative importance.

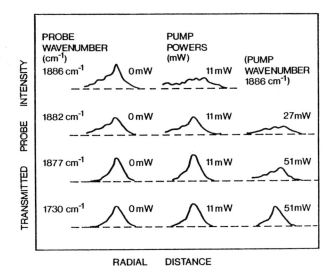

Fig.5. Transferred beam broadening in InSb. Two laser beams (spot size $\sim$ 150 μm) are coincident on the sample (5 mm thick, at 5 K), a pump and a probe. At 11 mW the pump far field profile is significantly broadened by self-defocusing; this broadening is observed transferred into the weak probe beam at the same wavelength. As the probe beam wavelength is moved from the pump wavelength progressively more pump power is required to induce beam broadening in the probe beam. At wavelengths much longer than the pump wavelength (e.g. 1730 cm^{-1}) no induced effect is observed with available beam powers

Following the knowledge that the nonlinear refraction induced by one beam can be transferred to another [3] we may attempt to observe induced de-focusing using two laser beams at various frequencies. If the origin of the nonlinearity was predominantly carrier plasma one would then expect the non-linearity to increase according to $1/\omega^3$. Thus pumping at 1886 cm^{-1} near the bandgap should therefore give a large effect at say 1730 cm^{-1}, the longest probe wavelength. As shown in Figure 5, probe laser beams at different frequencies are indeed broadened by the 1886 cm^{-1} pump but the effect falls off towards longer wavelengths in direct contradiction to this proposition. This experiment therefore favours 3 and 4.

In our original observations of nonlinear optical effects in InSb [5] we saw self-induced beam broadening at both 5 K and 77 K. Thus we predicted that optical bistability would be observable at 77 K as well as the original temperature regime at 5 K. As shown in Figure 6 we observe both a larger nonlinearity at 77 K ($\chi^{(3)} \sim 10^{-1}$ esu) and very clear optical bistability. This latter was observed in a crystal 130 μm thick coated to give a reflectivity of 0.7 with layers of Ge and ZnS. This bistability occurred in first order and required an onset power of only 7.9 mW. If the mechanism were excitonic we would expect a gross reduction in the strength of the non-linearity on increasing the temperature from 5 K to 77 K and a disappearance of the effect with impure samples. In fact we see a stronger nonlinearity at 77 K, and the bistability results shown here for 77 K were taken with inexpensive, impure polycrystalline InSb of the type normally used for monochromator order-blocking filters. This therefore argues strongly against excitonic effects (mechanism 1).

It should be possible to distinguish between the relative strengths of the refractive effects of direct saturation (i.e. population of states by direct optical excitation, mechanism (4)) and indirect saturation (i.e. population of states by scattering-in of excited carriers from other states, mechanism (3)) by making time-resolved measurements of nonlinear refraction. The total effect should be some combination of the two since one cannot exist without the other - indirect saturation can only result if there is some direct optical excitation, and intraband de-excitation of direct optical excitation must result in indirect population of other band states.

We have made preliminary measurements of time-resolved bistable switching in InSb using the observed bistability at 77 K. The laser beam intensity was modulated by $\sim$ 50% using an electro-optic modulator at a frequency of

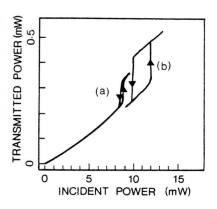

Fig.6. Optical bistability in InSb at 77 K. Crystal thickness 130 μm, surface reflectivity $\sim$ 70%, laser wavenumber 1827 cm^{-1}, spot size $\sim$ 150 μm. The onset of bistability, (a), is seen at 7.9 mW; clear, broad bistable regions are seen (e.g. (b)) as the cavity tuning is altered in the correct sense

∿ 1 KHz; bistable switching then occurred on a much faster timescale. The observations, currently limited by detector response times, show switching occurs in less than a microsecond and therefore conclusively eliminate any possibility of thermal origins for the process.

5. Conclusions

The small holding powers, switching energies, size, two-dimensional character and potentiality for fast operation give considerable promise for the application of narrow gap semiconductors to optically bistable devices and their various derivatives. The new bandgap-resonant mechanism discovered in this work seems likely to be applicable to a variety of materials, possibly at higher temperatures and shorter wavelengths.

6. Acknowledgement

This work was carried out with my colleagues, David Miller and Colin Seaton, and also reported at the First International Conference on Optical Bistability, Asheville, North Carolina, 1980.

References

[1]   D A B Miller, S D Smith and A M Johnston
      Appl. Phys. Lett., 35, 658 (1979).

[2]   H M Gibbs, S L McCall, T N C Venkatesan, S C Gossard, A Passner and
      W Wiegmann
      Appl. Phys. Lett., 35, 658 (1979).

[3]   D A B Miller and S D Smith
      Opt. Commun., 31, 101 (1979).

[4]   D A B Miller and S D Smith
      Appl. Opt., 17, 3904 (1978).

[5]   D A B Miller, M H Mozolowski, A Miller and S D Smith
      Opt. Commun., 27, 133 (1978).

[6]   D Weaire, B S Wherrett, D A B Miller and S D Smith
      Opt. Lett., 4, 331 (1979).

[7]   D A B Miller, R G Harrison, A M Johnston, C T Seaton and S D Smith
      Opt. Commun., 32, 478 (1980).

[8]   D G Seiler and L K Hanes
      Opt. Commun., 28, 326 (1979).

[9]   R K Jain and M B Klein
      Appl. Phys. Lett., 35, 454 (1979).

[10] P Lavallard, R Bichard and C Benoit a la Guillaume
      Phys. Rev., B16, 2804 (1977).

[11] A V Nurmikko
      Opt. Commun., 16, 365 (1976).

[12] D A B Miller, S D Smith and B S Wherrett
      to be published.

[13]   A Javan and P L Kelley
       IEEE J. Quantum Electron., QE-2, 470 (1966).

[14]   F Keilman
       IEEE J. Quantum Electron., QE-12, 592 (1976).

[15]   E Gornik, T Y Chang, T J Bridges, V T Nguyen and J D McGee
       Phys. Rev. Lett., 40, 1151 (1978).

Critical Behavior in Optical Phase-Conjugation

Chr. Flytzanis
Laboratoire d'Optique Quantique du C.N.R.S., Ecole Polytechnique
F-91128 Palaiseau, Cedex, France

G.P. Agrawal
Quantel, 17 Avenue de l'Atlantique, Z.I. F-91400 Orsay, France, and

C.L. Tang
School of EE, Cornell University, Ithaca, NY 14853, USA

1. Introduction

An externally driven nonlinear system can have more than one stable steady states under appropriate conditions [1]. The transition from the one stable state to the other occurs when the external parameter, such as the light intensity or light frequency reaches certain critical values. These critical values are consequences of a subtle balance of different competing mechanisms, the one overtaking the others stabilizes the system in a new state although the external parameter is smoothly changed.

Furthermore because of an inertia accompanying this balance of mechanisms the critical values are different as the external parameters increase or decrease and the system displays a *hysteresis* and *critical slowing down* reminiscent of nonequilibrium first-order phase transitions. It is therefore of outmost interest to sense the behavior of the nonlinear system at these very critical points.

Since bistable behavior can arise either as a consequence of the intrinsic anharmonicity of the material system, or because of the selfaction of the optical beam on its own optical path under appropriate boundary conditions, we shall consider these two situations separately by referring to them as *intrinsic* and *extrinsic* bistability, respectively. The first case will be illustrated with the behavior of an assembly of Duffing anharmonic oscillators driven by an intense electromagnetic field [2,3] and the second with that of a two-level medium in a Fabry-Perot cavity [4-7].

In order to sort out the essential features in both cases under similar conditions we have chosen the four-wave interaction process which encompasses many nonlinear processes of current interest and wide range of applications [8,9]. In both cases the material system is driven by a pump field E_p consisting of two counterpropagating plane waves of frequency ω

$$E_p(\underline{r},t) = E_p\cos(\omega t - \underline{k}_p\cdot\underline{r}) + E_p\cos(\omega t + \underline{k}_p\cdot\underline{r}) \qquad (1)$$

and its behavior will be probed with a weak field

$$E_s = E_s\cos(\omega t - \underline{k}_s\cdot\underline{r} + \Theta) \qquad (2)$$

of the same frequency ω and propagating in a different direction; the degenerate four-wave mixing of the pump and the signal will then generate a phase-conjugated field [9]

$$E_c = E_c\cos(\omega t + \underline{k}_s\cdot\underline{r} - \Theta) \qquad (3)$$

propagating oppositely to E_s to satisfy the phase-matching condition. As will be shown below the phase-conjugated field shows a very dramatic behavior when the critical condition for onset of bistability is approached.

2. Intrinsic Bistability. Driven Anharmonic Oscillator

We assume that the behavior of the material system in the presence of an intense electromagnetic field can be modelled by that of an assembly of driven Duffing anharmonic oscillators [2,3]

$$\ddot{Q} + \frac{1}{T}\dot{Q} + \omega_o^2 Q + \gamma Q^3 = \frac{e^*}{m}(E_p + E_s) \tag{4}$$

where T is a phenomenological relaxation time, ω_o is the oscillator frequency, γ is the anharmonicity, e^* and m are the effective charge and mass of the coordinate Q which can cover many situations (phonon, exciton, molecular vibration).

2.1 Mean-Field Approximation

The square of the pump field (1), which can also be written

$$E_p = 2E_p \cos \underline{k}_p \cdot \underline{r} \cos \omega t = 2E_p(\underline{r}) \cos \omega t, \tag{5}$$

contains a space-independent part which we take to give

$$P^2 = 2(\frac{e^*}{m})^2 E_p^2 \tag{6}$$

and we introduce the mean-field approximation for the pump which essentially implies that only the spatially independent term (6) drives the system. Then since we assume $|E_s| << |E_p|$ we may write

$$Q = \bar{Q}_o + \delta Q$$

where

$$\ddot{\bar{Q}}_o + \frac{1}{T}\dot{\bar{Q}}_o + \omega_o^2\bar{Q}_o + \gamma\bar{Q}_o^3 = P \cos \omega t, \tag{7}$$

and δQ satisfies the linearized equation ($S = e^* E_s/m$)

$$\delta\ddot{Q} + \frac{1}{T}\delta\dot{Q} + \omega_o^2\delta Q + 3\bar{Q}_o^2\delta Q = S \cos(\omega t - \underline{k}_s \cdot \underline{r} + \Theta). \tag{8}$$

The solution of (7) and (8) is given in [3]; here we only summarize the main results. If we neglect higher harmonics and make the ansatz

$$\bar{Q}_o = R \cos(\omega t + \phi), \tag{9}$$

one gets for the amplitude

$$R^2\left[\frac{\omega}{T^2} + (\omega^2 - \omega_o^2 - \frac{3}{4}\gamma R^2)^2\right] = P^2. \tag{10}$$

In Fig.1 we have plotted R as a function of E_p and ω for fixed ω and E_p, respectively. It is clear that the mode amplitude R exhibits bistability and hysteresis. The critical points (turning points in Fig.1a) are obtained using the condition $\partial E_p/\partial R = 0$. For $|\omega^2 - \omega_o^2| >> \omega/T$ we find

$$R_1^2 = \frac{4}{3\gamma}|\omega^2 - \omega_o^2| \quad \text{for} \quad P_1 \equiv \sqrt{2}(\frac{e^*}{m})E_1 = \frac{4\omega^2|\omega^2 - \omega_o^2|}{3\gamma T^2} \tag{11}$$

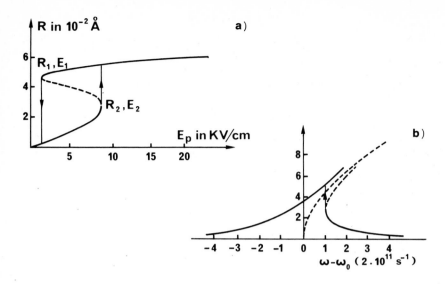

Fig.1 (a) Amplitude of the fundamental component of the anharmonic oscillator as a function of the driving field $E^2 = m^2 p^2/2e*^2$, $2e^2/m^2 = 10^{32}$ esu, $\omega_0 = 4\cdot10^{15}$ s^{-1}, $\omega - \omega_0 = 2\cdot10^{11}$ s^{-1}, $\tau = 10^{-10}$ s, and $\gamma = 10^{-46}$ esu. For this case, $E_1 \approx 0.6$ KV/cm and $E_2 \approx 8.5$ KV/cm. (b) Amplitude of the fundamental component as a function of frequency for $E_p = 8.5$ KV/cm

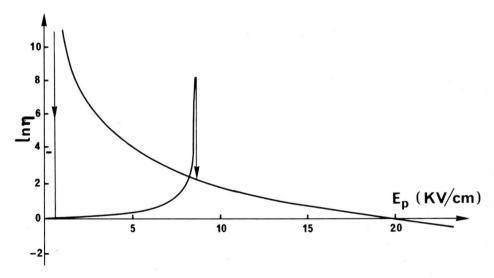

Fig.2 The enhancement factor showing critical behavior as a function of the pump field intensity. The horizontal axis, $\eta = 1$, corresponds to the usual perturbative result

$$R_2^2 = \frac{R_1^2}{3} \quad \text{for} \quad P_2 \equiv \sqrt{2}(\frac{e^*}{m})E_2 = \frac{16|\omega^2 - \omega_0^2|^6}{81\gamma} \tag{12}$$

Let us now derive the expression for the polarisation term responsible for the phase-conjugated signal [3]. In the mean-field approximation where the oscillators are taken to respond to the spatially independent part of the pump this polarisation term is simply given by

$$P_c = Ne \, \delta Q_c \tag{13}$$

where N is the number density of the oscillators and the expression of δQ_c is derived in [3]; it is that part of δQ which is phase conjugated to the signal field. When the oscillator is driven outside the resonance linewidth one gets

$$P_c(\omega) = \frac{Ne^{*2}\gamma R^2 E_s}{m(\omega^2 - \omega_0^2 - \frac{9}{4}\gamma R^2)(\omega^2 - \omega_0^2 - \frac{3}{4}\gamma R^2)} \tag{14}$$

which should be contrasted with the result

$$P_c^{(3)}(\omega) = \frac{Ne^{*2}\gamma E_s P^2}{m(\omega^2 - \omega_0^2)^4} \tag{15}$$

expected from the conventional perturbation analysis.
In Fig.2 we plot the ratio

$$\eta = \left|\frac{P_c(\omega)}{P_c^{(3)}(\omega)}\right| = \left|(1 - R^2/R_2^2)^{-1}(1 - R^2/R_1^2)^{-3}\right| \tag{16}$$

versus the pump field strength E_p; η expresses the enhancement of the phase-conjugated signal when a nonperturbative analysis is carried out over the result expected from the conventional perturbative approach. The important point to note is that η increases dramatically as R approaches the two critical points R_1 and R_2 and in a way that depends on which side they are approached.
More specifically, as P increases from zero and reaches P_2 from below, the phase-conjugated signal increases dramatically to drop abruptly to very low values once P crosses the critical value P_2 and further increases. When P starts decreasing after having crossed the value P_2 the phase-conjugated signal does not follow the same curve; it increases only slowly even when the point P_2 is crossed and then increases dramatically when the point P_1 is reached from above and drops to very low values for P below P_1. The magnitude of η as P approaches P_1 cannot be predicted with our analysis even when damping is included. The reason is when the critical point $R = R_1$ is approached by decreasing the pump, the inclusion of higher harmonics becomes necessary. It was recently pointed out [10] that in this region a driven anharmonic oscillator displays a set of cascading bifurcations until a chaotic state, characterized by the appearance of a strange attractor in physe space, is reached.

2.2 Rigorous Treatment. Multivalued Hysteresis

The mean-field approximation only gives a zero-order estimate to the actual behavior. The space-averaging process washes out a very subtle feature on which we wish to dwell on only qualitatively here as its more rigorous treatment requires some involved numerical calculation and its details are given elsewhere.

The new feature is a consequence of the standing wave nature of the pump. It arises because various oscillators within a spatial region of half wavelength are driven differently (pump field varies as $E_p \cos k_p \cdot r$) and because as stated above the enhancement of η occurs only at the transitional regions at P_1 and P_2 and this in a way that depends on from which side they are approached; as a matter of fact the pattern depicted in Fig.2 is not unique but depends on the history of the variations of E_p in a way reminiscent of the magnetization loops in magnetic substances because of the existence of domains [11].

To appreciate this crucial point let us return to (4) and still assume $|E_s| << |E_p|$ so that we may write

$$Q = Q_o + \delta Q \tag{17}$$

where now instead of (7) and (8) we have

$$\ddot{Q} + \frac{1}{T}\dot{Q}_o + \omega_o^2 Q_o + \gamma Q_o^3 = \frac{2e^*}{m}E_p \cos(\underline{k}_p \cdot \underline{r}) \cos \omega t \tag{18}$$

$$\delta\ddot{Q} + \frac{1}{T}\delta\dot{Q} + (\omega_o^2 + 3Q_o^2)\delta Q = \frac{e^*}{m}E_s \cos^2 \omega t - \underline{k}_s \cdot \underline{r} + \theta). \tag{19}$$

Using again an ansatz $Q_o = R \cos(\omega t + \phi)$, (18) can be solved as previously for each space point $\underline{r}$. If we neglect the damping, (19) becomes

$$\delta\ddot{Q} + (\omega_o^2 + 3R^2 \cos \omega t)\delta Q = S \cos(\omega t - \underline{k}_s \cdot \underline{r} + \theta) \tag{20}$$

where R is solution of

$$R^2\left[\frac{\omega^2}{T^2} + (\omega^2 - \omega_o^2 - \frac{3}{4}\gamma R^2)^2\right] = 2P^2 \cos^2(\underline{k}_p \cdot \underline{r}). \tag{21}$$

The mean-field result was obtained by taking the space-averaged part of (21). If instead the space dependence is kept the most general solution of (20) is of the form [12]

$$\delta Q(\underline{r},t) = \sum_n \delta Q^{(n)}(t) e^{in\underline{k}_p \cdot \underline{r}}. \tag{22}$$

Introducing this expression in (20) one obtains an infinite set of coupled equations for the various Fourier components $\delta Q^{(n)}$. A general solution requires numerical approximations. Here we only give a qualitative discussion based on physical arguments.

Once $\delta Q(\underline{r},t)$ is known, the induced polarization P_c, responsible for the phase-conjugated signal, is obtained using

$$P_c = \int dN(\underline{r})\delta Q_c(\underline{r}) \tag{23}$$

where $dN(\underline{r})$ is the number of oscillators around the point $\underline{r}$. Eq. (23) together with (15) is used to obtain the enhancement factor η of the phase-conjugated signal. The gross features depicted in Fig.2 are preserved but the integral in (23) depends on the history of the variation of E_p in a much more complicated way than the mean-field approximation predicts.

In order to illustrate the behavior of η when we go beyond the mean-field approximation, we note from Fig.2 that η peaks around the critical fields E_1 and E_2 (defined through (11) and (12), respectively) with some finite linewidth ΔE_j, $j = 1,2$. Let us consider the case when E_p starts from some value below E_1, increases monotonically up to a certain value E_{max} which can be above or below E_2 and then decreases monotonically to a value below E_1. Three cases are possible and they are schematically illustrated in Fig.3. In each case we show the spatial variation of the standing-wave pump field in a $\lambda/2$ region. The shaded region corresponds to the critical driving field which will produce an enhancement of the phase-conjugated reflectivity η.

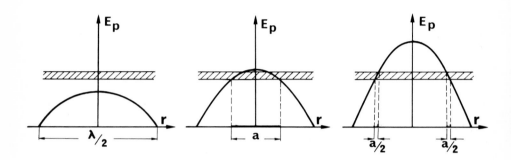

Fig.3 Schematic illustration of the behavior of η without using the mean-field approximation. The shaded area shows the region of the critical driving field (see text)

The number of oscillators which contribute to the enhancement of η is related to the length of the region where the sin-curve intersects the shaded area. The three possibilities are
  (a) if $E_{max} < E_2$, then the behavior of the phase-conjugated reflectivity derived from (23) does not differ appreciably from the mean-field result
  (b) if $E_2 < E_{max} < (E_2 + \Delta E_2)$, η will again show a peak but its height will be lower than the one in Fig.2 because now only the oscillators lying within the width $a < \lambda/2$ see the critical driving field
  (c) as $E_{max} < (E_2 + \Delta E_2)$, η falls off smoothly because a decreases (see Fig.3c). The rate of decrease of η will be slower than the mean-field case where a drops sharply to zero.

A new qualitative feature will arise when E_p is decreased from $E_{max} \gg E_2$ to a value below E_1. The resulting curve for η will not be unique but will depend on the magnitude of E_{max}. This behavior is explained as follows. The width a depends on E_{max} as is evident from Fig.3c. The oscillators within the width a, that reached the critical value R_2, and those outside this region of width $(\lambda/2 - a)$, that did not attain the amplitude R_2, will follow different paths and their contributions will be weighted differently in (23). This very remarkable behavior has some similarities with the nonuniqueness of the hysteresis curve in magnetic substances and there again this behavior arises because of domain formation [11].

3. Extrinsic Bistability. Two-Level Medium in a Fabry-Perot Cavity

We now consider the case of extrinsic bistability [4-7]. The nonlinear medium, modelled as a homogeneously broadened two-level atomic system, is placed inside a Fabry-Perot cavity which is driven by a strong pump field E_o. The signal field E_s is incident from the side of the cavity. We wish to obtain the phase-conjugated field as a function of E_o. The atomic system with resonance frequency ω_a may be in one-photon [6] or two-photon [7] resonance with the external driving field E_o of frequency ω. We consider both cases separately.

In the case of one-photon resonance the nonlinear susceptibility for the steady-state response is given by [5]

$$\chi(E) = \frac{\alpha}{k} \frac{\Delta + i}{1 + \Delta^2 + |E|^2/E_s^2} \tag{24}$$

where $k = \omega/c$ and α is the resonant absorption coefficient. Further $\Delta = (\omega_a - \omega_o)T_2$ is the atomic detuning parameter and $E_s^2 = \hbar^2 T_1 T_2/\mu^2$ is the saturation amplitude; T_1 and T_2 are the longitudinal and transverse times and μ is the atomic transition dipole moment. In (24), E is the total cavity field,

$$E = (E_1 + E_2) + (E_s + E_c) = E_p + \Delta E \tag{25}$$

where $E_p = E_1 + E_2$ consists of counterpropagating plane waves and ΔE is the sum of the signal and conjugated fields. Since $|\Delta E| << |E_p|$, $\chi(E)$ in (24) can be expanded about E_p [13]. The induced polarization is then given by

$$P = \varepsilon_o \chi_o E_p + \frac{\varepsilon_o \chi_o (1 + \Delta^2)\Delta E}{(1 + \Delta^2 + |E_p/E_s|^2)} - \frac{\varepsilon_o \chi_o (\Delta E)^* |E_p|^2/E_s^2}{(1 + \Delta^2 + |E_p/E_s|^2)} \tag{26}$$

where $\chi_o = \chi(E_p)$. Substituting (26) in Maxwell's equations one obtains a set of four equations satisfied by the pump fields E_1 and E_2 and the signal and conjugated fields E_s and E_c. We shall not go into details here which are available elsewhere [14]. The important point to note here is that the pump-field equations are uncoupled from the other two and can be solved first with proper boundary conditions. These equations are identical to those obtained in the context of optical bistability when $E_s = 0$ and no phase conjugation takes place. They may be readily solved in the mean-field approximation [5]. The pump intensity $I_p = |E_p|^2/E_s^2$ is constant inside the cavity and satisfies the relation

$$\frac{I_p T}{2}\left[\left(1 + \frac{2C}{1 + \Delta^2 + I_p}\right)^2 + \left(\phi - \frac{2C\Delta}{1 + \Delta^2 + I_p}\right)^2\right] = I_o \tag{27}$$

where $I_o = |E_o|^2/E_s^2$ is the incident intensity (in normalized units), T is the mirror transmittivity, $C = \alpha L/2T$ is the cooperativity parameter (L is the cavity length) and $\phi = 2(m\pi - kL)/T$ is the cavity detuning parameter (m is an integer).

For a given I_p two remaining coupled equations for E_s and E_c are linear and can be sovled exactly. The phase-conjugated reflectivity is found to be [13]

$$R_c = \left|\frac{E_c}{E_s}\right|^2 = \left|\frac{\kappa \sin g}{g \cos g + \beta \sin g}\right|^2 \tag{28}$$

where $g = (\kappa^2 - \beta^2)^{1/2}$ and

$$\beta = \frac{\alpha L'}{2} \frac{(1 + \Delta^2)}{(1 + \Delta^2 + I_p)^2} \quad , \qquad \kappa = \beta I_p / (1 + \Delta^2)^{1/2}. \tag{30}$$

Here L' is the length of the linear medium in the direction of the signal. Eq. (28) can be further simplified for the case of weak absorption $\alpha L' \ll 1$, which is the case in the mean-field approximation adopted here. We then obtain

$$R_c \approx \kappa^2 = (\alpha L' r_n / 8)^2 \tag{31}$$

where r_n is the normalized amplitude reflection coefficient,

$$r_n = \frac{4 I_p (1 + \Delta^2)^{1/2}}{(1 + \Delta^2 + I_p)^2} \, . \tag{32}$$

In Fig.4 we have plotted r_n as a function of the incident amplitude Y defined by

$$Y^2 = 2(E_0^2 / E_S^2 T) \tag{}$$

for several choices of the detuning parameters Δ and ϕ. In each case the cooperativity parameter was chosen $C = 20$. The unstable branch of the reflectivity curve is shown by a dashed line. Bistability and hysteresis of the phase-conjugated reflectivity is clearly displayed in Fig.4. At the upper critical field Y_{max}, R_c decreases by four orders of magnitude. Such a dramatic change in the strength of the phase-conjugated signal is a manifestation of the critical behavior of the bistable device and arises from an interplay between the nonlinearity of the medium and the cavity feedback mechanism.

We briefly consider the case of two-photon resonant bistability. The nonlinear susceptibility is given by [7]

$$\chi(E) = \frac{\alpha_2}{k} \left[\frac{(\Delta + i) I + I^2 \delta}{1 + \Delta^2 + 2 I \Delta \delta + (1 + \delta^2) I^2} \right] \tag{33}$$

where α_2 is the two-photon absorption coefficient, $I = |E|^2 / I_s$, I_s is the two-photon saturation intensity, $\Delta = (\omega_a - 2\omega) T_2$ is the two-photon detuning parameter and δ represents the optical Stark-shift effects associated with a two-photon transition. For the definitions of I_s, δ etc., see [10]. Details of the calculation of the phase-conjugated reflectivity are the same as in the one-photon resonant case and will be presented elsewhere. Once again we obtain an implicit equation for I_p which allows for multiple values of I_p for a given value of the driving field I_0. The phase-conjugated reflectivity is still given by (28) but the parameters α_R and κ are more complicated functions of I_0. In particular, they now depend on an additional parameter δ which denotes Stark-shift effects. The main point to note is that for a given I_0 multiple values of R are possible and bistability and hysteresis will be observed in the phase-conjugated reflectivity.

It should be mentioned that the use of the mean-field approximation is less critical in the case of extrinsic bistability than in the case of the intrinsic one. In the former case its use can be justified if the absorption $\alpha L \ll 1$ and the cavity transmission $T \ll 1$ while $\alpha L/T$ is kept finite. These conditions are easily met by the use of a low-density gas in a high-Q Fabry-Perot cavity. Even when these conditions are not satisfied, curves in Fig.4 remain qualitatively the same. This should be contrasted with the case of the intrinsic bistability

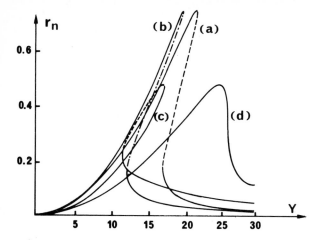

Fig.4 Variation of the phase-conjugated reflection coefficient r_n with the driving field Y (in normalized units). The dimensionless atomic and cavity detuning parameters, Δ and ϕ are as follows: (a) $\Delta = 1$, $\phi = 1$; (b) $\Delta = 1$, $\phi = -1$; (c) $\Delta = 2$, $\phi = 2$; (d) $\Delta = 2$, $\phi = -2$. In the last case the parameters are such that the bistability does not occur. Physically unaccessible part (unstable branch) of the reflectivity curve is shown by dashed line

where new features were found when we go beyond the mean-field approximation. The difference in the two cases comes from the fact that in the case of the extrinsic bistability the medium plays a passive role of providing the nonlinear susceptibility while in the case of the intrinsic bistability each anharmonic oscillator is driven independently and its behavior is very sensitive to the strength of the driving field it sees.

4. Applications

Optical bistability of the phase-conjugated fields should find many interesting applications. For instance, if the bistable switch is used as a memory element, a measurement of the phase-conjugated reflectivity by probing the device through a weak signal will allow to read the information without affecting the memory contents. When used in the context of intrinsic bistability, the phase-conjugated reflectivity provides a way to observe the critical behavior of collective modes in dielectrics. Bistability in the excitonic density in CuCl and Cu_2O comes in this category. The soft modes near a paraelectric to ferroelectric transition form another interesting system where bistability of the phase-conjugated signal may find applications.

References

1. See, for example, H. Haken: *Synergetics. An Introduction.* Nonequilibrium Phase Transitions and Self-Organization in Physics, Chemistry and Biology (Springer, Berlin, Heidelberg, New York 1978)
2. See, for example, G. Joos: *Theoretical Physics* (Black, London 1960) pp. 100-102
3. C. Flytzanis, C.L. Tang: Phys. Rev. Lett. 45, 441 (1980)
4. H.M. Gibbs, S.L. McCall, T.N.C. Venkatesan: Phys. Rev. Lett. 36, 1135 (1976)
5. G.P. Agrawal, H.J. Carmichael: Phys. Rev. A19, 2074 (1979)

6. R. Bonifacio, Lugiato: Opt. Commun. $\underline{19}$, 172 (1976)
7. G.P. Agrawal, C. Flytzanis: Phys. Rev. Lett. $\overline{44}$, 1743 (1980);
   F.T. Arecchi, A. Politi: Lett. Nuovo Cim. $\underline{23},\overline{65}$ (1978)
8. M.D. Levenson, C. Flytzanis, N. Bloembergen: Phys. Rev. $\underline{B6}$, 3962 (1972);
   J.J. Wynne: Phys. Rev. Lett. $\underline{29}$, 650 (1972)
9. R.W. Hellwarth: J. Opt. Soc. $\overline{Am}$. 67, 1 (1977);
   A. Yariv, IEEE J. $\underline{QE-14}$, 650 (1978)
10. B.A. Huberman, J.P. Crutchfield: Phys. Rev. Lett. $\underline{43}$, 1743 (1979);
    Y. Ueda: J. Stat. Phys. $\underline{20}$, 181 (1979)
11. See, for example, E. Hallén: *Electromagnetic Theory* (Chapman, London 1962)
    pp. 167-172
12. G.P. Agrawal, M. Lax: J. Opt. Soc. Am. $\underline{69}$, 1717 (1979)
13. R.L. Abrams, R.C. Lind: Opt. Lett. 2, 92 (1978)
14. G.P. Agrawal, C. Flytzanis: IEEE J. Quant. Electr. (1981) and
    G.P. Agrawal, R. Frey, F. Pradère, and C. Flytzanis: Applied Phys. Lett.
    (1981)

Transient Statistics in Optical Instabilities

F.T. Arecchi

Università di Firenze and Istituto Nazionale di Ottica, Firenze, Italy

Abstract

By photon statistics methods the statistical features of optical instabilities
have been measured with high accuracy over the past years. Their critical behavior can be classified with the language of equilibrium phase transitions.
Furthermore, in nonequilibrium optical systems it is possible to study the
transient build-up of an ordered state by a rapid passage through an instability. These transients are characterized by large fluctuations. An exact approach is described in terms of a stochastic time.

1. Introduction

Phase transitions in equilibrium systems are the result of a competition between the interparticle energy J and the thermal energy $k_B T$ which introduces
disorder. In quantum optics, even when interparticle interactions are negligible as in a very dilute gas, there may be particle correlations due to
common radiation field. The transition from disorder to order consists in a
passage from a regime where the atoms emit independently from one another,
to a regime where the atoms emit in a strongly correlated way.

Over the past fifteen years, the introduction of the photon statistics
method (for the theory see [1], for the experiments see [2]) has allowed a
careful investigation of optical instabilities.

From the first experiments [2] the threshold point, where the gain due to
the external excitation prevails over the internal losses, displayed the same
features of a continuous phase transition in an equilibrium system (large increase in fluctuations, slowing down).

Similarly, evidence of discontinuous jumps and hysteresis effects in a
laser with a saturable absorber suggested an analysis of this instability as
a first order phase transition [3]. Recently, injecting a laser field into
an interferometer filled with absorbing atoms, evidence of a hysteresis cycle
suggested a region of coexistence of two stable points, hence the name of optical bistability [4]. The corresponding theory, either when the instability
is due to the absorptive or the dispersive part of the atomic susceptibility,
shows the characters of a first-order phase transition [5].

Nonequilibrium systems can be driven through the instability point by a
rapid passage, that is, at a rate larger than the local relaxation rate of
steady state fluctuations. Such a transient situation was first observed for
the laser instability [6]. A phenomenological theory [7] has shown the uni-
versality character of the anomalous transient fluctuations.

Of course, since transient phenomena are not invariant for time translation,
there is no equivalent in equilibrium systems. The superfluorescence, that is
the spontaneous cooperative decay of N atoms all prepared in their excited

state in the absence of a classical field to drive them [8-10], is a transient collective behavior displaying a threshold.

Phase transitions in thermal systems, that is, in systems in contact with a thermal reservoir, have been recently shown also to display transients by induced temperature steps. The associated instability is called spinodal decomposition [11].

Other classes of transitions in nonequilibrium systems have been recently described in hydrodynamic and chemical systems [12].

The above introductory remarks are summarized in Table 1.

Table 1. Phase transitions in quantum optics

| | | | Nonequilibrium systems | |
		Thermal systems	Quantum optics	Others
steady state	2nd ord.	- order/disorder - para-ferromagnetic (H = 0)	- laser threshold	hydro-dynamic and chemical instabilities
	1st. ord.	- para-ferromagnetic (H = 0) - liquid gas	- laser plus saturable absorber - optical bistability	
transient		- spinodal decomposition	- laser transient - superfluorescence	

2. Cooperation and Phase Transitions in Radiative Interactions

A gas of N noninteracting particles in contact with a thermal reservoir is a single-phase system. If, however, there is a nonzero interparticle energy J, such as the intermolecular potential in a real gas or the exchange energy in a magnetic spin system, as soon as J prevails over the fluctuation energy $k_B T$, there is a transition from a disordered phase characterized by almost free particles to an ordered phase characterized by a collective or cooperative behavior.

A convenient classification of phase transitions is that of EHRENFEST depending on which thermal derivative of the free energy F is discontinuous at the transition point.

In some instances, when the interparticle interactions are long-range compared to the interparticle distance, the system is well described by a free energy which is a series expansion of an order parameter q. If, by symmetry arguments, the series has only even terms such as

$$F = U - TS = F_0 + a \cdot (T - T_c) \cdot q^2 + b q^4 \quad (a, b > 0) \quad ,$$

then it is an easy matter to show that at $T = T_c$, there is a 2nd-order (continuous q) transition. If an odd term has to be included, that is, if

$$F - F_0 = \xi q + Aq^2 + bq^4 \qquad [A = a(T - T_c)]$$

then the transition is 1st order (q discontinuous).

The probability $P(q)$ of the order parameter q is given by (N = normalization factor)

$$P(q) = N \exp^{(-F/k_B T)} \quad .$$

Both free energy F and probability P are plotted in Figs. 1 and 2.

In optics, even when interparticle interactions are negligible, there may be correlations due to the common radiation field giving rise to a transition from a disordered state (or single particle emission) to an ordered state. Generally, an optical device is an open system fed by a source of energy and radiating electromagnetic energy toward a sink.

Let us consider the equations for a field coupled with N two level atoms by a resonant transition at frequency ω. Writing field a, polarization S and population inversion Δ as slowly varying variables, in the interaction volume the coupled equations reduce to (at resonance a and S are real quantities)

$$\dot{a} = gS - ka$$

$$\dot{S} = 2ga\Delta - \gamma_\perp S$$

$$\dot{\Delta} = -2gaS - \gamma_{||} (\Delta - \Delta_0) \qquad\qquad (1)$$

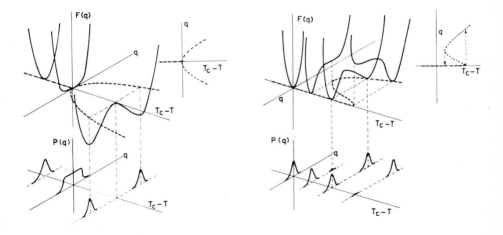

Fig. 1. Second-order phase transition.   Fig. 2. First-order phase transition.

Free energy F(q) and probability density P(q) versus the order parameter q at different temperatures T. The locus of equilibrium points is displayed separately in a q - T plane

where

$$g = \left(\frac{\omega\mu^2}{2\hbar\varepsilon_0 V}\right)^{1/2}$$

is the coupling constant and K, $\gamma_\perp$, $\gamma_{||}$ are loss rates. The normalization is such that $a^2 = n$ is the photon number and Δ the number of inverted atoms in the volume V. Δ_0 is a source term.

To give the order of magnitude, for a dilute gas of atoms with allowed transitions in the visible and for V 1 cm^3, it is

$$g \sim 10^4 \text{ s}^{-1} \quad , \quad \gamma_\perp \sim \gamma_{||} \sim \gamma \sim 10^8 \text{ s}^{-1} \quad ,$$

and

$$K \sim 10^7 \text{ s}^{-1} \quad \text{or} \quad \sim 10^{10} \text{ s}^{-1} \quad , \tag{2}$$

depending on whether the gas is in a laser cavity or distributed over a length of some centimeters, without mirrors at the ends.

For small deviations from ground or excited state ($\Delta_0 \sim \pm N/2$) we show by linearizations of (1) that the lossless atoms-field interaction has a rate [13]

$$\gamma_c^2 \equiv g^2 N \quad . \tag{3}$$

Losses introduce competing mechanisms with the rates as shown in Fig. 3.

The collective interaction will prevail on the separated uncorrelated dampings whenever

$$C \equiv \frac{g^2 N}{\gamma k} \equiv \frac{\gamma_c^2}{\gamma k} > 1 \quad . \tag{4}$$

It can be shown [14] that relation (4) rules all quantum-optical instabilities, namely,

I)   laser threshold, optical bistability,

II) superfluorescence,

III) optical turbulence.

These three cases correspond to different scales of damping times [(see relations (2)].

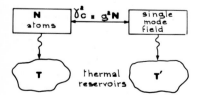

Fig. 3. Atoms-field interaction and dissipation to the thermal baths

In case I), $\gamma \gg k$, hence the fast atomic variables relax to a local equilibrium which is set by the slow field amplitude which then acts as the order parameter. The evolution equation for such order parameter is, in suitable adimensional units

$$x = (C - 1) x - C x^3 \quad .$$ (5)

In case II), $k \gg \gamma$, the field has a fast escape rate from the atomic medium and a detector witnesses a fast collective decay whenever atoms are prepared in an excited state in a short time.

In case III), the damping rates are comparable, (1) must be considered simultaneously, and their solutions show chaotic behavior [12,15,16].

The interaction with thermal reservoirs imposes to complete the equations for the collective variables [(1) in general, (5) for the laser case] with stochastic, or noise, sources, which can be taken in general as Gaussian processes with very short correlation times and correlation amplitudes that we call D.

The general statistical theory as well as the main experimental results are reviewed in [14]. Here we limit the discussion to the transient anomalous fluctuations, for which an exact approach is now available [17].

3. Transient Fluctuations in the Decay of an Unstable State

A nonequilibrium system, under the action of external parameters, may undergo transition in the sense that one (or a set) of its macroscopic observables have a sizable change. Usually these changes were studied by a slow setting at the external parameter, in order to measure the stationary fluctuations and their associated spectra around each equilibrium point.

More dramatic evidence, on the decay of an unstable state, can be obtained by applying sudden jumps to the driving parameter and observing the statistical transients [6]. The decay is initiated by microscopic fluctuations. In the first linear part of the decay process the fluctuations are amplified, hence during the transient, and until nonlinear saturation near the new stable point reduces them, fluctuations do not scale with the reciprocal of the systems size, as it is at equilibrium.

A first experiment on the photon statistics of the laser field during its switch on [6] has opened this investigation. Figures 4 to 6 give the transient photon statistics during a laser build up and the associated average photon number and variance.

Limiting to the case of one stochastic amplitude x, the most natural approach was to measure the probability density P (x, t) at a given time t after the sudden jump of the driving parameter. Under general assumptions, P (x, t) can be shown to obey a nonlinear FOKKER-PLANCK equation (FPE). A time dependent solution in terms of an eigenfunction expansion is unsuitable for the large number of terms involved, with the exception of small jumps near threshold [18] or the asymptotic behavior for long times [19].

Solving for the moments $\langle x^k(t) \rangle$ leads to an open hierarchy of coupled equations. A two-piece approximation first introduced for the laser [7] consists in first letting the system decay from the unstable point under the linearized part of the deterministic force, diffusing simultaneously because of the stochastic forces. This leads to a short time probability distribution of easy evalutation [20]. Then we solve for the nonlinear deterministic path and spread it over the ensemble of initial conditions previously evaluated in the linear regime.

A recent nonpiecewise treatment consisted in a 1/N expansion of the diffusion term (N being the system size) [21]. However, this approximation fails for small jumps above the threshold of instability or for nonlinear diffusion coefficients.

Another approach [22] was to trace back at any time a virtual ensemble of initial conditions, which inserted in the noise-free dynamic equations, would be responsible for the actual spread. This approach reduces the FPE to a diffusion equation, however, it fails for large deviations from the Gaussian as shown in a recent generalization [23].

Here we present an approach to transient statistics which overcomes the previous limitations. We consider the time t at which a given threshold z_F

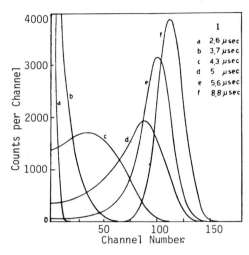

Fig. 4. Experimental statistical distributions with different time delays obtained on a laser transient

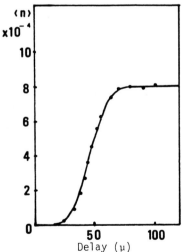

Fig. 5. Evolution of the average photon number $\langle n \rangle$ inside the cavity as a function of the time delay

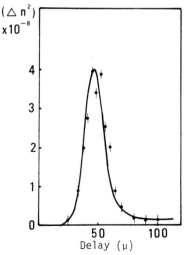

Fig. 6. Evolution of the variance $\langle \Delta n^2 \rangle$ of the statistical distribution of photons inside the cavity, as a function of the time delay

is crossed as the stochastic parameter, whose distribution Q (t, z, z_F) in terms of the interval between the initial position z and z_F must be assigned. Here, the time is no longer an ordering parameter but an interval limited by a start-stop operation. Let P (x, t) be the probability density for the amplitude x which gets unstable under a force F (x) and a noise delta-correlated with a correlation D (x). P (x, t) is solution of the FPE

$$\frac{\partial P}{\partial T} = - \frac{\partial}{\partial x} [F(x)P] + \frac{\partial^2}{\partial x^2} [D(x)P] \quad . \tag{6}$$

In order to develop an equation for the new density Q (t, z) the time must be assigned as a single value parameter of z. This amounts to consider the problem of the first passage time in the Brownian motion which is ruled by the KOLMOGOROV equation [24] (KE)

$$\frac{\partial Q}{\partial t} = F(z) \frac{\partial Q}{\partial z} + D(z) \frac{\partial^2 Q}{\partial z^2} \tag{7}$$

where z is the initial value $(t = 0)$, and the normalization is $\int_0^\infty Q dt = 1$.
Since we are studying the space evolution of the time distribution, (8) must be considered as a second-order differential equation and we need two boundary conditions, that is the final value (the threshold) z_F, and the value α above which the process has to remain limited during the evolution. Like for the usual FPE, the evaluation of the moments $T_m = <t^m>$ is formally equivalent to the solution of the equation, but in this case we have a simple recurrence formula as

$$F(z) T'_m + D(z) T''_m = - m T_{m-1} \tag{8}$$

(the apex denoting differentiation with respect to z).
In particular, we have for the mean time T_1

$$T_1(z) = \int_z^{z_F} \frac{dy}{W(y)} \int_{-\infty}^y dx \frac{W(x)}{D(x)} \tag{9}$$

where

$$w(x) \equiv 1xp \int_{x_0}^x d\xi \, F(\xi)/D(\xi) \quad .$$

(9) holds for $z_F > z$ and $\alpha = -\infty$. For a spread in the initial position z, $T_1(z)$ should still be averaged over the set of z. In a similar way we obtain $T_2(z)$, etc.
When we apply this formalism to the decay of unstable states, since D scales with the inverse system size, we can expand the above results in D series and display the first relevant correction to the deterministic solution. We find immediately that

$$T_1(z) = \int_z^{z_F} \frac{dy}{F(y)} + \int_z^{z_F} dy \, D \frac{dF}{dy} \frac{1}{F^3} \tag{10}$$

where the first term on the right-hand side is the deterministic part. Similarly, performing the approximation for T_2 we obtain for the variance $\Delta T \equiv T_2 - T_1^2$ the following relation

$$\Delta T = 2 \int_{Z}^{Z_F} dy \, \frac{D(y)}{F^3(y)} \quad . \tag{11}$$

In order to show the power of this approach, we have measured the crossing time probability distributions for an electronic oscillator driven from below to above threshold [25].

Figure 7 gives the mean oscillator amplitude and its variance versus time as in the usual stochastic treatment of transients, Figure 8 gives the variance of crossing times for increasing threshold as defined here.

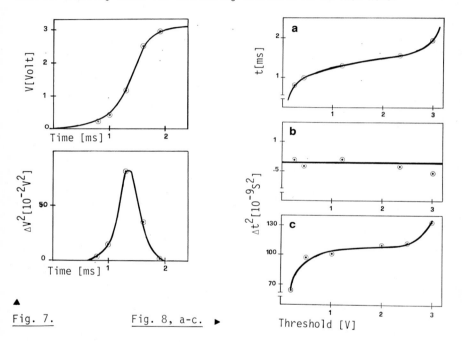

Fig. 7.　　　　Fig. 8, a-c. ▶

Fig. 7. Transient statistical evolution of an electronic oscillator driven from below to above threshold by a sudden jump. No external noise added. Average amplitude V and variance. (···) experiment; (——) theory

Fig. 8, a-c. Transient oscillator as in Fig. 7. Statistical distribution of the time intervals between the initial condition and the crossing time (a), variance under the action of the internal noise (b), variance for an added external noise (c). In (b) and (c) the scale is in units of 10^{-9} s^2. (···) experiment; (——) theory

The following comments convey some of the relevant physics: I) The first
term of (10) yields an average decay time which scales as $T_1 \sim \ln(N)$, that is,
a logarithmic divergence with the system size N; II) a constant variance for
increasing threshold means that the various trajectories are shifted versions
of the same deterministic curve, and the noise scaling as 1/N plays a role
only in spreading the initial condition; III) introduction of an external
noise D_0 adds a fluctuation peculiar for each path, giving a ΔT dependent
on s_f.

In conclusion, we have shown a clear separation between the role of the
initial spread and the noise along each path, and have introduced a new experimental characterization of a statistical transient which can be dealt
with in an exact way.

As here presented, the method seems limited to discrete variables. However, by a suitable mode expansion and selection of the lowest threshold
modes one can reduce field problems, (diffusive instabilities) to a set of
few discrete coupled variables which can be dealt with by our method.

References

1. R.J. Glauber: In *Quantum Optics and Electronics*, Proc. of 1964 les Houches
   School (McGraw-Hill, New York 1965) pp. 65-185
2. F.T. Arecchi: In *Quantum Optics*, Proc. of 1967 Varenna School (Academic
   Press, New York 1969) pp. 57-110
3. L.A. Lugiato, P. Mandel, S.T. Demlinski, A. Kossakowski: Phys. Rev. *A18*,
   238 (1978)
4. H.M. Gibbs, S.L. McCall, T. Venkatesan: Phys. Rev. Lett. *19*, 1135 (1976)
5. R. Bonifacio, L.A. Lugiato: Opt. Commun. *19*, 172 (1976)
6. F.T. Arecchi, V. Degiorgio, B. Querzola: Phys. Rev. Lett. *19*, 1168 (1967)
7. F.T. Arecchi, V. Degiorgio: Phys. Rev. *A3*, 1108 (1971)
8. R.H. Dicke: Phys. Rev. *93*, 99 (1954)
9. R. Bonifacio, P. Schwendimann, F. Haake: Phys. Rev. *A4*, 302 and 804 (1971)
10. H.M. Gibbs, Q. Vrehen, H. Hikspoors: Phys. Rev. Lett. *39*, 547 (1978)
11. W.G. Goldburg, J.S. Huang: In *Fluctuations, Instabilities, and Phase
    Transitions*, ed. by T. Riste (Plenum Press, New York 1975)
12. H. Haken: *Synergetics. An Introduction. Nonequilibrium Phase Transitions
    and Self-Organization in Physics, Chemistry, and Biology* (Springer, Berlin,
    Heidelberg, New York 1977)
13. F.T. Arecchi, E. Courtens: Phys. Rev. *A2*, 1730 (1970)
14. F.T. Arecchi: In *Dynamical Critical Phenomena and Related Topics*, ed. by
    C.P. Enz, Lecture Notes in Physics, Vol. 104 (Springer, Berlin, Heidelberg,
    New York 1979)
15. V. Benza, L.A. Lugiato: Zeit. Phys. *B35*, 383 (1979)
16. K. Ikeda, H. Daido, D. Akimoto: Phys. Rev. Lett. *45*, 709 (1980)
17. F.T. Arecchi, A. Politi: Phys. Rev. Lett. *45*, 1219 (1980)
18. H. Risken, H.D. Vollmer: Z. Phys. *204*, 240 (1967)
19. N.G. Van Kampen: J. Stat. Phys. *17*, 71 (1977)
20. M. Suzuki: J. Stat. Phys. *16*, 447 (1977)
21. F. Haake: Phys. Rev. Lett. *41*, 1685 (1978)
22. F. De Pasquale, P. Tombesi: Phys. Lett. *72A*, 7 (1979)
23. F.T. Arecchi, A. Politi: Lett. N. Cimento *27*, 486 (1980)
24. R.L. Stratonovich: *Topics in the Theory of Random Noise*, Vol. 1 (Gordon
    and Breach, London 1963)
25. F.T. Arecchi, M. Cetica, F. Francini, L. Ulivi: to be published

List of Contributors

Advances in Laser Chemistry

Proceedings of the Conference on Advances in Laser Chemistry,
California Institute of Technology, Pasadena, USA,
March 20–22, 1978
Editor: A. H. Zewail
1978. 242 figures, 2 tables. X, 463 pages
(Springer Series in Chemical Physics, Volume 3)
ISBN 3-540-08997-7

Contents: Laser-Induced Chemistry. – Picosecond Processes and
Techniques. – Non-Linear Optical Spectroscopy and Dephasing Pro-
cesses. – Multiphoton Excitation in Molecules. – Molecular
Dynamics by Molecular Beams.

Laser-Induced Processes in Molecules

Physics and Chemistry
Proceedings of the European Physical Society, Divisonal Conference
at Heriot-Watt University, Edinburgh, Scotland,
September 20–22, 1978
Editors: K. L. Kompa, S. D. Smith
1979. 196 figures, 31 tables. XIV, 367 pages
(Springer Series in Chemical Physics, Volume 6)
ISBN 3-540-09299-4

Contents: Study of Lasers and Related Techniques Suitable for
Applications in Chemistry and Spectroscopy. – Spectroscopic Studies
With and Related to Lasers. – Multiphoton Excitation, Dissociation
and Ionization. – Laser Control of Chemical Reactions. – Molecular
Relaxation.

Lasers and Applications

Proceedings of the Sergio Porto Memorial Symposium
Rio de Janeiro, Brasil, June 30 – July 3, 1980
Editors: W. O. N. Guimares, C. T. Lin, A. Mooradian
1981. 85 figures. Approx. 300 pages
(Springer Series in Optical Sciences, Volume 26)
ISBN 3-540-10647-2

Contents: Raman Spectroscopy. – Laser Spectroscopy. – Laser
Photochemistry. – New Laser Devices and Applications. – Laser
Biology and Medicine. – Picosecond Bistability.

Laser Spectroscopy III

Proceedings of the Third International Conference, Jackson Lake
Lodge, Wyoming, USA, July 4–8, 1977
Editors: J. L. Hall, J. L. Carlsten
1977. 296 figures. XI, 468 pages
(Springer Series in Optical Sciences, Volume 7)
ISBN 3-540-08543-2

Contents: Fundamental Physical Applications of Laser Spectro-
scopy. – Multiple Photon Dissociation. – New Sub-Doppler Inter-
action Techniques. – Highly Excited States, Ionization, and High
Intensity Interactions. – Optical Transients. – High Resolution and
Double Resonance. – Laser Spectroscopic Applications. – Laser
Sources. – Laser Wavelength Measurements. – Postdeadline Papers.

Springer-Verlag
Berlin
Heidelberg
New York

Laser Spectroscopy IV

Proceedings of the Fourth International Conference
Rottach-Egern, Fed. Rep. of Germany,
June 11–15, 1979
Editors: H. Walther, K. W. Rothe
1979. 411 figures, 19 tables. XIII, 652 pages
(Springer Series in Optical Science, Volume 21)
ISBN 3-540-09766-X

Contents: Introduction. – Fundamental Physical.
Applications of Laser Spectroscopy. – Two and
Three Level Atoms/High Resolution Spectroscopy. –
Rydberg States. – Multiphoton Dissociation, Multi-
photon Excitation. – Nonlinear Processes, Laser
Induced Collisions, Multiphoton Ionization. –
Coherent Transients, Time Domain Spectro-
scopy. – Optical Bistability, Superradiance. – Laser
Spectroscopic Applications. – Laser Sources. –
Postdeadline Papers. – Index of Contributors.

Picosecond Phenomena

Proceedings of the First International Conference
on Picosecond Phenomena, Hilton Head,
South Carolina, USA, May 24–26, 1978
Editors: C. V. Shank, E. P. Ippen, S. L. Shapiro
1978. 222 figures, 10 tables. XII, 359 pages
(Springer Series in Chemical Physics, Volume 4)
ISBN 3-540-09054-1

Contents: Interactions in Liquids and Molecules. –
Poster Session. – Sources and Techniques. – Biological Processes. – Poster Session. – Coherent Techniques and Molecules. – Solids. – High-Power
Lasers and Plasmas. – Postdeadline Papers.

Picosecond Phenomena II

Proceedings of the Second International Confe-
rence on Picosecond Phenomena, Cape Cod,
Massachusetts, USA, June 18–20, 1980
Editors: R. Hochstrasser, W. Kaiser, C. V. Shank
1980. 252 figures, 17 tables. XII, 382 pages
(Springer Series in Chemical Physics, Volume 14)
ISBN 3-540-10403-8

Contents: Advances in the Generation of Picosecond Pulses. – Advances in Optoelectronics. –
Picosecond Studies of Molecular Motion. – Picosecond Relaxation Phenomena. – Picosecond
Chemical Processes. – Applications in Solid State
Physics. – Ultrashort Processes/Biology. – Spectroscopic Techniques. – Index of Contributors.

A. B. Sharma, S. J. Halme, M. M. Butusov

Optical Fiber Systems and Their Components

An Introduction
1981. 125 figures, 12 tables. VIII, 246 pages
(Springer Series in Optical Sciences, Volume 24)
ISBN 3-540-10437-2

Contents: Introduction. – Generation, Modulation,
and Detection. – Light Propagation in Waveguides. – Components for Optical Fiber Systems. –
Fiber Measurement. – Fiber Optical Systems and
Their Applications. – Selected Problems. – References. – Subject Index.

Tunable Lasers and Applications

Proceedings of the Loen Conference,
Norway, 1976
Editors: A. Mooradian, T. Jaeger, P. Stokseth
1976. 238 figures. VIII, 404 pages
(Springer Series in Optical Sciences, Volume 3)
ISBN 3-540-07968-8

Contents: Tunable and High Energy UV-Visible
Lasers. – Tunable IR Laser Systems. – Isotope
Separation and Laser Driven Chemical Reactions. –
Nonlinear Excitation of Molecules. – Laser Photo-
kinetics. – Atmospheric Photochemistry and
Diagnostics. – Photobiology. – Spectroscopic Appli-
cations of Tunable Lasers.

Springer-Verlag
Berlin
Heidelberg
New York